OXFORD STUDIES IN EPISTEMOLOGY

OXFORD STUDIES IN EPISTEMOLOGY

OXFORD STUDIES IN EPISTEMOLOGY

Volume 5

Edited by

Tamar Szabó Gendler and John Hawthorne

OXFORD
UNIVERSITY PRESS

OXFORD
UNIVERSITY PRESS

Great Clarendon Street, Oxford, OX2 6DP,
United Kingdom

Oxford University Press is a department of the University of Oxford.
It furthers the University's objective of excellence in research, scholarship, and education by publishing worldwide. Oxford is a registered trade mark of Oxford University Press in the UK and in certain other countries

First Edition published in 2015

Published in the United States of America by Oxford University Press
198 Madison Avenue, New York, NY 10016, United States of America

British Library Cataloguing in Publication Data
Data available

Library of Congress Control Number: 2014941571

ISBN 978–0–19–872276–2

EDITORS' PREFACE

We are very excited about this fifth volume in the *Oxford Studies in Epistemology* series, which, like those before it, provides a showcase for some of the most exciting new work in the field of epistemology from throughout the English-speaking world. Published biennially under the guidance of a distinguished editorial board, each volume of *Oxford Studies in Epistemology* seeks to publish not only traditional work in epistemology, but also work that brings new perspectives to traditional epistemological questions, and that opens new avenues of investigation. Among the broad array of topics discussed in this issue are knowledge of abstracta, the nature of evidential support, epistemic and rational norms, fallibilism, closure principles, disagreement, the analysis of knowledge, and a priori justification. Papers make use of a variety of different methodologies, including those of formal epistemology and decision theory, as well as conventional philosophical analysis and argumentation, and suggest many new insights and perspectives that we hope our readers will find intriguing.

We are particularly pleased to announce that this volume of *Oxford Studies in Epistemology* includes the winner of the newly established Sanders Prize in Epistemology (a new award supported by the Marc Sanders Foundation open to scholars within fifteen years of receiving their PhD), Michael Titelbaum's "Rationality's Fixed Point (or: In Defense of Right Reason)," as well as the two runners-up, John Bengson's "Grasping the Third Realm" and Sarah Moss's "Time-Slice Epistemology and Action under Indeterminacy." Starting from the premise that *akrasia* is irrational, Titelbaum argues that it is always a mistake of rationality to have false beliefs about the requirements of rationality, and, from this conclusion, defends logical omniscience requirements, the claim that one can never have all-things-considered misleading evidence about what's rational, and the Right Reasons position concerning peer disagreement. Bengson invokes a form of naïve realism (according to which a successful intuition is constituted by the fact intuited) in order to explain how we could have knowledge of abstracta (a theory that he claims can be independently motivated by the possibility of acquiring knowledge of facts about colours and shapes via hallucinatory experience). And Moss defines and defends "time-slice epistemology," according to which there are no essentially diachronic norms of rationality, and highlights a more general moral about action under indeterminacy (namely, that time-slice theories are supported by strong analogies with ethical theories).

Several other papers in this volume make use of the tools of formal epistemology. In "Accuracy, Coherence, and Evidence," Kenny Easwaran and Branden Fitelson propose a new way of grounding formal, synchronic, epistemic coherence requirements for (opinionated) full belief that yields principled alternatives to deductive consistency, sheds new light on the preface and lottery paradoxes, and reveals novel conceptual connections between alethic and evidential epistemic norms. In "Fallibilism and Multiple Paths to Knowledge," Wesley Holliday argues that epistemologists should replace a "standard alternatives" picture of knowledge with a new "multipath" picture of knowledge and considers inductive knowledge and strong epistemic closure from this multipath perspective. In "New Rational Reflection and Internalism about Rationality," Maria Lasonen-Aarnio explores rational reflection principles and discusses a more general problem with any attempt to formulate such principles (which is that they invariably must take seriously certain kinds of uncertainty about what is rational, but not others). Finally, in "When Beauties Disagree," John Pittard presents a variant of the "Sleeping Beauty" case that shows that those who are "halfers" with respect to the original Sleeping Beauty problem are committed to holding that rationality can be perspectival in a rather extreme and surprising way—a conclusion that can be seen as calling into question a key principle often taken for granted in the disagreement literature, the proportionality principle.

Rounding out the collection are three papers that employ more traditional argumentative techniques to venture solutions to several significant and enduring epistemological puzzles. In "Evidence and Epistemic Evaluation," Jessica Brown argues against a popular version of the probability-raising conception of evidential support on the grounds that it has the problematic consequence that any proposition which is evidence for some hypothesis is evidence for itself, and contends that the probability-raising conception should be modified so as to avoid this consequence by appealing to the notion of warrant transmission. In "Knowledge Is Belief for Sufficient (Objective and Subjective) Reason," Mark Schroeder draws on more general principles about reasons, their weight, and their relationship to justification to offer answers to problems about defeat and the conditional fallacy that plagued early defeasibility analyses of knowledge and provides a sketch of a contemporary alternative. And, in "An Inferentialist Theory of the A Priori," Ralph Wedgwood proposes an account of the a priori based on a conception of inference that results from combining a generalization of the notion of degrees of conditional belief with the natural deduction approach, an account which he takes to have several advantages over its competitors (most notably, that it makes no appeal to any mysterious faculty of "intuition" or "rational insight").

Thanks are due our referees: Mark Balaguer, John Broome, Troy Cross, Jeremy Fantl, John Gibbons, Daniel Greco, Sophie Horowitz, Christopher Meacham, Richard Pettigrew, Stephen Schiffer, Brian Weatherson, Stephen Yablo, and two anonymous reviewers; and to our editorial board members Stewart Cohen (University of Arizona), Keith DeRose (Yale University), Richard Fumerton (University of Iowa), Alvin Goldman (Rutgers University),

Alan Hájek (Australian National University), Gil Harman (Princeton University), Frank Jackson (Australian National University and Princeton University), Jim Joyce (University of Michigan), Jonathan Vogel (Amherst College), and Tim Williamson (Oxford University). We would also like to welcome Jennifer Lackey (Northwestern University) and Jennifer Nagel (University of Toronto) to the board and extend a special thank you to the departing Scott Sturgeon (University of Birmingham) for his fine service over the years. Thanks also go out to all those who helped referee Sanders Prize submissions: David Christensen, Ram Neta, Matthias Steup, Fritz Warfield, Ralph Wedgwood, and two anonymous reviewers. We are also grateful to our managing editors Julianne Chung and Alex Worsnip for their invaluable assistance, and to Peter Momtchiloff at Oxford University Press for his outstanding support.

Tamar Szabó Gendler, Yale University
John Hawthorne, Oxford University

CONTENTS

CONTRIBUTORS

John Bengson
University of Wisconsin, Madison

Jessica Brown
Arché, University of St Andrews

Kenny Easwaran
Texas A&M University

Branden Fitelson
Rutgers University

Wesley H. Holliday
University of California, Berkeley

Maria Lasonen-Aarnio
University of Michigan, Ann Arbor

Sarah Moss
University of Michigan, Ann Arbor

John Pittard
Yale University

Mark Schroeder
University of Southern California

Michael G. Titelbaum
University of Wisconsin, Madison

Ralph Wedgwood
University of Southern California

1. Grasping the Third Realm

John Bengson

It by no means follows ... that [intuitions], because they cannot be associated with [causal] actions of certain things upon our sense organs, are something purely subjective. Rather ... their presence in us may be due to another kind of relationship between ourselves and reality.

Kurt Gödel, "What is Cantor's Continuum Problem?"

1. INTRODUCTION

Some things we know just by thinking about them: that identity is transitive, that Gettier's Smith does not know that the man who will get the job has ten coins in his pockets, that it is wrong to wantonly torture innocent sentient beings, that the ratio between two and six holds also between one and three, and various other things that simply strike us, intuitively, as true when we consider them. The question is *how*: how can we know things just by thinking about them?

Many philosophers have been attracted to a broadly platonistic or "third realm" conception of the entities—properties, relations, numbers, sets, norms, values, reasons, and other items—that such knowledge is about; I will use the label *realism*:[1]

> *Realism*: What are known are facts about mind-independent abstract entities (hereafter *abstract facts*).

To say that an entity is *abstract* is to say that it lacks spatiotemporal location and causal powers. To say that a fact is *mind-independent* is to say that it neither is nor holds in virtue of—it is not grounded in—a fact (or facts) about intelligent agents and their attitudes, languages, or practices. *Facts*, as I shall understand them, are distinct from true propositions. On a familiar view, a

[1] This realist thesis is endorsed at least locally (i.e. about some domains) by a diverse group, including Frege (1884/1953), Moore (1903), Quine (1960), Gödel (1964), Putnam (1971), Lewis (1983), Bealer (1993), Linsky and Zalta (1995), and Balaguer (1998). The expression "third realm" owes to Frege (1884/1953, 337), who famously held that some entities "are neither things in the external world nor ideas. A third realm must be recognized. Anything belonging to this realm has it in common with ideas that it cannot be perceived by the senses, but has it in common with things that it does not need an owner so as to belong to the contents of consciousness."

correspondence theory, true propositions are not identical to facts; rather, they *correspond to* facts. And, correlatively, according to a truth-maker theory, true propositions are *made true by* the corresponding facts. Viewed from another angle, while true propositions might be understood as quasi-sentential or intrinsically representational entities (they are about how the world is, or what is the case), the facts that correspond to true propositions or make those propositions true are themselves non-sentential, non-representational, worldly entities (they are how the world is, or what is the case).

Realism is often combined with a broadly rationalist epistemology:[2]

> *Rationalism*: The source of our knowledge of the relevant facts is a non-sensory, conscious mental state—e.g. a reflective or intellectual striking, or *intuition*.[3]

For example, when one reflects on whether Gettier's Smith has knowledge, it strikes one that he does not have knowledge; when one reflects on whether wantonly torturing innocent sentient beings is wrong, it strikes one that this action is wrong. According to rationalists, such facts are neither perceived nor inferred from what is perceived; rather, they are *intuited*.

When rationalism is combined with realism,[4] the resulting position may seem rather perplexing. Here, for example, is Paul Benacerraf:

> I find [the appeal to intuition] both encouraging and troubling. What troubles me is that [we lack] an account of the link between our cognitive faculties and the objects known... We accept as knowledge only those beliefs which we can appropriately relate to our cognitive faculties... [S]omething must be said to bridge the chasm... between the entities that form the subject matter of mathematics and the human knower... [T]he absence of a coherent account of how mathematical intuition is connected with the truth of mathematical propositions renders the over-all account unsatisfactory. (1973, 674–5)

Although Benacerraf here focuses on intuitive knowledge in the case of mathematics, the problem is far more general, arising wherever we find an appeal to intuition of facts regarding non-spatiotemporal, causally inert entities (i.e. abstracta). How does—or *could*—such intuition "work"?[5]

[2] The following rationalist thesis is historically popular; it has even been endorsed by philosophers (e.g. Locke 1689, IV.2.1) who have held broadly empiricist views regarding, say, the origins of ideas or the principle of sufficient reason. Rationalism has seen defense in recent years by Bealer (1992, 1998), BonJour (1998), Jackson (1998, ch. 3), Sosa (1998, 2007, ch. 3), Huemer (2005, ch. 5), and Ludwig (2007), among others.

[3] This is not to omit the possibility of knowledge of abstract facts via reasoning, which according to rationalism may be regarded as inference from intuition. Here and below I employ the broad use of 'state' familiar in contemporary philosophy of mind; on this use, properties, relations, and prima facie dynamic mental phenomena (e.g. events) may qualify as states even though they are not standing conditions.

[4] This combination is not compulsory: rationalism is also compatible with a nominalist, constructivist, or idealist (conceptualist, psychologistic) metaphysics (for discussion, see Bengson forthcoming, §6).

[5] Concerns of this type, which arguably have roots in the ancients, have been pressed by a number of philosophers in recent years: see, e.g., Hart (1977, 124ff.), Mackie (1977, 24 and 38),

The two main goals of this paper are to clarify the problem and to outline one way that realist rationalists might begin to address it. In my view the basic challenge centers on a question regarding "the link between our cognitive faculties and the objects known." The core of Benacerraf's worry is sometimes thought to rely upon a causal constraint on knowledge. Alternatively, the worry is commonly formulated in the language of "reliability," and it is sometimes construed so as to emphasize matters of etiology (or genealogy), in particular the evolutionary origins of human cognition or of particular attitudes. However, I believe that the basic challenge is not tied to a causal constraint on knowledge. It also remains distinct from, and is (in a sense that will be made clear) prior to, questions of reliability and etiology: what is needed is an explanation not simply of how the "link" is *reliable*, or how it *came to be* reliable, but of what the "link" even *is*, and of how it could ever obtain, in the first place.

The strategy will be to explore the possibility of a certain kind of non-causal explanation—what I will call a *constitutive explanation*—of the relevant knowledge. Two points of clarification about the project. First, I will not be arguing for either realism or rationalism in what follows: the aim is to explore how intuition could provide knowledge of abstracta, *if it does*. While this will require specifying a condition, namely, non-accidental correctness, that must be satisfied in order to have such knowledge, the reason for specifying this condition will be to engage the explanatory question of how it could be satisfied, *if it is*, rather than to address the skeptical question of whether any subject actually satisfies it. In effect, progress consists in, and can be measured by, clarification and development of the explanatory options.

Second, I will assume that the task—or, if you prefer, the burden—is to locate an explanation of intuitive knowledge of abstracta that is *dialectically adequate*, in the sense that it articulates resources that a proponent of realist rationalism can rationally use to respond to the challenge to account for the possibility of intuitive knowledge of abstracta. To achieve this end, the resources need not be accepted by (or even rationally acceptable to) the challenger. As Jim Pryor (2004) has emphasized in another context, a dialectically adequate answer can be usefully contrasted with a *dialectically persuasive* answer, which provides grounds for a rebuttal that are sufficient to rationally convince an opponent. In the present debate, in which more or less entire worldviews are at issue, it would appear that a dialectically adequate answer is the most that can be reasonably expected or demanded. Or so I shall assume in what follows.[6]

Bell (1979, §II), McDowell (1985, 111), Field (1989, 25–30 and 230–9; 2005), Rosen (1993), Hawthorne (1996, §§2–3), Boghossian (2001, 635), Cheyne (2001), Casullo (2003, §5.4), Peacocke (2004, 153), Wright (2004, 156–7), Devitt (2005, §§3–4), Goldman (2007, 6–8), Williamson (2007, 215), and Gibbard (2008, 20–1). Kitcher (1984, 59) does not speak only for himself when he concludes, "Benacerraf's point casts doubt on the ability of [intuition] to generate knowledge."

[6] Some philosophers will not be satisfied until realist rationalists identify subatomic particles, cellular groups, neural networks, or some other concrete entity or mechanism

The pursuit of a dialectically adequate explanation of intuitive knowledge of abstracta is not theoretically idle, but has potential significance for a wide variety of philosophical areas and debates, within epistemology as well as outside of it. For example, such an explanation would help to elucidate the contentious notion of "grasping," and other placeholders for mental-cum-epistemic achievement, to which friends of realist rationalism often appeal. It would also answer popular epistemological arguments against realist views in philosophy of mathematics, metaphysics of modality, and metaethics, and in effect undermine what is perhaps the primary motivation for the opposing, anti-realist views.[7] Likewise, it would enable a reply to influential objections, centering on accusations of "magic," "mystery," or "spookiness," against rationalist views in these areas, substantially reshaping debates regarding a priority (including the synthetic a priori), modal rationalism, and ethical intuitionism, among others. A second type of significance concerns philosophical methodology. To the extent that philosophical argumentation and theorizing employs intuitions about thought experiments and hypothetical examples (e.g. Gettier cases, Twin-earth scenarios), general principles and axioms (e.g. anti-coincidence theses, law of identity), or conflicting propositions (as in various puzzles and paradoxes), an explanation of intuitive knowledge may help to shed light on the character and scope of philosophical practice.

Here is a roadmap. §§2–3 clarify the challenge to realist rationalism, seeking to improve on the rough characterization of the question introduced above. §§4–5 begin to lay the foundations for an answer, which is then presented in §6. After, §7 responds to an important objection to the proposal; §8 discusses its broader explanatory implications; §9 concludes by identifying several of its theoretical virtues.

2. VERIDICAL INTELLECTUAL HALLUCINATION

I believe that we can begin to identify a clear, specific challenge to realist rationalism by focusing on a familiar distinction between accidentally correct and and non-accidentally correct conscious mental states. The challenge that

through which energy or information is "transferred" between thinkers and abstracta. These philosophers will find my approach unsatisfying, for I will not pursue the idea that somehow the behavior of leptons, bosons, cells, or neurons, through normal or quantum effects, provides the requisite philosophical explanation. I do not think that this is a difficulty for my project: to expect or demand an explanation in these terms is to import an extremely controversial perspective, to which a dialectically adequate response need not capitulate.

7 For example, the "access problem" for realism about numbers, attributed to Benacerraf, and Field's (1989, 25ff. and 230ff.) epistemological argument for mathematical fictionalism; Mackie's (1977, 24 and 38) queerness argument for an "error theory" of morality, at least on one way of understanding that argument, as well as cognate objections to ethical non-naturalism and objectivism about reasons; Hawthorne's (1996) "epistemological puzzle" for realism about modality. Cf. Peacocke (1999, ch. 1) on the "integration challenge," presented with reference to Benacerraf.

will then emerge is centered on the difficulty of constructing an explanation of non-accidentally correct intuitions, given a realist view of the nature or character of what they are about.

To fix attention, it is useful to begin with a case of perceptual experience. As is well known, one's experience might match one's environment, but only accidentally so (see, e.g., Grice 1961). For example, a capricious brain lesion might cause one to hallucinate that there is a red apple present; by a sheer coincidence, there is a red apple present: one got lucky. (At least since David Armstrong (1973, 171), this phenomenon has been referred to as 'veridical hallucination.') In such a case, even though one's experience is correct (true, accurate, veridical), one's experience is not related to the fact in such a way as to rule out accidentality. So the experience is not able to serve as a source of knowledge about one's environment.

Similarly, an intuition might get it right, but only accidentally so. For example, a capricious brain lesion might cause one to have the intuition that Goldbach's conjecture holds; suppose it does in fact hold: one got lucky. (I will call this phenomenon 'veridical *intellectual* hallucination.') In such a case, even though one's intuition is correct (true, accurate, veridical), one's intuition is not related to the fact in such a way as to rule out accidentality. So the intuition is not able to serve as a source of knowledge about this bit of mathematics.[8]

It will be useful to consider another example, based on a famous anecdote by the British mathematician G. H. Hardy regarding a hospital visit to the Indian mathematician Srinivasa Ramanujan:

> I remember once going to see him when he was ill at Putney. I had ridden in taxi cab number 1729 and remarked that the number seemed to me rather a dull one, and that I hoped it was not an unfavorable omen. "No," he replied, "it is a very interesting number; it is the smallest number expressible as the sum of two cubes in two different ways." (1940, 12)

Now suppose that Ramanujan (a mathematical genius) and I (a mathematical pedestrian) *both* have the intuition that 1729 is the smallest number expressible as the sum of two (positive) cubes in two different ways, upon considering the number for the very first time. Ramanujan has this intuition because he is a brilliant mathematician, whereas I have it because I have a capricious brain

[8] Related concerns about accidental correctness in the case of intuition, or a priori reflection, have been noted by Ayer (1946, 79), Tidman (1996, §1), BonJour (1998, 113), Kagan (2001), Linnebo (2006), Bedke (2009), Liggins (2010), and Setiya (2012, ch. 3). It should be clear that the possibility of accidentally correct intuition is neutral between various theories of what intuitions are: e.g. inclination to judge theories (Williamson 2007, 3 and 215ff.), attraction to assent theories (Sosa 2007, ch. 3), *sui generis* seemings theories (Bealer 1992, 1998; Huemer 2005, ch. 5), or quasi-perceptualist theories (Bengson, forthcoming). We need not abandon neutrality in order to ask the question, identified below, of how any such mental state could provide knowledge of abstracta, as realist rationalism maintains. (Cf. Goldman (2007, 7): "Whether intuitions are inclinations to believe, or a *sui generis* kind of seeming... it is still a puzzle why the occurrence of such a mental event should provide evidence for the composition of a Platonic form.") Of course, this does not imply that the *answer*—a specific account of successful intuition, pursued in later sections—must likewise be neutral.

lesion—or, instead, because I am so incompetent that nearly all sophisticated mathematical claims that are not obviously inconsistent strike me as true. Ramanujan's intuition may be non-accidentally correct, whereas mine is just lucky. My intuition is not able to serve as a source of knowledge about this interesting feature of the number 1729.

Some may protest that intuitions cannot be accidentally correct in this way. Presumably this would be the view of Cartesian rationalists who maintain that intuitions are infallible guarantors of knowledge. This view faces well-known objections: think, for example, of the intuition that every predicate defines a set, or the intuition that there is a set of all sets, or the intuition that there are more whole numbers than even numbers, or one of the intuitions that sustains your favorite paradox or conflicts with your favorite theory. A fallibilist view that accommodates this point might insist that, nevertheless, an intuition that has the phenomenology of "presenting as necessary"—a modally robust phenomenology that George Bealer (1992, §1; 1998, §I), in particular, has emphasized—is non-accidentally correct, if correct. This is the phenomenological correlate of David Lewis's (1986, 113) claim that attitudes about non-contingent matters are non-accidentally correct, if correct. But it is difficult to see why either claim should be granted. For example, a capricious brain lesion might cause me (a mathematical pedestrian) to suffer a series of intuitions that present as necessary forty extremely complicated mathematical propositions. It seems possible that thirty-nine of these intuitions are incorrect but, by a stroke of luck, I enjoy one intuition—perhaps the intuition regarding the non-contingent feature of the number 1729 mentioned above—that happens to be correct. In such a case, the single exception does not yield knowledge: the intuition might be correct, but only accidentally so.[9]

[9] Some theorists privilege intuition's connection to concept possession or understanding (e.g. Bealer 1998, §§3–4; Sosa 2002; Huemer 2005, §5.7; Ludwig 2007, 135ff.; Setiya 2012, §3.2), which might inspire a somewhat different fallibilist proposal: intuitions regarding propositions that one properly understands—that is, intuitions grounded in understanding—cannot be accidentally correct, if correct. However, it seems possible even for such understanding-based intuitions to be accidentally correct. Consider, for example, an aesthete whose intuitions tend to track mere elegance and beauty, not truth. The aesthete might fully understand a difficult mathematical hypothesis T (e.g. the four color theorem) whose elegance and beauty is revealed through and only through a full understanding of T. Suppose that, prompted by the aesthetic properties of T revealed through her understanding, the aesthete has the intuition that T is true. Suppose further that T is true. Although the aesthete may truly believe T on the basis of her intuition, which in turn is grounded in her understanding of T—no less here than when one's intuition is prompted by the truth of the proposition that is purportedly revealed through one's understanding—the aesthete cannot be said to know T. In such a case, it is a sheer coincidence that her aesthetically sensitive intuition is correct, so it does not yield knowledge, despite the fact that it is grounded in understanding. (Perhaps there is a natural mode of understanding—a type of "ideal alethic understanding"—that avoids this objection. But it is far from clear that there is such a mode, or if there is, that it can be identified in a way that avoids circularity or trivialization, as with, e.g., the mode *understanding-in-such-a-way-as-to-secure-non-accidental-correctness*.) A second, independent objection to understanding-based responses to worries about accidentality will be discussed in §3.2 and shown to generalize to appeals to infallibility, phenomenology, and various other mental or epistemic phenomena.

The basic point should be familiar. In order for one's mental state, whether perceptual experience or intuition, to be able to serve as a source of knowledge, *it must not be an accident that one's experience or intuition is correct.*[10] In both cases, correctness (truth, accuracy, veridicality) is not enough for success, which requires in addition that one's mental state be connected or related to the fact in question in such a way as to rule out accidentality. With this familiar point in hand, we are now in a position to formulate the powerful challenge it poses to realist rationalism.

3. THE NON-ACCIDENTAL RELATION QUESTION

What is perhaps the most forceful objection to realist rationalism concerns its apparent incapacity to render intelligible the relation between intuitions and abstract facts intuited. With successful perception, we seem to have some understanding of the relation that perceptual experiences bear to the facts perceived that could explain how those experiences are not cases of veridical hallucination—that is, how they can be non-accidentally correct, hence able to serve as sources of knowledge of those facts: namely, *some type of causal relation*. With successful intuition, we seem to lack any understanding of a relation between intuitions and the abstract facts intuited that could explain how a

[10] A comment on the relevant type of accidentality and its relation to other forms of epistemic luck. In veridical hallucination, sensory or intellectual, what is accidentally correct is a potential source of belief (a "source state"), such as a perceptual experience or intuition. Such *source accidentality* can be contrasted with *doxastic accidentality*, where what is accidentally correct is not the source state but a subsequent belief. The latter is illustrated by Goldman's (1976, 772–3) famous fake barn example, in which one believes that there is a barn present on the basis of a successful perceptual experience of a real barn in an area populated by many unperceived fake barns: here, there is a sense in which what is accidentally correct is *not* the source state but, rather, the belief (which is "defeated" by the presence of many unperceived fakes). While doxastic accidentality prevents a source state from resulting in knowledge (because of a problem downstream, in subsequent belief), there is a sense in which it leaves intact the ability of the source state to provide knowledge: one could come to know on the basis of the source state (e.g. were there no defeater for subsequent belief), even if one does not in fact do so. In cases of source accidentality, by contrast, there is a sense in which one could not come to know on the basis of the source state, since its correctness is accidental: the source state is unable to provide knowledge (regardless of whether, e.g., there is no defeater for subsequent belief) and is not merely prevented from resulting in knowledge. Throughout, our focus is the comparatively severe phenomenon of source accidentality (throughout, simply 'accidentality' or 'accidental correctness'). Such accidentality is not the same as unreliability, and its absence is not the same as reliability; it is also orthogonal to modal conditions such as safety, sensitivity, and adherence. Even if one's perceptual experiences or intuitions are reliable (tend to be correct) or satisfy such modal conditions (could not easily be mistaken or track truth), one's experience or intuition on a given occasion may still be accidentally correct, as in some cases of veridical hallucination (see, e.g., the examples in Davies 1983, §3; cf. Setiya 2012, 89–94). Hence the failure of attempts to explain the difference between successful perception and veridical hallucination in terms of reliability and modal conditions. The distinction between source and doxastic accidentality complements but is not intended to match the distinctions between "intervening" and "environmental" luck (Pritchard 2005) or "subject-directed" and "object-directed" accidental truth (Shafer 2014). For example, doxastic accidentality might be environmental and object-directed, as in the aforementioned fake barn example; or it might be non-environmental and subject-directed, as in the case of a defect in the subject's transition from a non-accidentally correct perceptual experience to subsequent belief.

thinker's intuitions are not cases of veridical intellectual hallucination—that is, how they can be non-accidentally correct, hence able to serve as sources of knowledge of abstract facts. In short, we seem to lack an answer to the following question about intuition:

> *The Non-accidental Relation Question*: What relation does a thinker's mental state—her intuition—bear to an abstract fact that explains how the state can be non-accidentally correct with respect to that fact, hence able to serve as a source of knowledge of it?

Let me make three comments about this question and the problem it poses.

3.1. *Comment 1: Benacerraf's Core Worry*

It should be plain that the problem is of a piece with Benacerraf's original worry, quoted in §1, regarding the absence of a "link," "relation," or "connection" between intuitions and abstracta that could account for how those intuitions can provide knowledge, given realism.[11] Although Benacerraf does not explicitly raise the issue of accidental correctness (at least not as such), a concern with this issue is highlighted in Hartry Field's self-described "reformulation" of Benacerraf's worry:

> The key point, I think, is that our belief in a theory should be undermined if the theory requires that it would be *a huge coincidence* if what we believed about its subject matter were correct... [Realism] postulate[s entities] that are mind-independent and bear no causal or spatiotemporal relations to us, or any other kinds of *relations* to us that would *explain* why our beliefs about them tend to be correct; it seems hard to give any account... that doesn't make the correctness of the beliefs *a huge coincidence*. (2005, 77, emphasis added)

What is needed to address the "key point," Field tells us, is a theory that identifies a "relation" between thinkers and abstracta "that would explain why" the correctness of a thinker's mental states about those abstracta is not "a huge coincidence." This is precisely what is required to answer the non-accidental relation question.[12]

Benacerraf's worry is sometimes interpreted or formulated in different terms. Field has pressed the point about coincidence, or accidentality, in a

[11] Other expressions can be found throughout the literature. For example, Fine (2001, 14, emphasis added) stresses "the problem of explaining how we can be in *appropriate contact* with an external realm of mathematical facts." It is safe to assume that *contact* is a type of relation, and that it is *appropriate* when it meets a further condition, viz. being such as to secure non-accidental correctness.

[12] That the non-accidental relation question captures the "key point" can be brought out further by considering Field's primary analogy, involving a subject with largely correct attitudes about the daily happenings in a remote Nepalese village (see Field 1989, 26–7). The unanswered question that generates the basic problem in this case—*what relation do the subject's mental states bear to the remote Nepalese village that explains how those mental states can be non-accidentally correct with respect to, hence able to serve as sources of knowledge of, the daily happenings in the remote Nepalese village?*—appears to be the precise analogue of the non-accidental relation question.

way that focuses on reliability and correlation, and I will critically examine this aspect of his reformulation in a moment (in §3.3). Famously, Benacerraf himself chose to develop his worry using a causal constraint on knowledge, according to which "for *X* to know that *S* is true requires some causal relation to obtain between *X* and the referents of the names, predicates, and quantifiers of *S*" (1973, 671). However, as many commentators have stressed (see esp. Hart 1977, 125–6 and Field 2005, 77), and as Benacerraf's original presentation indicates (1973, 667–8), the core of the worry is independent of any such constraint—which, perhaps unsurprisingly, has been accused of "begging the question" (Rosen 2001, 71). The present discussion goes beyond this familiar negative point by offering a positive, non-causal formulation of Benacerraf's core worry, in terms of the non-accidental relation question. That this formulation of the basic challenge does not rely on a causal constraint on knowledge, or any other such question-begging condition, might seem a narrow point so long as causation is the only (putative) candidate for a "non-accidental relation" between thinkers and mind-independent facts. But I will argue that there is an alternative, and I will explore its prospects as a response to the challenge.

3.2. *Comment 2: Constraint on an Adequate Answer*

Part of what makes the basic challenge—expressed by the non-accidental relation question—so difficult is that an adequate, non-mystical solution must identify a non-causal, explanatory relation, holding between a thinker and abstracta, that can be conceived in *non-epistemic, non-psychological* terms.

This constraint is operative in Benacerraf's (and Field's) discussion, and I believe it is fair. We have not yet resolved the problem, or answered the "key point," if we allege to explain fully how an intuition (a psychological phenomenon) can be non-accidentally correct, hence able to yield knowledge (an epistemic phenomenon), simply by citing more psychological or epistemic phenomena (e.g. attention, intelligence, understanding, evidence, double-checking, or following reliable "rules of inference" such as *believe that a is F when it is the case that a is G and a is H and if something is both G and H then it is F*), which seem to push the question back by invoking phenomena that raise the non-accidental relation question just as urgently as the particular instance of intuitive knowledge does. Compare: in the face of the possibility of veridical hallucination, it does not suffice to explain how a perceptual experience (a psychological phenomenon) can be non-accidentally correct, hence able to yield knowledge (an epistemic phenomenon), to simply cite more psychological or epistemic phenomena (e.g. attention, concentration, recognition, evidence, double-checking, or following reliable "rules of inference" such as *believe that there is something circular when it is the case that there is something circular in front of you*). Hence philosophers of perception have, rightly, felt the need to invoke a non-psychological, non-epistemic relation—normally, a type of causal relation—when theorizing about successful perception. A non-psychological, non-epistemic relation is similarly needed for an adequate theory of successful intuition.

Ernest Sosa shows recognition of this point when, at the end of a sustained discussion of the prospects of a response to Benacerraf's worry centered on understanding, he writes:

> Learning the tables by rote derives from drills that give you both understanding and belief in a single package. Through the numerals and the tables we gain access to the numbers and the cardinality properties and basic truths of elementary arithmetic. *But how can we bear even so much as this relation of understanding-plus-belief to a set of facts so far removed from us? This question does remain* ...[13] (2002, 383, emphasis added)

To invoke "understanding-plus-belief" (viz. belief grounded in understanding) is not yet to locate a non-psychological, non-epistemic relation, and thus it is not yet to address the non-accidental relation question.[14] This objection generalizes to any response to Benacerraf-style worries comprised by appeal to psychological or epistemic phenomena: for example, *self-evidence* (Audi 1999); *acquaintance* and *description* (Giaquinto 2001); *conscious directedness* (Tieszen 2002); *reflection, reasoning*, and *calculation* (Cassam 2007, ch. 6); *mental dispositions* (Wedgewood 2007, ch. 10); *reliable rules of inference* (Schechter 2010, 441–5); or *safety* and *sensitivity* (Clarke-Doane forthcoming). The problem is that such appeals fail to locate a non-psychological, non-epistemic explanatory relation that thinkers could bear to the abstract facts known, without which allegations of knowledge of abstracta look brute or mysterious.[15]

The indicated constraint is often overlooked or disregarded without argument. But it is an important element of the basic problem. While I will not try to prove that there is no room to oppose it, I believe it is worth pursuing an approach, ambitious as it may seem, which accepts the constraint and proceeds to locate an answer to the non-accidental relation question that does not simply appeal to further psychological or epistemic phenomena. At any rate, this is what I propose to do in what follows.

[13] Sosa's reply to this lacuna seems to be that here we may rest content knowing that we are partners in crime; the passage continues, "but is it *distinctively* a problem for [realism]?"

[14] Note 9 raised an initial worry that appeal to understanding does not secure non-accidental correctness. The present point is that, even if this initial worry were resolved, such an appeal would still not yet adequately answer the non-accidental relation question, hence Benacerraf's core worry. For the same reason, it will not do simply to cite elements of human nature (Setiya 2012, ch. 4) or social-historical phenomena (e.g. schooling, rote learning, interaction with experts), which tacitly invoke further psychological or epistemic phenomena. This yields a diagnosis of at least part of what is wrong with the "boring explanations" criticized by Linnebo (2006) and mentioned below.

[15] It should be clear that the basic problem is also not solved by simply opting for an epistemology of abstracta that favors: *abduction* (Quine 1960, Putnam 1971); *conceivability* (Yablo 1993, Chalmers 2002); *epistemic analyticity* (Boghossian 1996); *implicit definition* (Hale and Wright 2000); *conceptions* (Peacocke 2004, ch. 6); *postulation* (Fine 2005); or *imaginative simulation* (Williamson 2007, ch. 5). Such views fail to identify a non-psychological, non-epistemic relation between thinkers and abstracta, leaving Benacerraf's question regarding a "link" unaddressed. In this respect they are comparable to Plato's appeal to *anamnesis*, discussed by Benacerraf (1973, 675).

3.3. *Comment 3: Relation to other Explanatory Challenges*

As foreshadowed in §1, there are several questions that one can ask—and that have been asked—about intuition of abstracta. For example:

> *The Reliability Question:* How is it that a thinker's mental states—her intuitions—are non-accidentally correct with respect to abstract facts *sufficiently more often than not?*[16]
>
> *The Etiological Question*: How is it that a thinker *comes to have* a mental state—an intuition—that is non-accidentally correct with respect to an abstract fact?[17]

These questions can be combined; they can also be formulated for individual thinkers and populations of thinkers. For instance, one natural combination yields a population-level *etiological question about reliability*:

> *The Etiological Question about Reliability (population-level)*: How is it that we human beings came to have mental states—intuitions—that are non-accidentally correct sufficiently more often than not?[18]

However, while such questions regarding reliability and etiology are important, they are posterior to the non-accidental relation question in the following sense. Without any understanding of how a *given* intuition can successfully connect to the fact intuited, as required to answer the non-accidental relation question, it is impossible to make sense of the *recurrent* success of states of that type, as required to answer the reliability question, or of the *genealogy* of such successful states, as required to answer the etiological question(s). That is, we must first render intelligible how a thinker (or her mental state) can be appropriately related—or, to use Benacerraf's term, "linked"—in one case if we are to explain her being so linked *sufficiently often enough* to qualify as reliable, or to explain her *coming to have* a mental state so linked. In this sense, answers to the reliability and etiological questions, regarding the recurrence and genealogy of the link, already presuppose some answer to the non-accidental relation question, regarding the link itself. (We will return to these issues in §8.)

[16] Cf. Field (1989, 26 and 230–2; 2005, 278), who speaks of reliability, correlation, and regularity.

[17] Compare Sidgwick's (1907, 211) "psychogonical" question regarding the origin of moral intuition. Street (2006) has developed a specific etiological challenge (sometimes referred to as "genealogical critique"), centering on evolutionary considerations, to the thesis that values are mind-independent.

[18] This resembles what Schechter (2010, 444) names "The Etiological Question," which he privileges vis-à-vis various other "reliability challenges." Schechter does not articulate what I have called the non-accidental relation question, which I will now argue is the basic problem in the vicinity. Of course, none of this is meant to imply that the questions I have formulated are the only questions. For example, Chudnoff (2013) queries the *existence* and *ground* of awareness of abstract individuals (not facts, and not non-accidental relations thereto: see 710–11).

3.4. *A Lacuna in Realist Rationalism*

I believe that the foregoing considerations highlight the importance, as well as the legitimacy, of the demand that realist rationalists provide an answer to the non-accidental relation question. Put dramatically: at issue is the very possibility of successful intuition and hence intuitive knowledge of abstracta. Yet, no proponent of realist rationalism has yet ventured a plausible, non-mystical answer.

I have already mentioned (in §3.2) some proposals that fail to locate a non-psychological, non-epistemic relation between thinkers and abstracta. Other proposals perhaps manage to do this, but they fail to explain the relevant type of non-accidental correctness. To illustrate briefly, consider a plenitudinous or "full-blown" version of realism (see, e.g., Linsky and Zalta 1995; Balaguer 1998), which holds that there are abstract entities of all possible kinds. While such a view might manage to explain the correctness of our mental states, including our intuitions, by implying that these states could not have failed to be correct, it does not address a crucial epistemic division within the class of correct states: some are non-accidentally correct (like Ramanujan's intuition about 1729, in §2), others are accidentally correct (like mine about 1729), and the former remain unexplained. Consequently, a plenitudinous metaphysics, insofar as it treats all correct states on par, does not by itself provide a satisfactory answer to the non-accidental relation question. A similar limitation afflicts other popular attempts to engage Benacerraf's worry. Indeed, most responses, including partner-in-crime replies that appeal to the causal inertness of the propositional contents of beliefs and other attitudes (see, e.g., Plantinga 1993, 113ff.; cf. Burge 1990, 635n6), as well as those standard, deflationary replies that flat-footedly reject a causal constraint on knowledge, perhaps in favor of what Øystein Linnebo (2006) aptly calls a "boring explanation" of knowledge of abstract facts in terms of subjects' backgrounds and educational histories (see, e.g., Burgess and Rosen 1997, §I.A.2.c), or in quietistic fashion supply no explanation whatsoever (Katz 1998, ch. 2; cf. Lewis 1986, §2.4 and Pust 2004), simply do not venture an answer to the non-accidental relation question.[19] None of these approaches addresses, let alone solves, the basic problem—which, as the foregoing indicates, is both genuine, arising out of reflection on a simple contrast between accidentally correct and non-accidentally correct conscious mental states (recall §2), and exceedingly difficult, unyielding to familiar forms of explanation.

[19] BonJour (1998, §6.7) has, by contrast, proposed a substantive answer in terms of the *instantiation* of abstracta by a thinker's mental states. However, as Boghossian (2001, 636–7) has observed, this proposal is not acceptable. Given that thoughts do not literally *instantiate* the abstracta they are about (a thought about the color red is not itself red, i.e. it is not an instance of red), it is unclear what the proposal is. Even waiving that problem, it is difficult to see how the proposal is meant to provide the requisite explanation: correct intuitions as well as mistaken intuitions (what BonJour refers to as "apparent rational insights"), which *cannot* yield knowledge, are said by BonJour to equally instantiate the abstracta they are about.

It is widely thought that extant discussions, including those mentioned in the preceding paragraph and those described in §3.2, do not provide satisfactory replies to the challenge raised by Benacerraf. The present discussion explains this verdict: simply put, those responses do not adequately address the non-accidental relation question. Indeed, it is fair to say that a workable answer to this question remains to be found. The rest of this paper strives to fill this lacuna in the realist rationalist position by developing such an answer.

4. A PECULIAR CASE OF NON-CAUSAL KNOWLEDGE

Having clarified the question, we are now ready to try to answer it. We will proceed gradually and somewhat indirectly. The first step, taken in this section, will be to show that an analogous question arises for a certain type of experience-based knowledge without causation. This may seem like a detour. But it will point the way towards an explanatory framework, developed in the next section, which enjoys substantial precedent while providing the resources needed to answer Benacerraf's challenge.

Standard examples of knowledge without causation, such as knowledge about the future, might perhaps be explained by citing principles of inference that somehow extend antecedent knowledge (e.g. of various lawlike regularities), which is itself ultimately explained by citing causal relations to the facts antecedently known. However, some instances of knowledge without causation seem to involve knowledge that is not a mere extension of such antecedent knowledge; consequently, it cannot be explained in the same way.

It will be useful to think through an example. Consider:

Trip has never before encountered the colors red, orange, or blue. Nor has he ever encountered any elliptical, circular, or hexagonal shapes. Then, one evening, Trip has an experience with the phenomenal character of the experience had when viewing a red ellipse labeled 'I,' an orange circle labeled 'II,' and a blue hexagon labeled 'III.'

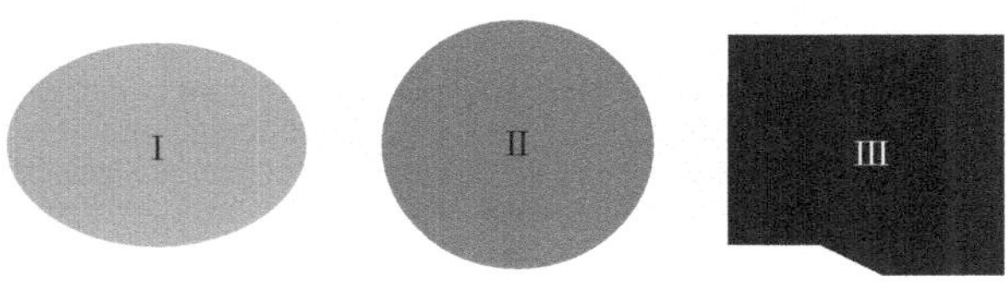

As it happens, Trip is not actually viewing these things: rather, he is unwittingly the subject of a spontaneous, vivid, hallucinatory experience. On the basis of this experience, Trip—a smart, attentive fellow—comes to believe the following:

[α] The color of I resembles the color of II more than the color of III.

[β] The shape of I resembles the shape of II more than the shape of III.

The question on which I would like to focus concerns the epistemic status of these attitudes.[20]

Notice, to begin, it is extremely plausible to think that Trip has, or could have, knowledge of α and β. After all, his beliefs are true, and his experience gives him excellent reason to believe α and β. Of course, Trip could have acquired such knowledge even if he had been enjoying a successful perceptual experience with the same phenomenal character, that is, even if he had been successfully perceiving objects in his environment that possess the relevant colors and shapes, rather than suffering a vivid hallucination. The lesson is not that hallucination provides knowledge that successful perception cannot, but rather that hallucination can provide knowledge: for instance, knowledge of α and β.[21]

Now suppose we embrace a widely accepted realist view of colors and shapes, according to which colors and shapes are mind-independent entities that comprise mind-independent facts concerning such entities. What is then needed is a relation between Trip's hallucinatory experience and the relevant facts about colors and shapes that explains how his experience can be non-accidentally correct with respect to those facts, hence able to provide Trip with knowledge of α and β. Here we encounter the following explanatory question about hallucination:

> *The Non-accidental Relation Question about Hallucination:* What relation does Trip's hallucinatory experience bear to the relevant facts about colors and shapes that explains how his experience can be non-accidentally correct with respect to those facts, hence able to serve as a source of knowledge of them?

Answering this question can be regarded as the problem of hallucinatory knowledge.

[20] Similar cases have been discussed in another context by Johnston (2004) and Pautz (2007). It is not clear how best to articulate the relevant truths. One might worry that the formulation in the text suffers from reference-failure, given the plausible supposition that hallucination does not involve a relation to a mental or non-mental individual (e.g. Pricean sense datum or Meinongian non-existent object). Hawthorne and Kovakovitch (2006, 158) demonstrate that worries about reference-failure in hallucination can be accommodated by focusing on the *properties* that hallucinatory experience presents: redness, circularity, and so forth. (Perhaps they are also "topic-neutral.") α and β could then be reformulated as follows: *red resembles orange more than blue* and *ellipticality resembles circularity more than hexagonality*, respectively. Such resemblance may be understood in terms of proximity on a spectrum (see Jackson's suggestion in the next note), co-determination of a common determinable (or the absence thereof), or in some other way. For example, Byrne (2003, §7) proposes to articulate such truths in terms of magnitudes, defined as sets of particular color or shape properties together with a ratio scale.

[21] This is compatible with representational and non-representational theories of experience. For instance, one might hold that the experience that is the basis of Trip's beliefs is best understood as involving a sensory representing relation to a propositional content concerning the relevant resemblance relations. See, e.g., Jackson's (1998, 111) suggestion that "color experience . . . represents [color] properties as occupying certain places in the three-dimensional color array (red is opposite green, orange is nearer red than green, etc.)." See also Johnston (1997, 173), Tye (2000, 164–5), Byrne (2003, §7), and Pautz (2007, 508ff.).

Let us pause to highlight why the problem is a problem. Recall that Trip is hallucinating: there need not be anything with, or any instances of, those colors and shapes causing his experience. (Compare: a hallucinatory experience as of there being a pink, mouse-shaped item present does not require that one's experience be caused by anything with, or any instances of, the indicated color or shape.) So, it is possible that Trip's experience does not stand in any causal relation to the relevant colors and shapes—though it is, nevertheless, able to provide knowledge about them.

One might protest that the case of Trip is impossible. However, it is not clear what could justify this response. The case seems internally coherent, and it is not obviously incompatible with either common sense or "received" theories, such as physicalist or materialist views of the metaphysics of mind. As Michael Tye observes, such views can (and should) allow that, "given the right causal proximal stimulations, a brain that grows in a vat—a brain that is never properly embodied—has perceptual experiences of features to which it bears no causal connections" (2000, 64). Modulo the vat, this is the situation in the case at hand: Trip is enjoying an experience of features to which his experience bears no causal connections.[22]

To be sure, although Trip's experience does not stand in a causal relation to the relevant colors and shapes, there may still be some proximal cause,[23] and thus some causal explanation, of Trip's experience. In addition, Trip's experience may stand in a causal relation to his belief. I do not wish to dispute these observations. On the contrary, I propose to accept them. The point is simply that, compatibly with those observations, it is possible that the colors and shapes do not cause Trip's experience, so causation cannot be the relation that his experience bears to the colors and shapes that explains how his experience can serve as a source of knowledge about them.

Herein lies the difficulty. Given the absence of a causal relation between Trip's experience and that which it is about, what is needed is a *non-causal relation* that explains how Trip's experience can serve as a source of knowledge. What could this relation be?[24]

[22] Some disjunctivists will object to Tye's use of "perceptual" to characterize such experience; fortunately, the adjective is inessential for present purposes. So, too, is Trip's vat-less status: we could have placed Trip in a vat; or he could be a corporeal, ancestor-less victim of an evil demon or malicious scientist (these possibilities, too, are compatible with physicalism). Or perhaps he is simply congenitally blind. Research in cognitive psychology concerning the dreams of congenitally blind individuals has been said to indicate that visualization or imaging of colors is possible without prior successful perception of anything with, or any instances of, the visualized or imaged colors (see, e.g., Goldstein 2009, 893).

[23] Recall Tye's "causal proximal stimulations."

[24] Some may wish to appeal to reliability. But, first, as explained in note 10, reliability does not suffice for non-accidental correctness. Second, as explained in §3.2, what is needed is a non-psychological, non-epistemic explanatory *relation* that a subject could bear to mind-independent colors in the absence of a causal relation. Put differently, insofar as reliability is a feature of a relation, it simply pushes the question back: what is the non-causal relation between Trip's experience and the relevant facts? Obviously, "some reliable relation" will not do.

Let me offer a preview of what is to come. I will suggest that we need the notion of a mental state that is not causally, but is rather *constitutively*, related to what it is about to make sense of how Trip's experience is able to serve as a source of knowledge about the colors and shapes. It is not that the colors and shapes are constituted by Trip's experience, but the other way around: Trip's experience constitutively depends on the (mind-independent) colors and shapes, and that is why it can serve as a source of knowledge about them. Then, after observing the role of constitutive dependence in a well-known theory of perceptual knowledge, I will return to our original explanatory question in order to explore how this approach might help with intuitive knowledge as well. But first we will need to say more about constitutive dependence and its role in explanation.

5. CONSTITUTION AND CONSTITUTIVE EXPLANATION

A plausible constraint on explanation in general, which I propose to accept, is that the explanandum must bear an asymmetric dependence relation to (i.e. asymmetrically depend on) the explanans. The exact asymmetric dependence relation, as well as the modal force of the dependence (nomological, metaphysical, etc.), may differ in different cases. In putative causal explanations, the relation will be one of asymmetric *causal* dependence. In putative non-causal explanations, such as (a)–(d) below, the relation will be one of asymmetric *non-causal* dependence.[25]

a. The planetary orbits are stable because space-time is four-dimensional.
b. The vase is fragile because it is constituted by a piece of glass.
c. Torture is wrong because it does not maximize utility.
d. The set {Obama, Biden}exists because Obama exists and Biden exists.

Let us focus on the type of non-causal explanation found in (b), which invokes the traditional metaphysical notion of constitution (constitutive dependence). There are two main questions to ask. First, what is constitution? Second, how does it yield explanation? We will consider each question in turn.

The notion of constitution has applications in a wide variety of areas (see, e.g., Johnston 2005). Consider some potentially familiar—even if not uncontroversial—examples of constitution, material and otherwise:

e. The vase is constituted by the piece of glass.
f. {Obama, Biden} is constituted by Obama and Biden.[26]
g. The event of Derek running is constituted by Derek, the property of running, and a time (on a Kimian theory of events; see Kim 1973).

[25] For useful discussion of non-causal explanation, see especially Kim (1974) and Ruben (1990, ch. 7) and the citations therein.

[26] Currie (1982, 69, emphasis added) summarizes Frege's view as follows: "Sets are *constituted* by their members, while classes are not."

h. The speech act of assertion is constituted by the norm *assert only what you know* (on a Williamsonian view of assertion).[27]

While there are many differences between them, all of (e)–(h) can be understood as exemplifying in one way or another the non-causal dependence of one entity—the *constituted* entity—upon another entity—the *constituting* entity.

This is no place to try to engage skepticism regarding constitution. Nor is it the place to attempt a comprehensive account or reductive analysis of this rich and intricate phenomenon. Nevertheless, it is natural—and, given what is to come, important—to say something about the character of this relation.

Constitution, as I will understand it, is distinct from mereological (part-whole) composition, containment, supervenience, and identity. Take, for example, (e): the vase is constituted by the piece of glass. But the piece of glass is not one of the vase's *parts*: it is not a part that the vase has. Nor does the vase *contain* the piece of glass. Likewise, the vase cannot be said to (merely) *supervene on*—modally covary with—the piece of glass. Nor are the vase and piece of glass *identical*. Rather, the vase *constitutively depends* on the piece of glass.

Positively, such dependence has several interesting features. It is *irreflexive*: the vase is constituted by the piece of glass; but the vase is not constituted by itself. And it is *asymmetric*: the vase is constituted by the piece of glass; but the piece of glass is not constituted by the vase. However, it is *non-causal*: the vase is constituted by the piece of glass; but the vase is not caused by the piece of glass. At the same time, constitutive dependence is not strictly logical (or merely "formal"), but *ontological*, insofar as it concerns the *being* or *existence* of the entity in question: what it is for the entity to be or exist.[28]

In general, as I will understand the notion, to specify the constitution of a given entity is to say *what it is*, or *part of what it is*, for that entity to be or exist:

> *a* is (partly) constituted by *b* iff (part of) what it is for *a* to exist is for *b* to exist.

For example:

e′. *What it is* for the vase to exist is for there to be an object that is thus-and-so (in the example above, the piece of glass).

f′. *What it is* for the set {Obama, Biden} to exist is for Obama to exist and Biden to exist.

[27] Williamson (2000, 238, emphasis added): "[T]he speech act [of assertion], like a game and unlike the act of jumping, is *constituted* by rules." Those skeptical of Williamson's claim may replace (h) with another constitution claim, perhaps one regarding the game of chess: The game of chess is constituted by the norm (or rule) *the bishop moves any number of vacant squares in any and only a diagonal direction*. See also the gloss below.

[28] This observation might be traced, if not wholly credited, to Aristotle. Cf. Fine (1994) and, e.g., Koslicki (2004, 340).

g′. *What it is* for the event of Derek running to exist is for Derek to instantiate the property of running at a particular time.
h′. *What it is* for an assertion to exist is for there to be an illocutionary act and a norm (namely, the norm *assert only what you know*), and for the norm to govern the act.[29]

In each case, the constituted entity stands in a *non-causal, irreflexive, asymmetric ontological dependence relation*—the constitution relation—to the constituting entity.

These four expedients—examples (e)–(h), contrasts with nearby metaphysical relations, specification of features, and general characterization—serve jointly to elucidate the notion of constitution, thus answering our first question about the constitution relation. Let us turn now to our second question, concerning the role that this relation may play in a type of non-causal explanation. For example, recall (b) above, in which the fragility of the vase is *explained* in terms of the fragility of the piece of glass of which the vase is *constituted*. What is the "mechanism" underwriting such constitutive explanation?

Notice that, in (b), the property (being fragile) possessed by the subject in the explanandum (the vase) is one and the same as the relevant property (being fragile) possessed by the subject in the explanans (the piece of glass). Call this *property inheritance*: the vase *inherits* the property of being fragile from the piece of glass, of which it is constituted. It is this that appears to make (b) a successful constitutive explanation.

In one sense, this appearance is correct; but there is another sense in which it is not. To see why, consider that property inheritance is a special case of a more general phenomenon, which I will call *property ensurance* (or simply "ensurance").

> *Property Ensurance*: a's having F **ensures** b's having $G =_{def}$: (i) a constitutes b, and (ii) necessarily, if a is F and a constitutes b, then thereby b is G,

where "thereby" indicates that the consequent of the conditional holds in virtue of the antecedent.[30]

Inheritance occurs when $F = G$. But ensurance does not require inheritance. Perhaps the simplest example occurs when an entity b is constituted by an entity a, F is the property of constituting b, and G is the property of being constituted by a; since a's being F then ensures b's being G even though F and

[29] Those wishing to focus on chess (recall note 27) may replace (h′) with the following: *What it is* for a game of chess to exist is for there to be a series of events involving a board and a set of pieces (or their equivalent), including a bishop, and a set of norms (including the norm *the bishop moves any number of vacant squares in any and only a diagonal direction*), and for the set of norms to govern the events.

[30] More formally: Fa **ensures** $Gb =_{def}$: $aCb \;\&\; \Box\,((Fa \;\&\; aCb) \Rightarrow Gb)$, where 'C' designates the constitution relation and '⇒' designates that what follows it holds in virtue of—is grounded in—what precedes it. For an initial treatment of the logic of ground, see Fine (2011).

G are distinct, the result is ensurance without inheritance. (It should be clear why inheritance is the wrong model for such a case: given the irreflexivity of constitution, in such a case it is impossible for *a* to be *G* and it is impossible for *b* to be *F*. Yet it is clear that in such a case *b* could not fail to be *G* given that *a* is *F*: thus, *a*'s being *F* guarantees *b*'s being *G*. This guarantee is property ensurance.)

While ensurance is a broader or more general notion than inheritance, it is nevertheless *selective* in the sense that it holds only in certain cases. To illustrate, let *F* be the property of being fragile and *G* be the property of being priceless. Since the piece of glass might have *F* and constitute the vase although the vase lacks *G*, the piece of glass's having *F* does not ensure the vase's having *G*—even if, as it happens, the vase does have *G*. In short: the piece of glass's fragility does not ensure the vase's pricelessness.

The notion of ensurance picks out a qualified modal connection that is important insofar as it may underwrite a general theory of constitutive explanation. A hypothesis is this:

> *Hypothesis*: Whenever *a*'s having *F* ensures *b*'s having *G*, the fact that *b* is *G* is explained by citing its constitution: b *is* G *because* b *is constituted by* a (*which is* F).

While various refinements might be pursued, the central idea that this hypothesis seeks to express is that it is sometimes possible to explain why an entity has a certain property by citing its constitution, insofar as its constitution ensures that it has that property. The hypothesis articulates a valuable schema with a host of concrete applications. Here are a few illustrations.

First, recall (b): if the vase is constituted by the piece of glass, then we can explain the fragility of the vase by citing a certain fact about its constitution, namely, *the vase is fragile because the vase is constituted by a piece of glass (which is fragile)*. Here, the fragility of the piece of glass (*F*) *ensures* the fragility of the vase (*G*). This enables the fragility of the piece of glass to *explain* the fragility of the vase. By contrast, as we saw above, the fragility of the piece of glass does *not* ensure the pricelessness of the vase: there is no such qualified modal connection between a piece of glass being fragile and a vase being priceless. Hence, in this case, the relevant ensurance claim is false. So the fragility of the piece of glass does *not* explain the pricelessness of the vase. This is, I assume, the correct verdict—one that is correctly predicted by the hypothesis above.

Second, recall the example in which *b* is constituted by *a*, *F* is the property of constituting *b*, and *G* is the property of being constituted by *a*. In such a case, *a*'s being *F* ensures *b*'s being *G*. This allows us to explain why *b* has *G* by citing a certain fact about its constitution, namely, *b has G because b is constituted by a.* This is perhaps the simplest case of constitutive explanation in the absence of inheritance. And it illustrates why the mechanism underwriting constitutive explanation must be ensurance rather than the narrower inheritance, as in the hypothesis above.

As a third example, which also illustrates the potential philosophical significance of the hypothesis, recall the (Williamsonian) constitution claim in (h). Suppose, for the purpose of illustration, that this constitution claim is true. It might then be used to answer the question: Why are false assertions inappropriate? The explanation might proceed as follows: *a false assertion is inappropriate* *because assertion is a speech act constituted by the norm* assert only what you know.[31] Here, the norm *assert only what you know* has the property (*F*) of being satisfied only under certain conditions, e.g. when the assertion is true. This *ensures* that assertion, which is a speech act constituted by this norm, has the property (*G*) of being inappropriate when false. As predicted by the hypothesis above, this ensurance claim, together with the constitution claim in (h), delivers a constitutive explanation of the explanandum.[32]

6. EXPLAINING KNOWLEDGE

The preceding section located a non-causal relation—constitution—that allows a type of non-causal explanation, namely, *constitutive* explanation. It also identified a mechanism underwriting such explanation, a qualified modal connection that I labeled "property ensurance," and described how it works in a few putative examples. This section shows that the resulting explanatory framework may have wide application in epistemology.

6.1. *Explaining Knowledge via Hallucination*

Recalling the case of Trip (from §4), suppose that the following constitution claim is true:

i. Trip's hallucinatory experience is constituted by the relevant colors and shapes.

That is, part of *what it is* for Trip's hallucinatory experience to exist is for those very colors and shapes, replete with certain resemblance relations that hold between them, to exist. This constitution claim, which seems to enjoy substantial prima facie plausibility,[33] might be used to explain the ability of Trip's hallucinatory experience to serve as a source of knowledge about those

[31] This is basically Williamson's explanation (2000, 249ff.); he suggests that it is a significant virtue of his constitutionalist view of assertion that it enables such an explanation. However, Williamson himself does not attempt to identify the mechanism underwriting such explanation, as I have done (through the notion of ensurance).

[32] The present treatment could instead be applied to the case of chess (recall notes 27 and 29). Suppose that White attempts to move her bishop horizontally. We may explain the impermissibility of her attempt as follows: *moving the bishop horizontally in the course of a game of chess is impermissible* *because chess is a game constituted by the norm (or rule)* the bishop moves any number of vacant squares in any and only a diagonal direction. As we say, that is simply part of *what it is* to play chess.

[33] For instance, it is arguably part of "naïve common sense" that red is "a *constituent* of experience when one hallucinates [something] red" (Hawthorne and Kovakovitch 2006, 178, emphasis added). It is also part of some popular contemporary theories of consciousness: Tye (2000, 48, emphasis in original) writes, "qualities [i.e. the qualities represented by

colors and shapes. The explanation would proceed as follows: *Trip's hallucinatory experience is non-accidentally correct, hence able to serve as a source of knowledge about the relevant colors and shapes, because it is partly constituted by those colors and shapes.*

The explanation is made possible by ensurance. The colors and shapes have the property (*F*) of standing in certain resemblance relations. This *ensures* that Trip's hallucinatory experience, which is constituted by those colors and shapes, has the property (*G*) of being non-accidentally correct with respect to those resemblance relations, hence able to serve as a source of knowledge about them. In other words, Trip's hallucinatory experience has a certain constitution: part of *what it is* for this mental state to exist is for the relevant colors and shapes, replete with certain resemblance relations that hold between them, to exist. Accordingly, this mental state will have a certain property, namely, the property of being non-accidentally correct with respect to, hence able to provide knowledge about, the resemblance relations that hold between those colors and shapes.[34]

One virtue of this constitutive explanation is that it seems to pinpoint what makes it the case that Trip is in a position to know what he does. To see this, it is useful to contrast the case of Trip with the following example, involving Lucky:

Lucky receives an anonymous email with the following text: "After viewing the three colored shapes in the attached document, determine whether the color of I more closely resembles the color of II or the color of III, and whether the shape of I more closely resembles the shape of II or the shape of III." Lucky attempts to open the attached document but is unable to do so: every time he tries, an error message appears. The attached document, which he fails to open, includes a red ellipse (labeled 'I'), an orange circle (labeled 'II'), and a blue hexagon (labeled 'III'). Although Lucky is unable to open the document, and thus cannot view the colored shapes in question, he decides to go ahead and guess the answers. He settles on α and β:

[α] The color of I resembles the color of II more than the color of III.
[β] The shape of I resembles the shape of II more than the shape of III.

Lucky's guess is correct, but only accidentally so; hence, Lucky does not know α and β. Consider the following explanation of why Lucky's guess does not provide knowledge: *Lucky's guess is not able to serve as a source of knowledge about the relevant colors and shapes because it is not constituted by the relevant colors and shapes.* To say that Lucky's guess is not constituted by the relevant colors and shapes is to say that it is not the case that part of *what it is* for the guess to exist

an experience] at least partly *constitute* phenomenal character," including the phenomenal character of hallucination (cf. Tye 2009, 82–3). See also Johnston (2004).

[34] As indicated in note 10, the focal point is source accidentality: to the extent that Trip's hallucinatory experience is constituted by the relevant colors and shapes, it is plausible that Trip's experience could not fail to be *able* to serve as a source of knowledge about those colors and shapes, even if it does not in fact yield knowledge on a particular occasion (perhaps, e.g., because of doxastic accidentality).

is for those colors and shapes to exist. Consequently, Lucky's mental state—his guess—does not have the property of being able to provide knowledge about those colors and shapes. This is the desired result. In this way, constitution may serve as the relation that (constitutively) explains the difference between Trip's *knowledge* about the colors and shapes and Lucky's *mere opinion* about them.

The next two subsections consider two further applications of this explanatory strategy. One is our original problem case: intuitive knowledge of abstract facts. Before, we will first consider another case, involving ordinary perceptual knowledge of empirical facts.

6.2. *Explaining Perceptual Knowledge of Empirical Facts*

The following question confronts us once we accept a broadly realist view of the empirical world, according to which what are known through perception are facts about mind-independent concrete entities ("empirical facts"):

> *The Non-accidental Relation Question about Perception*: What relation does a perceiver's perceptual experience bear to an empirical fact that explains how the perceptual experience is non-accidentally correct with respect to that fact, hence able to serve as a source of knowledge of it?

Answering this question can be regarded as the problem of perceptual knowledge.[35]

There have arisen a variety of theses about perception that aim to "bridge the chasm" between the mental states of subjects, their perceptual experiences, and the external world. One of these invites us to go naïve:

> *Naïve Realism about Perception*: Those mind-independent items that are successfully perceived partly constitute one's perception.

On this increasingly popular view, successful perceptual experiences (e.g. states of seeing) are intentional states partly constituted by the mind-independent items (individuals, properties, events, facts, etc.) that they are about. Michael Martin summarizes: "Some of the objects of perception—the concrete individuals, their properties, the events these partake in—are *constituents* of the experience. No experience like this ... could have occurred had no appropriate candidate for awareness existed"[36] (2004, 273, emphasis added).

[35] This problem has perplexed most of the figures in the canon, realists and non-realists alike. For instance, on one prominent reading, Kant thought that realism cannot explain the possibility of perceptual experience of empirical facts non-dogmatically; his solution was to reject realism in favor of transcendental idealism.

[36] This counterfactual should, I think, be read as partly elucidating, rather than fully analyzing, the constitution claim in the previous sentence. Besides Martin, those sympathetic to naïve realism include McDowell (1982), Putnam (1994), Campbell (2002), Johnston (2004, 229), Snowdon (2005, 136–7), and Hellie (2007). The position is not new, however; for example, the early twentieth-century Oxonian Cook Wilson (1926, 70, emphasis added) wrote, "what we apprehend ... is *included* in the apprehension as a part of the activity or reality of apprehending."

On this view, when one successfully perceives the fact that, for example, there is a red apple present, that very fact partly constitutes the mental state one is in. The idea is not simply that successful perceptual experience (e.g. seeing) is "direct," an unmediated relation between the perceiver and the external world, but that successful perceptual experience has a chunk of the external world as a constituent. In perception, one quite literally has the world in mind.

Of course, accidentally correct perceptual experiences, in which one merely seems to see, and non-accidentally correct perceptual experiences, in which one sees, may be subjectively indistinguishable: in some cases, they cannot be told apart from the inside. But it does not follow that they are one and the same mental state. According to naïve realism, the crucial difference is that the presence of *this kind of mental state*—demonstrating a non-accidentally correct perceptual experience as if p (e.g. a mental state of seeing)—constitutively depends on the fact that p; not so for unsuccessful perceptual experience: that is, part of *what it is* for a successful perceiver's perceptual experience as if p to exist is for it to be a fact that p; but it is not the case that part of *what it is* for a capriciously lesioned (or, say, envatted) subject's perceptual experience as if p to exist is for it to be a fact that p. So, although the successful and unsuccessful states may be subjectively indistinguishable, they are distinct kinds of mental state, for they are differently constituted.[37]

It may be helpful to view this as a modest version of disjunctivism about perceptual experience: when one has a perceptual experience as if p, *either* one has an unsuccessful perceptual experience as if p *or* one has a successful perceptual experience as if p, where these are different kinds of mental state. As Jonathan Dancy (1995) has observed, the indicated disjunction need not be viewed as an analysis of perceptual experience; nor must its proponent deny that instances of the two kinds share some "common factor."[38] For instance, it is compatible with the relevant, modest form of disjunctivism that, as proponents of the view labeled "intentionalism" maintain,[39] in both a successful and an unsuccessful perceptual experience as if p, the experiencing subject stands in the *merely perceptually experiencing as if* relation to the *proposition* that p;[40]

37 Compare the view that part of what it is to know that p is to stand in a relation to the fact that p (we know *facts*), whereas this is not so for belief that p (we believe *propositions*); see Vendler (1967), Harman (2003), and Moffett (2003). Consider also Williamson (2000, 47, emphasis added): "To know is not merely to believe while various other conditions are met; it is to be in *a new kind of state*, a factive one." On this view, knowing and merely believing are differently constituted, despite being subjectively indistinguishable. This may provide a useful analogy for the present approach.

38 See Byrne and Logue (2008a; 2008b, esp. xi) for helpful discussion of some varieties of disjunctivism.

39 Cf. Pautz (2007).

40 My use of the term 'mere' should not be misinterpreted as deflating the justificatory status of unsuccessful perceptual experience. While one might deny that mere perceptual experience and perceptual awareness justify equally (cf. McDowell 1982), I believe that both states provide prima facie justification for corresponding beliefs (see Bengson forthcoming, §5). Such egalitarianism about *justification* is compatible with the thesis, pursued in the text, that perceptual awareness alone is able to provide *knowledge*, because it alone is non-accidentally correct.

however, in a successful perceptual experience as if p, the experiencing subject also stands in the **perceptual awareness** relation to the FACT that p.[41]

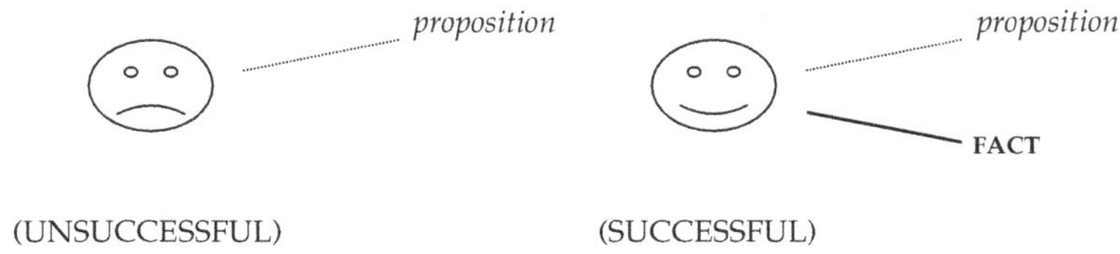

Successful perceptual experience and unsuccessful perceptual experience may thus be regarded as distinct determinates of the same common determinable, or as two distinct sub-types of the type perceptual experience. (I reserve "perceptual experience" for this common determinable or type.)

Recent years have seen an array of arguments in defense of naïve realism (see, e.g., the citations in note 36). Rather than appraise these arguments, however, the question on which I would like to focus is how naïve realism is meant to address the problem of perceptual knowledge. I believe that the explanatory framework introduced above provides the answer.

Suppose that naïve realism about perception is true, and successful perceptual experiences are constituted by facts perceived:

j. A successful perceptual experience as if p is constituted by the fact that p.

How would this constitution claim be used to explain how perceptual experience can be non-accidentally correct, hence able to serve as a source of knowledge of empirical facts? In light of our discussion of constitutive explanation, we can formulate the explanation as follows: *a perceiver's perceptual experience as if p is non-accidentally correct, hence able to serve as a source of knowledge that p, because it is partly constituted by the fact that p.*

Our discussion also provides insight into how this explanation works. The fact that p has the property (*F*) of being how the world is. This *ensures* that a successful perceptual experience as if p, which is constituted by the fact that p, has the property (*G*) of being not merely accidentally correct, hence able to serve as a source of knowledge that p. That is, a successful perceptual experience as if p has a certain constitution: part of *what it is* for this mental state (demonstrating such an experience) to exist is for it to be a fact that p. Accordingly, this mental state has a certain property, namely, the property of

[41] The rightmost picture in the diagram represents a case of seeing in which one enjoys perceptual awareness of a fact. It is an open question whether such fact-perception, which I have been referring to as "successful perceptual experience as if p," can always be forced into the mold of "perceiving-that." Suppose I see Derek running; that is, I have a successful perceptual experience as if this is so. I thus see a certain fact, namely, the one consisting of Derek's running. Do I thereby *see that* Derek is running? One might say: perhaps not, if I do not realize (or do not subsequently believe) that it is *Derek*, rather than (say) a very tall person, who is running.

being non-accidentally correct. Consequently, it is able to serve as a source of knowledge of the fact that p.[42]

One virtue of this approach is its identification of a feature of successful perception that is relevant to its status as a potential source—albeit not a guarantor—of knowledge. To see this, recall our veridical hallucinator (from §2) who forms a true belief about the external world on the basis of a capricious-lesion-induced accidentally correct perceptual experience. How are we to explain the veridical hallucinator's subsequent lack of knowledge? The naïve realist about perception may offer the following explanation: *a veridical sensory hallucination as if p is unable to serve as a source of knowledge that p because it is not constituted by the fact that p*. To say that a veridical hallucination as if p is not constituted by the fact that p is to say that it is not the case that part of *what it is* for that mental state to exist is for it to be a fact that p. Consequently, a veridical hallucinator's accidentally correct perceptual experience does not have the property of being able to provide knowledge of the fact that p—the desired result. Constitution may thus serve as the relation that (constitutively) explains the difference between those perceptual experiences that can, and those that cannot, provide perceptual knowledge of empirical facts.

6.3. *Explaining Intuitive Knowledge of Abstract Facts*

We are now in a position to answer our original explanatory question about intuition, repeated below:

> What relation does a thinker's mental state—her intuition—bear to an abstract fact that explains how the state can be non-accidentally correct with respect to that fact, hence able to serve as a source of knowledge of it?

I propose that the answer lies in the notions of constitution and constitutive explanation.

Naïve realism about successful perception entails that some mental states are partly constituted by the mind-independent items that they are about.

[42] The focal point, again, is source accidentality: the explanandum is how a perceptual experience is *able* to serve as a source of knowledge, even if it does not entail the presence of such knowledge (perhaps, e.g., because of doxastic accidentality). To illustrate, recall a standard fake barn example, in which one believes that there is a barn present on the basis of a successful perceptual experience of a real barn in an area populated by many unperceived fake barns. Naïve realism allows us to make sense of a familiar dual reaction to such a case. On one hand, something has gone wrong: due to the presence of a defeater for the belief that there is a barn present (viz. the many unperceived fakes in the area), one does not know that there is a barn present. On the other hand, something has gone right: one is in a perceptual state constituted by the fact that there is a barn present: part of *what it is* for that state to exist is for it to be a fact that there is a barn present; as such, the state is non-accidentally correct with respect to this fact, hence it is *able* to serve as a source of knowledge of it, even though it does not do so (given the many unperceived fakes). Such a perceiver is importantly different from the veridical hallucinator described next.

Consider the following application of this general idea to the case of intuition:

> *Naïve Realism about Intuition*: Those mind-independent items that are successfully intuited partly constitute one's intuition.

To illustrate, when you successfully intuit the fact that identity is transitive, that very fact partly constitutes the mental state you are in. You are *grasping* the *fact* itself. A successful intuition thus has a chunk of the world—a fact—as a constituent; in this sense, present in the intuition is the reality intuited.[43]

A prominent motivation for naïve realism about perception is that in successful perception, we seem to enjoy an "openness to the world": through perception, we seem to directly grasp facts about the empirical realm—for instance, that there is a red apple present.[44] In perception, this fact is made manifest, just by looking. In successful intuition, we seem to enjoy a similar sort of "openness to the world": through intuition, we seem to directly grasp facts about the intellectual realm—for instance, that identity is transitive. This fact is made manifest, just by thinking. Indeed, in many cases, rationalists and empiricists alike have found it natural to describe the state we are in when we have such an intuition by saying that we can just *see* or *grasp* that things are thus-and-so. In successful intuition as in perception, our mental state does not "fall short" of the world, for it is partly constituted by the fact grasped. Hence its success.

Cartesian rationalists might be tempted to employ this thought in service of the ambition to secure the transparency of successful intuition to the reflective mind. However, such a transparency thesis is no part of naïve realism about intuition, which allows, to the contrary, that there may be no failsafe guide, mark, or method that could be used to ascertain whether one enjoys an intuition constituted by the fact intuited. This is not to say that successfully intuiting is in all respects the same exact mental state as its unsuccessful counterpart. Recall the contrast (from §2) between Ramanujan's intuition and mine with respect to the very interesting number 1729: we may say that Ramanujan *grasps* the fact that 1729 is the smallest number expressible as the sum of two positive cubes in two different ways, whereas I do not. According to naïve realism about intuition, the difference consists in this: the presence of *this kind of mental state*—demonstrating a non-accidentally correct intuition that p (i.e. a mental state of grasping), such as Ramanujan's—constitutively depends on the fact that p; not so for an unsuccessful intuition, such as mine. In other words, part of *what it is* for a successful intuiter's intuition that p to exist is for it to be a fact that p; but it is not the case that part of *what it is* for a

[43] Chudnoff (2013, §3) distinguishes three naïve realist views about awareness of abstract individuals ("primitive," "material," and "formal"), none of which are equivalent to the thesis pursued here (which, given the existential characterization of constitution in §5, together with our focus on facts, might be labeled "existential factual").

[44] Cf. Crane (2006, 134; 2008) and the citations in note 36. Though alluded to below, I prescind from detailed treatment of the phenomenology of intuition here. I treat this topic in a way that complements—and fills out—the present discussion in Bengson (forthcoming).

capriciously lesioned intuiter's intuition that p to exist is for it to be a fact that p. So, although the two intuitions may be subjectively indistinguishable, they are distinct kinds of state, for they are differently constituted.

The result, once again, is a type of disjunctivism: when one has the intuition that p, *either* one has an unsuccessful intuition that p *or* one has a successful intuition that p, where these are different kinds of mental state. Compatibly with this disjunction, in both a successful and an unsuccessful intuition that p, the thinking subject may be regarded as standing in one and the same *merely intuiting* relation to the *proposition* that p; however, in a successful intuition that p, the thinking subject also stands in the **intuitive awareness** relation to the FACT that p.[45] Successful intuition and unsuccessful intuition may thus be regarded as distinct determinates of the same common determinable, or as two distinct sub-types of the type intuition. (I reserve "intuition" for this common determinable or type.)

In these respects, naïve realism about intuition is the intellectual mirror-image of naïve realism about perception. And just as the latter addresses the problem of perceptual knowledge, the former addresses the problem of intuitive knowledge. We have already seen the core explanatory strategy at work in the previous subsection. I will briefly summarize the main points.

According to naïve realism about intuition:

k. A successful intuition that p is constituted by the fact that p.[46]

This constitution claim might be used to explain the ability of intuition to serve as a source of knowledge of abstract facts, as follows: *a thinker's intuition that p is non-accidentally correct, hence able to serve as a source of knowledge that p, because it is partly constituted by the fact that p*. The mechanism underwriting this explanation is, once again, ensurance: the fact that p has the property (F) of being how the world is *ensures* that a successful intuition that p, which is constituted by the fact that p, has the property (G) of being not merely accidentally correct.[47]

Contrast a veridical intellectual hallucinator (such as myself in the case above), who *lacks* intuitive knowledge. The naïve realist about intuition may offer the following explanation: *a veridical intellectual hallucination that p is unable to serve as a source of knowledge that p because it is not constituted by the fact that p*. In this way, constitution marks an epistemically significant contrast, serving as the "the link between our cognitive faculties and the objects known," which explains why the correctness of a thinker's mental states about them is not merely coincidental. The result is a (constitutive)

45 Recall the diagram from §6.2. The rightmost picture can be used to represent a case of grasping, in which one enjoys intuitive awareness of a fact.

46 It is natural to object to this thesis that mental states cannot be constituted by abstracta. I respond to this concern in §7.

47 As before, this explanation focuses on source accidentality: the explanandum is how a given mental state is *able* to serve as a source of knowledge, even if it does not entail the presence of such knowledge (perhaps, e.g., because of doxastic accidentality).

explanation of how intuition works—how successful intuitions are able to provide knowledge of abstract facts.

6.4. *Summary*

There is a question about how certain mental states are able to serve as sources of knowledge. Being correct—i.e. being related to true propositions—is not enough. What is needed, in addition, is to be appropriately related to the facts; only this can ensure a given state's non-accidental correctness. Enter the idea that some mental states are *constituted* by mind-independent facts—that is, they stand in a non-causal, irreflexive, asymmetric ontological dependence relation to the way the mind-independent world is. Because these mental states are constitutively related to the facts, they are not merely accidentally correct. They may be logically independent of the facts; but they are not metaphysically independent of them: on the contrary, that there is such a fact is part of *what it is* for such a mental state to exist. Of course, enjoying such a mental state does not by itself *entail* having knowledge of that fact (e.g. there may still be defeaters). Nevertheless, such a state is, given its constitution, non-accidentally correct, hence *able* to serve as a source of knowledge of that fact. The result is a constitutive explanation for why some mental states, including some hallucinatory experiences, some perceptual experiences, and some intuitions, are potential sources of knowledge of what they are about.[48]

In the case of intuition, this proposal is usefully thought of as broadly Gödelian, at least inasmuch as it might enable us to make good theoretical sense of Gödel's evocative remark, quoted at the outset, about the objectivity of intuition despite the absence of causal effects on our sense organs. If naïve realism about intuition is correct, then when we come to know about mind-independent abstracta via intuition, "their presence in us" is indeed due, not to causation, but to "another kind of relationship between ourselves and reality," namely, a *constitutive* relationship.[49]

The proposed explanation of intuitive knowledge has the following two key elements (where "| p |" designates the fact that p):

[48] Peacocke (2009, 731) has independently registered the plausibility of the idea that sometimes "the explanation of why a cognitive state is a means of acquiring knowledge has to mention the constitutive nature of that state." However, Peacocke's approach differs from mine in several respects. For instance, he does not identify the mechanism underwriting such explanation, as I have done (through the notion of ensurance); he also tends to focus on epistemic entitlement and does not pursue the point, at center stage in the present discussion, that the constitution of a state may secure its non-accidental correctness.

[49] Perhaps something like this thought lies also behind Frege's (1884/1953, 115, emphasis added) comment that "In arithmetic we are not concerned with objects which we come to know as something alien from without through the *medium* of the senses, but with objects *given directly* to reason and, as its nearest kin, utterly *transparent* to it." On a naïve realist view of intuition, our grasp of arithmetical facts does not involve any "medium" at all, and what we come to know is not something "alien" to the grasp by which we come to know it, for the arithmetical facts are to be understood as *constituents* of such grasp.

I. Naïve Realism (about intuition): A non-accidentally correct intuition with content p is a mental state constituted by |p|, where |p| has the property of being how the world is (recall §1 on the nature of facts); not so for an accidentally correct or false intuition with content p.

II. Ensurance Claim: For a particular fact, |p|, and a particular mental state σ with content p, necessarily, if |p| has the property of being how the world is and |p| constitutes σ, then thereby σ has the property of being non-accidentally correct with respect to |p|, hence able to serve as a source of knowledge that p.[50]

I submit that if these two elements are in place, then an explanation of intuitive knowledge of abstracta is achieved. While I have at various points indicated some of the motivations for these elements, or (in the case of I) correlates thereto, including their central role in explaining Trip's knowledge via hallucination (§6.1) and enabling an important style of response to the problem of perceptual knowledge (§6.2),[51] both elements remain to be given comprehensive defense. In this sense, the present proposal is programmatic.

But this is not to say that it is unsupported. In addition to the aforementioned motivations, an abductive argument can be given on behalf of the two elements: if the present proposal is the best explanation of our knowledge regarding domains such as mathematics, logic, morality, and modality, realistically conceived, as it seems to be (arguably, it is the only viable candidate; recall §§3.2–3), then realist rationalists have excellent reason to accept elements I and II, insofar as both figure ineliminably in that explanation.

I will call the conjunction of realism, rationalism, and element I *naïve realist rationalism*. Element II marks the general explanatory mechanism that enables naïve realist rationalism to answer the non-accidental relation question about intuition. The next section responds to what I take to be the most pressing objection to naïve realist rationalism; the subsequent section discusses its broader explanatory potential, specifically with respect to questions regarding reliability and etiology.

7. ON BEING CONSTITUTED BY ABSTRACTA

There is a simple but important objection to the idea, central to the proposed explanation, that a thinker's intuition, a non-abstractum, may be constituted by an abstractum, such as the abstract fact that identity is transitive or that wantonly killing innocents is wrong. It might be allowed that non-abstracta can be constituted by other non-abstracta. However, the objection

50 As should be clear, II is not the universal claim that mental states are *always* able to serve as sources of knowledge of facts about their constituents—to which the physicalist view that mental states are constituted by neural states would, if true, provide myriad counterexamples. As explained in §5, ensurance is *exclusive*, and it clearly does not relate all state-fact pairs.

51 Additionally, theses corresponding to I together with II allow constitutive explanations of, for example, a fragile vase's fragility and a false assertion's inappropriateness (§5).

goes, non-abstracta (causally efficacious denizens of space and time) cannot be constituted by abstracta (causally inert denizens of the third realm), for this would require an abhorrent style of commingling.

The objection raises a host of interesting metaphysical issues. For present purposes, I will focus on two points that provide reason to think that the constitution claim in question does not face any obvious logical, conceptual, or metaphysical obstacles.

The first point highlights a difference between constitution and causation that is significant in the present context. Recall that the reason that non-abstracta cannot be *caused* by abstracta is that this directly follows from the nature of abstracta: they are non-spatiotemporal *and causally inert*. It does not similarly follow that non-abstracta cannot be *constituted* by abstracta. In short, there is a principled reason, pursuant to the nature of the entities in question, why there could not be causal relations between mental states and abstracta. However, it is not clear that there is a similarly principled reason, pursuant to the nature of the entities in question, why there could not be constitutive relations between mental states and abstracta. Additional, and in all likelihood controversial, assumptions would be needed to establish such a result—assumptions that are not compulsory, and that the naïve realist about intuition need not share. In this sense, there is no "structural" problem in the idea that non-abstracta may be constituted by abstracta.

Nor is there a lack of precedent for this idea. This is the second point. Traditionally, abstracta, though *causally inert*, are not by any means *constitutionally inert*. For example, abstracta may be (and have been) said to be constituted by other abstracta, as in (l) and (m) below. Similarly, abstracta may be (and have been) said to be constituted by non-abstracta, as in (n), (o), and (p) below. Finally, given the prominent—and in this dialectical context admissible—view that properties, norms, universals, and numbers are abstract entities, non-abstracta may be (and have been) said to be constituted by abstracta, as in (g), (h), (q), and (r).

l. The set of whole numbers between one and ten is an abstract entity constituted by its abstract members, e.g. the number three.
m. Fregean propositions are abstract entities constituted by abstract entities, namely, Fregean senses.
n. The set {Obama, Biden} is an abstract entity constituted by its non-abstract members, namely, Obama and Biden.
o. Russellian propositions are abstract entities partly constituted by concrete entities, namely, material objects.
p. Lewisian properties are abstract entities constituted by concrete entities, namely, concrete individuals in concrete worlds (Lewisian possibilia).
g. The event of Derek running is constituted by Derek, the property of running, and a time (on a Kimian theory of events).
h. The speech act of assertion is constituted by the norm *assert only what you know* (on a Williamsonian view of assertion).

q. The game of chess is constituted by the norm (or rule) the bishop moves any number of vacant squares in any and only a diagonal direction.
r. The thick particular which is the red apple is a concrete entity that is constituted by the universal redness (on an Armstrongian view of particulars; see Armstrong 1997).[52]

It is not my intention to suggest that each and every one of these constitution claims is true; the point is simply that they do not all state claims that are out of bounds, as it were. To the contrary, in some cases, the claims enjoy substantial initial plausibility and argumentative support. This runs counter to the objection at hand, which is based on the idea that reality abhors commingling. A general prohibition against commingling does not, however, offer secure footing from which to object to the (dialectically adequate) explanation proposed by naïve realist rationalism, which from this vantage point can be seen as merely exploiting a type of possibility that is already exploited in, and recognizable from, other areas of philosophy.

8. OTHER EXPLANATORY QUESTIONS

The proposed explanation of intuitive knowledge of abstracta seeks to answer a particular type of explanatory question, namely, the non-accidental relation question. There are other explanatory questions, some of which were discussed in §3.3. It is a limitation of the foregoing answer to the non-accidental relation question that it does not answer all of these questions. But, in a way, that is only to be expected: different questions require different answers. And it is a virtue of the proposed answer to the non-accidental relation question that it might help eventually to make headway on these other questions as well.

To appreciate the sense in which this might be so, it may help to outline a strategy for addressing the reliability question, repeated below.

> How is it that a thinker's mental states—her intuitions—are non-accidentally correct with respect to abstract facts *sufficiently more often than not*?

Consider, first, the case of perceptual experience. How might we explain the reliability of competent perceivers' perceptual experiences, as expressed by the following reliability principle?

> [RELIABILITY$_{PE}$] If competent perceivers have a perceptual experience as if p, then p (for most propositions p regarding standard middle-sized dry goods).

[52] Perhaps another example from metaphysics: according to a bundle theory that accepts Campbell's (1990) trope theory, an individual horse is a concrete entity constituted by a bundle of abstract particulars.

Presumably an explanation may take the following two-step form:

> [STEP 1] When a perceiver has a successful perceptual experience as if p, she has a mental state that bears relation **R** to the fact that p.
>
> [STEP 2] For the most part, competent perceivers' perceptual experiences bear **R** to the facts (regarding standard middle-sized dry goods).

What is needed for STEP 1 is to supply a plausible value for **R**. This, together with a vindication of the anti-skeptical STEP 2, would yield the desired explanation of the reliability of competent perceivers' perceptual experiences.

A similar strategy might allow progress in the case of intuition. To see this, we can simply replace all reference to competent perceivers and their perceptual experiences with competent thinkers and their intuitions (e.g. competent mathematicians and their mathematical intuitions) to yield the following reliability principle (cf. Field 1989, 26 and 230–1):

> [RELIABILITY$_{INT}$] If competent thinkers have an intuition that p, then p (for most propositions p regarding familiar abstracta).

Here, too, an explanation may take the following two-step form:

> [STEP 1*] When a thinker has a successful intuition that p, she has a mental state that bears relation **R*** to the fact that p.
>
> [STEP 2*] For the most part, competent thinkers' intuitions bear **R*** to the facts (regarding familiar abstracta).

What is needed for STEP 1* is to supply a plausible value for **R***. This, together with a vindication of the anti-skeptical STEP 2*, would yield the desired explanation of the reliability of competent thinkers' intuitions.

The primary theoretical question here concerns **R** and **R***. What could **R** and **R*** be? The present proposal, which pursues a naïve realist approach to both perceptual experience and intuition, enables the following answer: the relevant relation is the *being constituted by* relation.

What about questions regarding the etiology of intuition? These raise many complex issues, but they, too, may prove tractable for realists once the constitution relation is recognized. Take, for instance, the etiological question articulated in §3.3:

> How is it that a thinker *comes to have* a mental state—an intuition—that is non-accidentally correct with respect to an abstract fact?

A natural strategy is to ascertain (i) the relation **R** that one's mental state—an intuition—bears to the facts when it is non-accidentally correct (cf. STEP 1/1*), and (ii) the process by which one comes to enjoy a state bearing non-accidental relation **R** to the facts. Regarding (ii), it is plausible to hold that, just as one typically comes to have a perceptual experience that bears a non-accidentally correct relation to the facts perceived via *a process of looking* (in suitable perceptual conditions: alertness, proper lighting, etc.), one typically comes to have an intuition that bears a non-accidental relation to the facts intuited via *a process of reflection* (in suitable intellectual conditions: attentiveness, proper intelligence, etc.), where a process of reflection consists in, at minimum, entertaining a proposition with the intention of determining whether it (or some suitably related proposition) is true.[53] What remains is to specify the non-accidental relation **R** cited in (i). The present proposal enables the following treatment: the relevant relation is the *being constituted by* relation.

I have outlined a uniform strategy for answering questions about the reliability and etiology of intuition. (There are of course additional questions beyond these, but I lack the space to pursue them here.) The central point, however, is not that such questions are easily answered—there is undoubtedly more to be said—but rather that the present proposal helps make progress towards this end.

9. CONCLUSION

My primary aims have been, first, to clarify Benacerraf's worry about intuitive knowledge of abstracta, concerning "the link between our cognitive faculties and the objects known," and, second, to develop an explanatory framework with the resources to address it. Regarding the first, I have articulated an explanatory question, prompted by reflection on cases of veridical hallucination, about how intuitions can be related to abstract facts so as to preclude coincidence—accidental correctness. Regarding the second, the central idea I have sought to develop, in response to this question, is that we need the notion of a mental state that is constitutively related to the fact that it is of or about.

This notion has a variety of applications. It allows us to make sense of how hallucinatory experience is able to serve as a source of knowledge about colors and shapes. A philosopher of perception may make use of this notion to make sense of how successful perception puts us "in touch" with a realm of mind-independent empirical facts. The notion may be put to similar use in the philosophy of intuition to make sense of how successful intuition puts us "in touch" with a realm of mind-independent abstract facts—and thus, for the naïve realist rationalist, to render intelligible the possibility of grasping the third realm.[54]

This approach affords a picture of how intuition works that responsibly engages an important explanatory challenge. The proposed theory of intuitive

[53] Realist rationalists might pursue (ii) in a way that sometimes privileges character or affect (see Fitzpatrick forthcoming), much as in some accounts of the etiology of certain perceptions. But, obviously, this is not obligatory. At any rate, in both cases—perceptual and intuitive—there will also be a corresponding series of subvenient non-mental, physical events, if physicalism is true. But it is far from obvious that citing such phenomena is either necessary or sufficient for answering the relevant type of etiological question in either case.

[54] I have focused on the explanation of intuitive knowledge of mind-independent abstract facts, both analytic and synthetic. Intuition might also be said to provide knowledge of facts partly about *concreta* (e.g. that a particular material object is self-identical) or about mind-*dependent* abstracta (see, e.g., Thomasson 1999). I believe that the proposal can be extended to cover these cases as well.

knowledge has several virtues. First, it is *general*, in at least the following two ways: it uncovers a uniform, as opposed to piecemeal, answer to Benacerraf-style worries across a priori domains, and it introduces an epistemological perspective that simultaneously addresses non-accidental relation questions regarding knowledge of abstracta, knowledge via hallucination, and ordinary perceptual knowledge. Second, it is *sober*: while it is unabashedly philosophical, it is not religious or mystical, and at no point does it invoke spiritual powers or supernatural forces, nor does it indulge in mystery or superstition. Third, and relatedly, it is *conservative*: it employs extant tools, such as the notion of constitution and the phenomenon of non-causal explanation, that draw upon resources familiar from other areas of philosophy (e.g. metaphysics and the theory of explanation). Fourth, it is *independently motivated*: it requires no special pleading, for it already seems to be called for by the case of Trip. Furthermore, it develops and extends a historically noteworthy and increasingly popular approach in the philosophy of perception, naïve realism, whose theoretical interest and explanatory virtues by no means narrowly or *ad hoc*-ly rely on the specific case of intuition. Fifth, it is *non-skeptical*: while it does not by itself purport to convince skeptics or vindicate any claims to knowledge (i.e. it is not *anti*-skeptical),[55] it allows for knowledge in just those places where it seems correct to allow it. In particular, it makes room for the knowledge—of properties, relations, numbers, sets, norms, values, reasons, and various other items, realistically conceived—that we seem to acquire just by thinking.[56]

REFERENCES

Armstrong, D. 1973. *Belief, truth, and knowledge*. Cambridge: Cambridge University Press.

Armstrong, D. 1997. *A world of states of affairs*. Cambridge: Cambridge University Press.

Audi, R. 1999. Self-evidence. *Philosophical Perspectives*, 13: 205–9.

[55] As explained in §1, the project is not the anti-skeptical one of defending claims to intuitive knowledge of abstracta. Nor is the project to ascertain a guide or method that could be used to check that one has such knowledge. The project is explanatory: to articulate, in a dialectically adequate way, an explanation of such knowledge, if any there be.

[56] I have been exploring the ideas in this paper for many years and have benefited tremendously from discussion with numerous individuals and groups during that time, including audiences at Arché Philosophical Research Centre at St. Andrews University, Australian National University, University of British Columbia, University of California at San Diego, University of Melbourne, University of Texas at Austin, University of Wisconsin at Madison, University of Wyoming, Yale University, and a session at the Pacific APA. Individually, I am particularly grateful to Ralf Bader, Derek Ball, George Bealer, Matt Bedke, Chad Carmichael, Eli Chudnoff, Louis de Rossett, Sinan Dogramici, David Enoch, Enrico Grube, Shelly Kagan, Rob Koons, Dan Korman, Anna-Sara Malmgren, Elliot Paul, Adam Pautz, Raul Saucedo, Anat Schechtman, Daniele Sgaravatti, David Sosa, Ernie Sosa, Zoltan Szabo, Leslie Wolf, Crispin Wright, and two anonymous reviewers for *Oxford Studies in Epistemology*.

Balaguer, M. 1998. *Platonism and anti-platonism in mathematics*. Oxford: Oxford University Press.

Bealer, G. 1992. The incoherence of empiricism. *Proceedings of the Aristotelian Society*, 66: 99–138.

Bealer, G. 1993. Universals. *The Journal of Philosophy*, 90: 5–32.

Bealer, G. 1998. Intuition and the autonomy of philosophy. In M. Depaul and W. Ramsey (eds.), *Rethinking intuition: The psychology of intuition and its role in philosophical inquiry*. New York: Rowman & Littlefield Publishers, Inc.

Bedke, M. 2009. Intuitive non-naturalism meets cosmic coincidence. *Pacific Philosophical Quarterly*, 90: 188–209.

Bell, D. 1979. The epistemology of abstract objects. *Proceedings of the Aristotelian Society*, 53: 135–52.

Benacerraf, P. 1973. Mathematical truth. *The Journal of Philosophy*, 70: 661–79.

Bengson, J. Forthcoming. The intellectual given. *Mind*.

Boghossian, P. 2001. Inference and insight. *Philosophy and Phenomenological Research*, 63: 633–40.

BonJour, L. 1998. *In defense of pure reason*. Oxford: Oxford University Press.

Burge, T. 1990. Frege on knowing the third realm. *Mind*, 101: 633–50.

Burgess, J. P. and G. Rosen. 1997. *A subject with no object: Strategies for nominalistic interpretation of mathematics*. Oxford: Oxford University Press.

Byrne, A. 2003. Color and similarity. *Philosophy and Phenomenological Research*, 66: 641–65.

Byrne, A. and H. Logue. 2008a. Either/or. In A. Haddock and F. Macpherson (eds.), *Disjunctivism: Perception, action, knowledge*. Oxford: Oxford University Press.

Byrne, A. and H. Logue. 2008b. Introduction. In A. Byrne and H. Logue (eds.), *Disjunctivism*. Cambridge, MA: MIT Press.

Campbell, J. 2002. *Reference and consciousness*. Oxford: Oxford University Press.

Campbell, K. 1990. *Abstract particulars*. Oxford: Blackwell.

Cassam, Q. 2007. *The possibility of knowledge*. Oxford: Oxford University Press.

Casullo, A. 2003. *A priori justification*. Oxford: Oxford University Press.

Cheyne, C. 2001. *Knowledge, cause, and abstract objects: Causal objections to platonism*. Dordrecht: Kluwer Academic Publishers.

Chudnoff, E. 2013. Awareness of abstract objects. *Noûs*, 47: 706–26.

Clarke-Doane, J. Forthcoming. What is the Benacerraf problem? In F. Pataut (ed.), *New perspectives on the philosophy of Paul Benacerraf: Truth, objects, infinity*.

Crane, T. 2006. Is there a perceptual relation? In T. S. Gendler and J. Hawthorne (eds.), *Perceptual experience*. Oxford: Clarendon Press.

Crane, T. 2008. The problem of perception. *The Stanford Encyclopedia of Philosophy* (Fall Edition).

Currie, G. 1982. Frege's letters. *The British Journal for the Philosophy of Science*, 33: 65–76.

Dancy, J. 1995. Arguments from illusion. *Philosophical Quarterly*, 45: 421–38.

Davies, M. 1983. Function in perception. *Australasian Journal of Philosophy*, 61: 409–26.

Devitt, M. 2005. There is no a priori. In M. Steup and E. Sosa (eds.), *Contemporary debates in epistemology*. Oxford: Blackwell.
Field, H. 1989. *Realism, mathematics, and modality*. Oxford: Basil Blackwell.
Field, H. 2005. Recent debates about the a priori. In T. S. Gendler and J. Hawthorne (eds.), *Oxford studies in epistemology, volume 1*. Oxford: Oxford University Press.
Fine, K. 1994. Essence and modality. *Philosophical Perspectives*, 8: 1–16.
Fine, K. 2001. The question of realism. *Philosophers' Imprint*, 1: 1–30.
Fine, K. 2005. Our knowledge of mathematical objects. In T. S. Gendler and J. Hawthorne (eds.), *Oxford studies in epistemology, volume 1*. Oxford: Oxford University Press.
Fine, K. 2011. The pure logic of ground. *Review of Symbolic Logic*, 5: 1–25.
Fitzpatrick, W. Forthcoming. Ethical realism in a Darwinian world. In T. Cuneo and D. Loeb (eds.), *Empirical dimensions of metaethics*.
Frege, G. 1884/1953. *The foundations of arithmetic*. Translated by J. L. Austin. Oxford: Basil Blackwell.
Giaquinto, M. 2001. Knowing numbers. *The Journal of Philosophy*, 98: 5–18.
Gibbard, A. 2008. *Reconciling our aims: In search of bases for ethics*. Oxford: Oxford University Press.
Gödel, K. 1964. What is Cantor's continuum problem? In P. Benacerraf and H. Putnam (eds.), *Philosophy of mathematics: Selected readings*. Englewood Cliffs, NJ: Prentice-Hall.
Goldman, A. 1976. Discrimination and perceptual knowledge. *Journal of Philosophy*, 73: 771–91.
Goldman, A. 2007. Philosophical intuitions: Their target, their source, and their epistemic status. *Grazer Philosophische Studien*, 74: 1–26.
Goldstein, B. 2009. *Encyclopedia of perception*. Thousand Oaks, CA: Sage Publications, Inc.
Grice, H. P. 1961. The causal theory of perception. *Proceedings of the Aristotelian Society*, 35: 121–53.
Hardy, G. H. 1940. *Ramanujan*. Cambridge: Cambridge University Press.
Harman, G. 2003. Category mistakes in M&E. *Philosophical Perspectives*, 17: 165–80.
Hart, W. D. 1977. Review of Steiner's *Mathematical knowledge*. *The Journal of Philosophy*, 74: 118–29.
Hawthorne, J. 1996. The epistemology of possible worlds: A guided tour. *Philosophical Studies*, 84: 183–202.
Hawthorne, J. and K. Kovakovitch. 2006. Disjunctivism. *Proceedings of the Aristotelian Society*, 80: 145–83.
Hellie, B. 2007. Factive phenomenal characters. *Philosophical Perspectives*, 21: 259–306.
Huemer, M. 2005. *Ethical intuitionism*. New York: Palgrave MacMillian.
Jackson, F. 1998. *From metaphysics to ethics: A defence of conceptual analysis*. Oxford: Clarendon Press.
Johnston, M. 1997. Postscript: Visual experience. In A. Byrne and D. Hilbert (eds.), *Readings on color, volume 1*. Cambridge, MA: MIT Press.

Johnston, M. 2004. The obscure object of hallucination. *Philosophical Studies*, 120: 113–83.
Johnston, M. 2005. Constitution. In F. Jackson and M. Smith (eds.), *The Oxford handbook to contemporary philosophy*. Oxford: Oxford University Press.
Kagan, S. 2001. Thinking about cases. *Social Philosophy and Policy*, 18: 44–63.
Katz, J. 1998. *Realistic rationalism*. Cambridge, MA: MIT Press.
Kim, J. 1973. Causation, nomic subsumption, and the concept of an event. *Journal of Philosophy*, 70: 217–36.
Kim, J. 1974. Noncausal connections. *Noûs*, 8: 41–52.
Kitcher, P. 1984. *The nature of mathematical knowledge*. Oxford: Oxford University Press.
Koslicki, K. 2004. Constitution and similarity. *Philosophical Studies*, 3: 327–64.
Lewis, D. 1983. New work for a theory of universals. *Australasian Journal of Philosophy*, 61: 343–77.
Lewis, D. 1986. *On the plurality of worlds*. Oxford: Basil Blackwell.
Liggins, D. 2010. Epistemological objections to platonism. *Philosophy Compass*, 5: 67–77.
Linnebo, Ø. 2006. Epistemological challenges to mathematical platonism. *Philosophical Studies*, 129: 545–74.
Linsky, B. and E. Zalta. 1995. Naturalized platonism vs. platonized naturalism. *Journal of Philosophy*, 92: 525–55.
Locke, J. 1689. *Essay concerning human understanding*.
Ludwig, K. 2007. The epistemology of thought experiments: First versus third person approaches. *Midwest Studies in Philosophy*, 31: 128–59.
Mackie, J. L. 1977. *Ethics: Inventing right and wrong*. London: Penguin.
Martin, M. G. F. 2002. The transparency of experience. *Mind & Language*, 4: 376–425.
Martin, M. G. F. 2004. The limits of self-awareness. *Philosophical Studies*, 120: 37–89.
McDowell, J. 1982. Criteria, defeasibility, and knowledge. *Proceedings of the British Academy*, 68: 456–79.
McDowell, J. 1985. Values and secondary qualities. In T. Honderich (ed.), *Morality and objectivity*. London: Routledge & Kegan Paul.
Moffett, M. 2003. Knowing facts and believing propositions: A solution to the problem of doxastic shift. *Philosophical Studies*, 115: 81–97.
Moore, G. E. 1903. *Principia ethica*. Cambridge: Cambridge University Press.
Pautz, A. 2007. Intentionalism and perceptual presence. *Philosophical Perspectives*, 21: 495–541.
Peacocke, C. 1999. *Being known*. Oxford: Oxford University Press.
Peacocke, C. 2004. *The realm of reason*. Oxford: Oxford University Press.
Peacocke, C. 2009. Means and explanation in epistemology. *Philosophical Quarterly*, 59: 730–9.
Plantinga, A. 1993. *Warrant and proper function*. New York: Oxford University Press.
Pritchard, D. 2005. *Epistemic luck*. Oxford: Oxford University Press.
Pryor, J. 2004. What's wrong with Moore's argument? *Philosophical Issues*, 14: 349–78.
Pust, J. 2004. Explaining our knowledge of necessity. *Dialectica*, 58: 71–87.

Putnam, H. 1971. *Philosophy of logic*. New York: Harper and Row.
Putnam, H. 1994. Sense, nonsense, and the senses: An inquiry into the powers of the mind. *The Journal of Philosophy*, 91: 445–517.
Quine, W. V. O. 1960. *Word and object*. Cambridge, MA: The MIT Press.
Rosen, G. 1993. The refutation of nominalism (?). *Philosophical Topics*, 21: 149–86.
Rosen, G. 2001. Nominalism, naturalism, epistemic relativism. *Philosophical Perspectives*, 15: 69–91.
Ruben, D. H. 1990. *Explaining explanation*. London: Routledge.
Schechter, J. 2010. The reliability challenge and the epistemology of logic. *Philosophical Perspectives*, 24: 437–64.
Setiya, K. 2012. *Knowing right from wrong*. Oxford: Oxford University Press.
Shafer, K. 2014. Knowledge and two forms of non-accidental truth. *Philosophy and Phenomenological Research*, 89: 373–93.
Sidgwick, H. 1907. *Methods of ethics*, 7th edition. London: Macmillan.
Snowdon, P. 2005. The formulation of disjunctivism: A response to Fish. *Proceedings of the Aristotelian Society*, 105: 129–41.
Sosa, E. 1998. Minimal intuition. In M. Depaul and W. Ramsey (eds.), *Rethinking intuition: The psychology of intuition and its role in philosophical inquiry*. New York: Rowman & Littlefield Publishers, Inc.
Sosa, E. 2002. Reliability and the a priori. In T. S. Gendler and J. Hawthorne (eds.), *Conceivability and possibility*. Oxford: Oxford University Press.
Sosa, E. 2007. *A virtue epistemology: Apt belief and reflective knowledge, volume 1*. Oxford: Oxford University Press.
Street, S. 2006. A Darwinian dilemma for realist theories of value. *Philosophical Studies*, 127: 109–66.
Thomasson, A. 1999. *Fiction and metaphysics*. Cambridge: Cambridge University Press.
Tidman, P. 1996. The justification of a priori intuitions. *Philosophy and Phenomenological Research*, 56: 161–71.
Tieszen, R. 2002. Gödel and the intuition of concepts. *Synthese*, 133: 363–91.
Tye, M. 2000. *Consciousness, color, and content*. Cambridge, MA: MIT Press.
Tye, M. 2009. *Consciousness revisited: Materialism without phenomenal concepts*. Cambridge, MA: MIT Press.
Vendler, Z. 1967. *Linguistics in philosophy*. Ithaca, NY: Cornell University Press.
Wedgewood, R. 2007. *The nature of normativity*. Oxford: Oxford University Press.
Williamson, T. 2000. *Knowledge and its limits*. Oxford: Oxford University Press.
Williamson, T. 2007. *The philosophy of philosophy*. Oxford: Blackwell.
Wilson, C. 1926. *Statement and inference*. Oxford: Clarendon Press.
Wright, C. 2004. Intuition, entitlement, and the epistemology of logical laws. *Dialectica*, 58: 155–75.

2. Evidence and Epistemic Evaluation

Jessica Brown

1. INTRODUCTION

In this paper, I examine the viability of the probability-raising conception of evidential support according to which a proposition is evidence for a hypothesis for a subject if and only if 1) it is part of the subject's evidence; and, 2) it raises the probability of that hypothesis. Although this approach has widespread popularity, I will argue that it has a problematic consequence, namely that any proposition that is evidence for some hypothesis is evidence for itself. In order to see whether the probability-raising conception can avoid this problematic consequence I focus on each of the two conditions it places on a proposition being evidence for a hypothesis for a subject. First, I examine whether the probability-raising conception can avoid this problem by being combined with some specific invariantist account of what it is for a proposition to be evidence for a subject. Second, I examine whether it can avoid the problem by appeal to contextualism either about evidence or evidential support. Finally, I defend a solution that appeals to an invariantist modification of the evidential support condition, more specifically one that appeals to the notion of warrant transmission.[1]

2. EVIDENTIAL SUPPORT

In this paper, I focus on evidence that takes a propositional form. This leaves it open whether there is evidence that is non-propositional. Whatever the outcome of the latter debate, propositions are clearly one important kind of evidence. From now on, by 'evidence' I mean propositional evidence.

Under what conditions does evidence support a hypothesis? One popular answer to this question is provided by the probability-raising conception of evidential support.[2] On this view, some evidence is evidence for a hypothesis if it raises its probability. A proposition is not evidence for a subject for a hypothesis unless it is part of her evidence. So, on a probability-raising conception, there are two conditions for a proposition to be evidence for a subject for a hypothesis: that it is part of her evidence and that it raises the probability of

[1] Thanks to two anonymous referees for extremely helpful comments, and to Heddon, McGrath, Schechter, Sgaravatti, and Weatherson for comments on earlier drafts.

[2] Kelly (2008a) describes this view of evidential support as 'more popular than any competitor' and 'widely held' (941).

the relevant hypothesis. Thus, the probability-raising approach would define evidential support as follows:

> *Probability-raising (PR)*: e is evidence for h for S if and only if S's evidence includes e and P(h/e) >P(h). (Williamson 2000: 187)

It is important to note that this principle is a static not a dynamic principle. It explains evidential support in terms of the relation between two quantities: the probability of a hypothesis conditional on evidence, e, and the prior probability of that hypothesis. It does not explain evidential support in terms of the way one's credence in a hypothesis ought to shift consequent upon acquiring evidence. Relatedly, the relevant notion of probability that features in *PR* should not build the relevant evidence, e, into the background information (Williamson 2000: 187). For, if it did, then the probability of e would be 1, and the probability of h given e would just be identical to the probability of h, so that e would not count as evidence for h. Instead, Williamson suggests that the relevant probability is 'intrinsic probability prior to investigation' (211). Thus, according to *PR*, e is evidence for h for S if and only if e is part of S's evidence and the probability prior to investigation of h given e is greater than the probability prior to investigation of h.

Despite the popularity of the probability-raising conception, it has problematic consequences. In particular, as one of its defenders notes, it has the consequence that if a proposition is evidence for any hypothesis, it is evidence for itself (Williamson 2000: 187). By the definition above, if e is evidence for h, then it follows that the probability of h given e is greater than the probability of h. That in turn requires that the probability of e is neither zero (otherwise P(h/e) is ill defined) nor 1 (otherwise P(h/e) = P(h)). So, the probability of e is positive, but less than 1. Since the probability of e given e is 1, the probability of e given e is greater than the probability of e. Thus, e is evidence for e, substituting e for h in *Probability-raising*.[3]

To illustrate the consequences of the probability-raising approach, suppose that a detective, Morse, is collecting evidence relevant to his investigation into the recent burglary at the Central Jewellery Store. By questioning eyewitnesses, he acquires the evidence that the notorious burglar, Burglar Bill, was in the vicinity of the Central Jewellery Store just before the theft. This proposition is evidence for various hypotheses, including the hypothesis that Burglar Bill was near enough to the scene of the crime to have perpetrated the burglary. It is a consequence of the probability-raising conception of evidential support that the proposition that Burglar Bill was in the vicinity of the Central Jewellery Store just before the theft is also evidence for itself.

However, this result is in tension with the way we use evidence to evaluate belief. Suppose that we are trying to critically evaluate Morse's belief that Burglar Bill was in the vicinity of the Central Jewellery Store just before the theft. We ask Morse, 'What evidence do you have for the claim that Burglar

[3] Williamson says, 'one consequence of EV [our *Probability-raising*] is that e is evidence for h only if e is evidence for itself' (2000: 187).

Bill was in the vicinity of the Central Jewellery Store just before the theft?'[4] In reply, it seems inappropriate for Morse to say, 'Burglar Bill was in the vicinity of the Central Jewellery Store just before the theft.' But, on the probability-raising account of evidential support, it's hard to explain why that is so. On the proposed account, it is literally true to cite the proposition that Burglar Bill was in the vicinity of the Central Jewellery Store just before the theft as evidence for itself. Indeed, the probability that Burglar Bill was in the vicinity of the Central Jewellery Store just before the theft on that very proposition is 1. So, if Morse were to reply to the request for evidence for the claim that Burglar Bill was in the vicinity of the Central Jewellery Store just before the theft by citing that very claim itself, then he not only cites evidence for that claim, but also excellent evidence; no evidence could give that hypothesis any higher probability. I call this problem, 'the evaluative problem'.

It is useful to note that the problem focused on here is distinct from some other problems raised for the probability-raising conception of evidential support. Objections have been raised to both the necessity and sufficiency directions of the probability-raising account. For instance, against the necessity direction, Achinstein (2001) argues that some evidence may be evidence for a proposition even without raising its probability (70–1). However, it is the sufficiency direction that is central to our discussion. Achinstein criticizes the sufficiency direction of the probability-raising account on the grounds that some evidence might raise the probability of a proposition yet leave its probability so low that it is counterintuitive to describe that evidence as evidence for the proposition. For example, information that some of the original tickets in a large lottery have been destroyed raises the probability that my ticket will win but might leave its probability at such a low level that it seems incorrect to say that the information about the destruction of tickets is evidence that I will win (69–70). As a result, Achinstein suggests that e is evidence for p only if the probability of p/e meets a certain threshold. However, the evaluative problem cannot be addressed by imposing a threshold such that e is evidence for p only if the probability of p/e meets a certain threshold. After all, the probability of any proposition on itself is 1, so meets any possible threshold. Thus, the evaluative problem arises not only for the probability-raising conception of evidential support, but also for the main probabilistic alternative, the threshold conception according to which an evidence proposition is evidence for a hypothesis if and only if it raises its probability to a specified threshold.

[4] One might worry that the question posed ('What evidence do you have ...?') implicitly asks for all of Morse's evidence, or perhaps the strongest evidence, and this is what explains the impropriety of the reply. However, the reply seems equally inappropriate as a reply to the different question 'Give me some evidence ...' In addition, as pointed out in the main text, on *PR*, a proposition p is not only evidence for itself but excellent evidence in the sense that the probability of p given p is 1.

In the rest of the paper, I investigate a number of possible ways of modifying the probability-raising conception of evidential support to avoid the evaluative problem. Before doing so, I motivate these modifications by considering, but rejecting, the suggestion that the infelicity of citing a proposition as evidence for itself can be explained pragmatically.

3. PRAGMATIC EXPLANATIONS

A defender of the probability-raising account of evidential support may hope to offer a pragmatic explanation of the infelicity of citing a proposition as evidence for itself. For instance, Williamson suggests that the impropriety of citing p as evidence for p can be explained by appeal to familiar Gricean mechanisms (2000: 187–8). On this view, although it would be true to cite p as evidence for p, it is conversationally inappropriate. However, a Gricean explanation of the felt infelicity faces two serious problems. First, a pragmatic account of the infelicity of citing p as evidence for p seems unlikely to deal with all aspects of the evaluative problem. For, at best, such a pragmatic explanation would explain why it is inappropriate to cite p as evidence for p in conversation. But, the evaluative problem does not merely concern what it is felicitous to say. Even in solitary thought, it seems problematic to cite a proposition as evidence for itself. As a result, pragmatic accounts do not deal with all aspects of the evaluative problem.

Second, problems arise even for the pragmatic explanation of the infelicity of citing a proposition as evidence for itself in conversation. In standard Gricean examples, we can recognize that a proposition is literally true even though conversationally inappropriate. For example, consider Grice's example in which a passerby responds to a stranded motorist who asks, 'Is there a garage nearby?' by saying, 'Yes, there is one just round the corner,' when the passerby knows that that garage is closed today. Grice argues that although the reply states a literal truth, it is infelicitous since it conveys the falsehood that the garage is open. Note that we can recognize that what the passerby says is literally true, even though it conveys a falsehood. However, it is hard for us to recognize that it is true to cite a proposition as evidence for itself. If a subject is asked to list all her evidence for a proposition, p, no matter how obvious, it is very unlikely that she would include p itself. Further, if it is suggested to her that she should include p itself, she would likely be dumbfounded or perhaps deny that p is evidence for itself. For example, if you ask a detective to list all her evidence for the claim that the suspect is guilty no matter how obvious, she is very unlikely to include the claim that the suspect is guilty. Furthermore, if you suggest to the detective that she should include the claim that the suspect is guilty, she would likely reject that suggestion, and deny that the claim that the suspect is guilty is evidence for the claim that the suspect is guilty. On one mainstream view, the intuitive truth conditions of statements are strong evidence for their semantic content (e.g. DeRose 1998, and Stanley 2005). On this view, the intuitive truth conditions of evidence statements are strong evidence against the pragmatic story. DeRose (1998) would take the relevant denial as especially telling evidence against the pragmatic story since he argues that a literally false utterance cannot seem true by conveying a truth. While this mainstream view is controversial,[5] it is widely agreed that the cancellation test is one of the most important tests for pragmatic implication. I now turn to argue that it is difficult for the purported pragmatic explanation of the infelicity of citing a proposition as evidence for itself to pass the cancellation test.

Consider the following attempted pragmatic explanation of the infelicity of citing p as evidence for itself. It may be suggested that in many, if not all, contexts, a request for evidence for p is a request for evidence for p consisting of propositions other than p. Suppose that, despite this, Anne replies to Bill's request for evidence that climate change is occurring by simply saying, 'Climate change is occurring.' On the assumption that Anne is being cooperative, her reply generates the implicature that she has no evidence for the claim that climate change is occurring consisting of propositions distinct from the proposition that climate change is occurring. If she does in fact have further evidence, then her reply is infelicitous since it conveys a falsehood. It is a problem for this suggested explanation that it seems hard to cancel the relevant implication. For instance, Anne might attempt to cancel the relevant implication by saying, 'Climate change is occurring but I don't mean to imply that I don't have any other evidence for the claim that climate change is occurring.' However, it's hard to make sense of this attempted cancellation. In particular, it's hard to make sense of the first conjunct, which, in the context, amounts to the claim that the fact that climate change is occurring is evidence that it is occurring. But, it seems false to claim that the fact that climate change is occurring is evidence that it is occurring. Since the cancellation test is one of the best diagnostic tests of pragmatic implication, this is a serious problem for any pragmatic attempt to explain the infelicity of citing a proposition as evidence for itself.[6]

I conclude that there is little prospect of a pragmatic explanation of the infelicity of citing a proposition for itself. In the rest of the paper, I turn to consider attempts to defend the probability-raising conception of evidential support by focusing on the two conditions it places on a proposition, p, being evidence for a subject for a hypothesis, namely that p be part of the subject's evidence and that p raise the probability of the hypothesis. I start by examining invariantist modifications of the first condition.

[5] For instance, Bach (2002) denies that intuitive truth conditions are strong evidence for semantics. Further, some argue against DeRose that even a literally false claim can seem felicitous if it conveys a truth (e.g. Brown 2006).

[6] Brown (2013) criticizes attempts to explain the infelicity of citing p as evidence for p by appealing to norms governing conversation other than Grice's principles, including the idea that assertion is governed by a challenge-retract norm, or a norm of dialectical effectiveness.

4. INVARIANTIST ACCOUNTS OF EVIDENCE

According to the probability-raising account of evidential support, a proposition is evidence for a subject for a hypothesis only if it is part of that subject's evidence. A defender of the probability-raising account might suggest that she can avoid the problem facing that account by combining it with an appropriate account of what it takes for a proposition to be part of a subject's evidence. So far, we have said very little about what's required for a proposition to be part of a subject's evidence. But, perhaps, a restrictive account would help defend the probability-raising approach?

To consider the suggestion, let us start with Williamson's suggestion that it is sufficient for a proposition to be part of a subject's evidence that she knows it, or *Sufficiency*. (He also endorses the reverse conditional: that it is necessary for a proposition to be part of a subject's evidence that she knows it.) Combined with the probability-raising conception of evidential support, *Sufficiency* has the consequence that any known proposition that raises the probability of some hypothesis is evidence for itself. As a result, *Sufficiency* offers no way for the probability-raising conception of evidential support to avoid the evaluative problem. To see this, let us reconsider the original problem posed by the case of Morse who learns from eyewitness testimony the proposition that Burglar Bill was near the scene of the crime just before it happened, or b. If knowledge is sufficient for a proposition to be part of a subject's evidence, then as soon as Morse learns b, this proposition is part of his evidence. Plausibly, b does raise the probability of some hypothesis or other, say the hypothesis that Burglar Bill was responsible for the crime. Thus, by the probability-raising account of evidential support, b is evidence for itself. But, as we have seen, it would be infelicitous for Morse to reply to a request for evidence for the claim that Burglar Bill was near the scene of the crime just before it happened by saying, 'Burglar Bill was near the scene of the crime just before it happened.'

Clearly, a defender of the probability-raising conception of evidential support cannot avoid the evaluative problem by holding that some condition weaker than knowledge is sufficient for a proposition to be part of one's evidence. Instead, she may hope to avoid the problem by making the conditions for a proposition to be evidence tougher than knowledge. For instance, some suggest that one's evidence is restricted to what is known in some particular kind of way, say what can be non-inferentially known.[7] Alternatively, it may be suggested that a proposition is part of a subject's evidence only if she knows that she knows it.

However, it is not clear that the proposed restrictions on evidence avoid the problematic result.[8] Rather, the problematic result still seems to apply to the

[7] Compare Goldman (2009), who restricts evidence to what is non-inferentially justified.

[8] To the extent that the class of known propositions meeting the relevant restriction is much smaller in size than the class of known propositions then these views may face the objection that they overly restrict evidence. In particular, on the suggestion that evidence is

more restricted notion of evidence. For example, suppose that one has non-inferential knowledge that one has hands so that the proposition that one has hands is part of one's evidence. The probability that one has hands given that one has hands is 1, and so plausibly higher than the prior probability that one has hands. So, the proposition that one has hands meets the conditions for being evidence for the proposition that one has hands. But, if one is asked for one's evidence for the claim that one has hands, it seems infelicitous to reply by saying that one has hands.

The same point applies to the suggestion that a proposition is part of a subject's evidence only if she knows that she knows it. Even in a case in which a subject knows that she knows that p, it still seems problematic for her to reply to a request for evidence for p merely by saying p. For instance, perhaps a scientist not only knows that increased alcohol consumption increases the risk of cancer, or i, but knows that she knows this. The probability of i given i is 1 and so plausibly higher than the prior probability of i. Nonetheless, it would seem inappropriate for her to reply to a request for evidence that increased alcohol consumption increases the risk of cancer merely by restating that claim.

The idea that any proposition that is part of one's evidence is evidence for itself may seem less problematic if such propositions are restricted to propositions concerning how things seem experientially to one. For instance, if one is asked to cite one's evidence for the claim that it seems to one as if one is in pain, one might respond to this request simply by saying that it seems to one as if one is in pain. The proposed restriction might be motivated by the restriction of evidence to what is non-inferentially known and the claim that one has non-inferential knowledge only of propositions concerning how things seem experientially to one.

This traditional view is, of course, controversial. However, even setting aside the correctness of this traditional view, it faces another objection. On the suggested view, one's evidence is restricted to propositions concerning how things seem to one because only such propositions are non-inferentially known. As a result, the suggested view must hold that all other propositions are inferentially known. In particular, it must hold that propositions about the external world are inferentially known. For example, it might be suggested that I gain knowledge that I have a hand by inference to the best explanation from the proposition that it seems to me as if I have a hand. Alternatively, it might be suggested that I gain knowledge that I have a hand by inference from my non-inferential knowledge that it seems to me as if I have a hand and my knowledge that such seemings are typically reliable. But, the history of such attempts to provide inferential knowledge of propositions about the external world is not reassuring. It is hard to see why the best explanation of our sensory experiences is that they typically represent the world veridically.

what you know that you know, it seems that subjects who know things, but lack the concept of knowledge, don't have any evidence.

It is hard to provide an account of how one knows that seemings are reliable prior to having any knowledge whatsoever about the external world. Thus, restricting one's evidence to propositions that are non-inferentially known understood as propositions concerning how things seem experientially to one is unlikely to deliver a satisfactory epistemology of the external world.

In conclusion, it seems implausible that a defender of the probability-raising conception of evidential support can avoid the evaluative problem by imposing invariant conditions for a proposition to be evidence stronger than that the proposition be known. However, it might be suggested instead that a contextualist approach would help the defender of the probability-raising conception. I consider that suggestion in the next two sections, by first looking at contextualism about evidential support, and then considering contextualism about evidence.

5. CONTEXTUALISM ABOUT EVIDENTIAL SUPPORT

According to contextualism about evidential support, what it takes for a proposition that is evidence to count as supporting a hypothesis varies with context.[9] On this contextualist view, a proposition that is evidence in two distinct contexts might count as evidence for a certain hypothesis in one but not the other context. Contextualism about evidential support can be combined with invariantism about evidence. On this view, what it takes for a proposition to be part of a subject's evidence is invariant across contexts, even though what it takes for an evidence proposition to support a hypothesis varies across contexts. This combination of views will be my focus here.

Just as contextualist accounts of knowledge come in several main varieties, so do contextualist accounts of evidential support. On a contrastivist approach to knowledge, whether it is true to ascribe knowledge that p to a subject depends on the contrasts to p (e.g. Lewis 1996, Schaffer 2004). Analogously, one could propose a contrastivist view of evidential support, on which whether it is true to say 'p is evidence for q' depends on the contrasts to q. For example, the proposition that Burglar Bill left his fingerprints all over the safe might count as evidence in two contexts, but count as evidence for the proposition that Burglar Bill stole the rubies in the first context, in which the question is whether Burglar Bill rather than some other thief stole the rubies, but not in the second context, in which the question concerns what was stolen.

On a standards-shifting version of contextualism about knowledge (e.g. DeRose 1995), context affects the standard required for a knowledge attribution to be true. Similarly, on a standards-shifting version of contextualism about evidential support, context affects the standard required for an evidential support statement to be true. For instance, the proposition that it

[9] Contrastivism about justification is defended inter alia by Schaffer (2005: 255) and Sinnott-Armstrong (2008).

seems to me as if I have hands might count as evidence in both an ordinary and a sceptical context, but count as evidence for the hypothesis that I have hands only in the ordinary, but not the sceptical, context.

Let us now consider whether either version of contextualism about evidential support could help with the evaluative problem, starting with the contrastivist version. Suppose that p is part of one's evidence and is also evidence for q. We might try to explain why it is nonetheless not evidence for p by appealing to the fact that the relevant contrasts to p and q are distinct. Perhaps, p is evidence for q rather than the contrasts to q, whereas p is not evidence for p rather than the contrasts to p. However, in fact, it is hard to see how appeal to the fact that p and q have different contrasts helps in the least. p is incompatible with any not-p possibility whatsoever. So, if p is part of one's evidence, it is conclusive evidence for p, regardless of what not-p possibilities we consider.

Even if a contrastive account of evidential support does not seem to help, a contextualist about evidential support might instead appeal to a standards-shifting version of contextualism about evidential support. This would most naturally fit with a threshold account of evidential support rather than a probability-raising one. On a threshold account, e is evidence for p only if the probability of p/e meets a certain threshold. Such a threshold account of evidential support lends itself to a contextualist view on which the relevant threshold varies with conversational context. Thus, we can imagine that evidence that makes a certain proposition p probable to a degree n might count as evidence for p in a context in which the threshold for evidential support is lower than n, but not count as evidence for p in a context in which the threshold for evidential support is higher than n.

However, this standards-shifting contextualist account of evidential support is no more successful than a contrastive account of evidential support in explaining why a proposition that is evidence is not evidence for that very proposition itself. The probability of a proposition p given p is 1. And, of course, the probability of p given p cannot be higher than 1. Thus, there is no prospect of explaining why p is not evidence for p by appeal to the claim that the probability of p given p does not meet the threshold required for evidential support.

In conclusion, contextualism about evidential support doesn't seem to offer a solution to the evaluative problem. Instead, in the next section, I consider contextualism about evidence.

6. CONTEXTUALISM ABOUT EVIDENCE

According to contextualism about evidence, what counts as evidence depends on the conversational context (e.g. Neta 2003). For example, it might be suggested that in a conversational context in which we are interested in the evidence for p, p no longer counts as evidence. This contextualist suggestion might seem to help with the evaluative problem. In particular, it would explain why, in a context in which we are interested in the evidence for p, it is false to cite p as evidence for p.

We may worry that the proposed contextualist view makes it too easy to change the truth value of attributions of evidence. On the proposed contextualist view, one can change the truth value of an evidence attribution merely by raising the question, 'What is the evidence for p?' In particular, one needn't provide any evidence against p, or any evidence that the process by which the belief that p was formed was unreliable. But surely it's not so easy to change the truth value of evidence attributions, or their related normative consequences? The truth of ascriptions of evidence appears to have normative consequences for belief. An invariantist might put this by saying that one should proportion one's belief to the evidence. A contextualist could accept a suitably modified version of this claim, perhaps saying that S should believe in accordance with whatever S can truly self-ascribe as 'evidence'. For instance, suppose that it is initially true for Morse to say, 'Part of our evidence that Scarlet is the murderer is that Scarlet's gun was used in the murder.' The truth of this self-ascription has consequences for what Morse ought to believe. However, on the proposed contextualist view, the truth of claims about evidence can be changed merely by raising the question, 'What is the evidence for p?' For instance, if someone asks what is the evidence that Scarlet's gun was used in the murder, then it is no longer true for Morse to say, 'Part of our evidence that Scarlet is the murderer is that Scarlet's gun was used in the murder.' This is so whether or not the enquirer provides any evidence against the claim that her gun was used in the murder, or that the process by which this belief was formed was unreliable.

A more specific worry for the proposed contextualist view is that it fails to offer a good answer to the evaluative problem. According to the probability-raising conception, if a proposition is part of a subject's evidence and raises the probability of some hypothesis or other, then it is evidence for itself. But, in general, it is infelicitous to cite a proposition as evidence for itself. According to the proposed contextualist solution, a request for evidence for p so affects the conversational context that it is no longer true to cite p as evidence. In a context in which it is not true to cite p as evidence, it is not true to cite it as evidence for any proposition whatsoever. In this way, contextualism about evidence might be offered as an explanation of why it is typically infelicitous to cite a proposition as evidence for itself. However, there is a mismatch between the proffered contextualist explanation and the data to be explained. What we wanted was not an explanation of why, in some contexts, a proposition p cannot be truly cited as evidence and so cannot be truly cited as evidence for any proposition whatsoever. Instead, we wanted an explanation of why it may seem infelicitous to cite a proposition p as evidence for p itself even if it is felicitous to cite p as evidence for other propositions.

To see the problem in more detail, consider a context in which a subject initially truly claims that a proposition, p, is part of his evidence. For example, suppose that Holmes has been carrying out an investigation into a recent murder. He discovers that Scarlet's gun was used as the murder weapon, where this proposition raises the probability that Scarlet was the murderer. Thus, initially, it is true for Holmes to say, 'That Scarlet's gun was used as the murder weapon is evidence that Scarlet was the murderer.' Now, suppose that Watson asks Holmes what is the evidence that Scarlet's gun was used as the murder weapon. On the proposed contextualist view, after Watson asks this question, Holmes can no longer truly say, 'Part of my evidence is that Scarlet's gun was used as the murder weapon.' Thus, the contextualist view under discussion predicts that, after Watson's question, the following two claims made by Holmes should both seem infelicitous for the same reason, namely that they both involve him falsely citing a proposition as evidence:

1. That Scarlet's gun was used as the murder weapon is evidence that she is the murderer.
2. That Scarlet's gun was used as the murder weapon is evidence that Scarlet's gun was used as the murder weapon.

However, it seems that 1) and 2) are in fact infelicitous in very different ways. After Watson's request for evidence for the claim that Scarlet's gun was used as the murder weapon, 1) doesn't seem false but instead irrelevant though true. It seems irrelevant since 1) doesn't offer evidence for the claim that Scarlet's gun was used as the murder weapon, but rather uses the latter claim as evidence for some further claim. By contrast, 2) seems false. For, it is hard to understand how the claim that Scarlet's gun was used as the murder weapon is evidence that her gun was used as the murder weapon. But, on the proposed contextualist view, 1) and 2) should both seem false in precisely the same way, namely that they both involve Holmes falsely citing a proposition as evidence.

A further worry arises for the proposed contextualist view from cases in which no explicit question is raised about the evidence for some proposition. Suppose that, as before, Holmes has discovered that Scarlet's gun was used as the murder weapon where the latter proposition raises the probability that Scarlet was the murderer. But, this time, Watson doesn't ask for evidence that Scarlet's gun was used as the murder weapon. Even if Watson fails to ask this question, it would seem true for Holmes to utter 1) but false for him to utter 2). The contextualist might attempt to explain this by saying that the very statement 2) implicitly raises the question of what is the evidence that Scarlet's gun was used as the murder weapon. Thus, it so changes the context in such a way that it is no longer true for Holmes to cite as part of his evidence the proposition that Scarlet's gun was used as the murder weapon. On the proposed view, it is true for Holmes to cite as evidence the proposition that Scarlet's gun was used as the murder weapon when he makes statement 1) but not when he makes statement 2). In this way, the contextualist might hope to explain why 1) seems felicitous and true, even though 2) seems infelicitous and false. However, this reply is equally unsuccessful. For, it fails to explain felicity differences in 1) and 2) even when their order is reversed. In particular, even if statement 2) is made before statement 1), 1) still seems felicitous and true. In addition, the suggested contextualist view has difficulties with conjunctive claims such as the following:

3. That Scarlet's gun was used as the murder weapon is not evidence that Scarlet's gun was used as the murder weapon but is evidence that Scarlet was the murderer.

3) seems true, but of course involves consideration of the evidence for the claim that Scarlet's gun was used as the murder weapon. By the suggested contextualist view, when we are interested in the evidence for the claim that Scarlet's gun was used as the murder weapon, it is no longer true to cite that proposition as evidence. But, 3) involves one and the same proposition, namely that Scarlet's gun was used as the murder weapon, being cited as evidence for a distinct proposition, but not for itself.[10]

The contextualist approach to evidence considered so far fails to deal with the evaluation problem. However, an alternative version of contextualism about evidence might seem to fare better. On this alternative, the pool of evidence is implicitly relative to the hypothesis to be evaluated, so that 'evidence' has an implicit argument place for a hypothesis, and h is never included in the pool of evidence relevant to the evaluation of h itself. This second form of contextualism about evidence seems to overcome the main objection to the formulation of contextualism about evidence considered above, namely the mismatch problem. For this second style of contextualism about evidence would allow that, in a single context, it may be true for a subject to say, 'p is evidence for q,' but not 'p is evidence for p.'[11]

However, to the extent that our aim is simply to deal with the evaluative problem, it seems that we could do so by adopting the suggested rule about the relation of evidence and hypothesis within a non-contextualist framework. For instance, we could simply modify *PR* by adding a third condition ruling it out that a proposition can be evidence for itself:

> *PR*: e is evidence for h for S if and only if i) S's evidence includes e; ii) P(h/e) >P(h); and iii) e is not h.

This suggests that what is doing the work in the alleged contextualist solution to the evaluative problem is not contextualism, but simply the rule that a proposition cannot be evidence for itself.

[10] The contextualist might try to deal with 3) by appeal to the idea that there is a context shift within sentence 3). But this is problematic for reasons familiar from the debate concerning contextualism about knowledge. Contextualism is standardly motivated by closure. Thus, contextualists about knowledge hold that, in no context is it true for a subject to say, 'I know I have hands but I don't know that I'm not a BIV.' Similarly, contextualism about evidence is motivated by closure about evidence (Neta 2003: 20). But if context can shift within a sentence, then even on contextualism about evidence it might be true to say, 'I have evidence that I have hands but I don't have evidence that I'm not a BIV.' Thus, contextualists would not be able to avoid the 'abominable conjunctions' that they charge affect closure deniers.

[11] Thank you to one of the anonymous referees for making this suggestion.

Further, we will soon see that the simple rule so far considered fails to deal with a generalized version of the evaluative problem. In the next section we will see that just as it is infelicitous to cite p as evidence for p, so it is also often infelicitous to cite as evidence for p, the conjunction of p with some other proposition. If that is right, then, we need a more complicated rule governing the relation between evidence and hypothesis, whether that is embedded within a contextualist or an invariantist account of evidence.

Within the next section I pursue the question of what rule could solve the generalized evaluative problem within an invariantist account of evidence, suggesting that the relevant rule should appeal to the notion of warrant transmission. This solution should appeal to those who reject the contextualist view on which what counts as evidence is relative to what hypothesis is under evaluation, and instead take it that what one's evidence is doesn't depend on what hypothesis one is considering. For example, it should be congenial to those who accept the popular idea that some doxastic condition, perhaps knowledge, or some stronger condition, is sufficient for a proposition to be part of one's evidence. However, someone who thinks there are independent motivations for the contextualist idea that 'evidence' is relative to the hypothesis under consideration could combine their contextualism with the warrant-transmission rule I later recommend. But, we should be clear that it is the rule that is central to the solution to the evaluative problem, rather than the contextualism with which it might be combined.

7. INVARIANTIST ACCOUNTS OF EVIDENTIAL SUPPORT

As we have seen, an invariantist defender of the probability-raising conception of evidential support might hope to avoid the evaluative problem by adding some condition to the account of *Probability-raising* which has the effect that a proposition is never evidence for itself. Most simply, she could modify *Probability-raising* by simply adding the simple stipulation that e is evidence for S for h only if e is not h.

However, the simple stipulation doesn't offer a sufficiently general solution to the evaluative problem. So far, we have concentrated on the fact that when asked for one's evidence for p, it is infelicitous to reply with 'p'. But, it is also infelicitous to reply with 'p or p', or 'p and p'. For instance, if Morse is asked for his evidence that Burglar Bill committed the burglary, it is not only infelicitous for him to reply by saying, 'Burglar Bill committed the burglary' but also infelicitous for him to reply by saying either, 'Burglar Bill committed the burglary or Burglar Bill committed the burglary,' or 'Burglar Bill committed the burglary and Burglar Bill committed the burglary.'

Notice that the probability-raising conception of evidential support seems to have the result that the propositions (p or p) and (p and p) are evidence for p. Assuming that p is part of the subject's evidence and is evidence for some hypothesis or other, the probability of p is less than 1. Given a minimum of logical acuity, a subject who knows that p can deduce and so know that (p or p) or (p and p). Assuming that knowledge is sufficient for evidence, the propositions (p or p) and (p and p) are part of the subject's evidence. Furthermore, the probability of p/(p or p) equals the probability of p/(p and p) equals 1. Thus, (p and p) and (p or p) are evidence for p by the probability-raising conception of evidential support. So, the question arises whether adding the simple stipulation would rule it out that (p and p) and (p or p) are evidence for p.

Whether it can do so depends on the controversial issue of the individuation of propositions. On Stalnaker's coarse-grained account, p is the same proposition as q if and only if p and q are true in all the same possible worlds. Thus, combined with Stalnaker's account, the simple stipulation rules it out that any of the following are evidence for p: p, (p or p), or (p and p). The same result follows even on one fine-grained account, namely Frege's. On Frege's account, roughly, p is the same proposition as q if a rational thinker couldn't take conflicting attitudes to p and q. A subject couldn't rationally take conflicting attitudes to p, (p or p), or (p and p). However, on the increasingly popular structured propositions view, p, (p or p), and (p and p) are distinct propositions. Thus, on the structured propositions view, the simple stipulation rules out neither of the following propositions from being evidence for p: (p or p), (p and p).

Someone holding a structured propositions view might attempt to rule it out that (p or p) and (p and p) are evidence for p by introducing a distinct stipulation according to which p is evidence for q only if it is not the case that p is logically equivalent to q. While this would certainly rule out the problematic propositions, this stipulation seems too strong. For, there may be cases in which one first knows one proposition, p, and then comes to know a distinct proposition, q, by a complicated deduction from p, where it seems correct to say that p is one's evidence for q, even though p and q are logically equivalent. Instead, a defender of a structured propositions view might appeal to a Gricean explanation of the relevant infelicity, suggesting that saying (p and p) or (p or p) is a more prolix way of saying p and so violates Grice's Maxim of manner. I won't consider these options any further since it is controversial how to individuate propositions. Rather, I will set aside worries arising from (p or p) and (p and p) to focus on other cases.

When Morse is asked for his evidence that Burglar Bill committed the burglary, it is not only infelicitous for him to reply with 'Burglar Bill committed the burglary,' but also infelicitous for him to reply by saying, 'Burglar Bill committed the burglary and Obama is president.' Yet, neither of the two stipulations so far considered helps. The conjunction (that Burglar Bill committed the burglary and Obama is president) is neither the same proposition as the proposition that Burglar Bill committed the burglary, nor logically equivalent to it. Further, on the probability-raising conception of evidential support, the conjunction (Burglar Bill committed the burglary and Obama is president) may count as evidence for Morse for the claim that Burglar Bill committed the burglary. To see this, suppose that when Morse comes to know that Burglar Bill committed the burglary, or b, he also independently knows that Obama is president, or o. Further, let us suppose that b is evidence for Morse for some hypothesis, say that Burglar Bill is still active in the area. So, the probability of b is less than 1. From Morse's knowledge of b and his knowledge of o he deduces, and so comes to know, the conjunction (b and o). Assuming that knowledge is sufficient for evidence, the conjunctive proposition (b and o) is part of his evidence. The probability of b given (b and o) is 1. Thus, the probability of b given (b and o) is higher than its prior probability, and so the conjunction counts as evidence for b on the probability-raising conception.

A defender of the probability-raising conception of evidential support might try to deal with this problem by adding a further stipulation to her account.[12] For instance, she might suggest that if p is evidence for q, then it is not the case that p entails q. But, this stipulation seems too strong. Surely, one way to have evidence for a proposition is for it to be the conclusion of a valid argument whose premises one knows. But the suggested stipulation would rule this out. Instead, a defender of a probability-raising conception might suggest that p is not evidence for q if p is the conjunction of q with some other proposition. Thus, on the proposed view, the account of evidential support now includes two distinct stipulations: 1) p is not evidence for q if p just is q; and 2) p is not evidence for q if p is the conjunction of q with some other proposition.

One potential worry with this suggestion is that it makes the conditions for evidential support look disjunctive and ad hoc. More seriously, the proposed stipulation 2) seems too strong. For, it's not clear that the evidence for p can never take the form of the conjunction of p and another proposition. To see this, consider the following inference:

4. The murderer was female and left-handed.
5. So, the murderer was female.

Imagine that Holmes comes to know 4) by combining his knowledge that the murderer was female and his knowledge that the murderer was left-handed. For instance, perhaps he knows that the murderer was left-handed by analysing the strangulation marks on the victim, and knows that the murderer was female by eyewitness testimony. In such a case, it seems correct to rule out 4) as being evidence for 5). Rather, it seems that 5) is part of Holmes' evidence for 4).

12 Notice that one cannot deal with the problem by pointing out that the conjunction contains irrelevant information, namely that Obama is president. It would seem infelicitous to cite as evidence for b the conjunction of b with another proposition that is relevant to b. For instance, it would seem infelicitous to cite as evidence that Burglar Bill is the burglar the conjunctive proposition (Burglar Bill is the burglar and CCTV camera footage caught him entering the bank just before the heist).

However, there may be circumstances in which 4) is evidence for someone for 5).[13] Suppose that Watson asks Holmes what he has learnt about the murderer. Holmes utters 4) but without telling Watson how he learnt 4). Watson comes to know 4) from Holmes' utterance and then in his dull way infers 5) from 4). As he does so, he says in shock, 'So the murderer was a woman. Who would have thought a woman could do such a dastardly deed!' In this case, there seems nothing wrong epistemically with Watson inferring from 4) to 5), nor with saying that Watson's evidence for 5) is 4). After all, Watson does not know the more detailed evidence from which Holmes inferred that the murderer is a left-handed woman. Further, we needn't think of the evidence that Watson gains from Holmes as merely being that Holmes said, 'The murderer was female and left-handed.'

Cases in which a conjunction may be evidence for one of its conjuncts may be provided not only by testimony, but also by inference to the best explanation. When one first comes to know some conjunction by inference to the best explanation, it's arguable that one can infer from the conjunction to one of its conjuncts, and that the conjunction is evidence for its conjuncts. An example might be provided by bird identification. Standardly, one comes to know that the bird one is looking at is, say, a female eider duck as the best explanation of the bird's observed plumage, behaviour, and location combined with one's background knowledge of the appearance of various birds and their seasonal migration patterns. One doesn't first have the separate pieces of knowledge that the duck is female and that the duck is an eider, and then combine them to know that the duck is a female eider. Given that male and female birds of the same species typically look quite different, one's knowledge that a bird is of a particular species, say an eider, does not standardly precede one's knowledge that it is, say, a female eider. Further, while a zoologist could know by an anatomical investigation that a bird is female before knowing its species, this is not the standard procedure by which a birdwatcher comes to know that some bird is female. Rather, a birdwatcher first knows, say, that the bird is a female eider duck, and can then deduce and so come to know that it is female. In such a case, the proposition that the duck is a female eider is arguably evidence that it is female.

In conclusion, it seems that we cannot solve the evaluative problem for the probability-raising conception of evidential support by the simple stipulation that p is evidence for q only if p is not the same proposition as q. For, it is not only infelicitous to reply to a request for evidence for p by saying, 'p', but also typically infelicitous to reply by saying 'p and q'. It would be preferable if we could add a single condition that explains both why it is infelicitous to cite a proposition as evidence for itself, and why, in many cases, it is infelicitous to cite as evidence for p, the conjunction of p with some other proposition. What we are looking for, then, is a condition on evidential support which seems both philosophically motivated and provides a unified account of why

[13] Thanks to Sgaravatti for this example.

it is infelicitous to cite a proposition as evidence for itself and (often) infelicitous to cite as evidence for p the conjunction of p and another proposition. I want to end the paper by tentatively exploring a potential solution, which appeals to the role or function of evidence. The solution appeals to the idea that one central function of evidence is to enable us to gain justified belief or knowledge of propositions for the first time. For example, if we think of the role of evidence in enquiry, the paradigmatic situation of interest is one in which evidence enables us to extend the range of hypotheses we justifiably believe.

The role of evidence in enquiry might seem to offer an explanation of both why it is infelicitous to cite p as evidence for itself, and why, in many cases, it is infelicitous to cite as evidence for p a conjunctive proposition one of whose conjuncts is p. The suggestion would be that both of these are infelicitous because what is cited as evidence doesn't provide a route to first-time justified belief or knowledge of the relevant proposition. For example, consider the following simple circular argument:

6. p.
7. Therefore, p.

One cannot use this argument to acquire first-time justified belief in its conclusion. Since the premise is the conclusion, in having justified belief in the premise one already has a justified belief in the conclusion. Nor could the inference from p to p strengthen one's justified belief in the conclusion. As it is sometimes put, such a circular argument is question-begging or exhibits a 'failure of transmission of warrant'.[14]

Similarly, consider the following argument:

8. (p and q).
9. Therefore, p.

One cannot gain first-time justified belief in p by the inference from 8) to 9), at least where one has a justified belief in the premise (p and q) only because one infers (p and q) from one's justified belief that p and one's justified belief that q.

As we have already seen, there are arguably some cases in which one has a justified belief in the premise (p and q) in some way other than by inference from one's justified belief that p and one's justified belief that q. In such a case, there seems no reason to deny that warrant transmits across the argument from 8) to 9). In our earlier example, Watson comes to know the conjunction that the murderer was female and left-handed from Holmes' assertion

[14] Appealing to the notion of transmission of warrant not only helps rule it out that a proposition can be evidence for itself, but offers a defender of the structured propositions view the promise of arguing that neither (p and p) nor (p or p) are evidence for p. One cannot gain first-time justified belief in p by the inference from (p or p) to p, at least where one has a justified belief in the premise (p or p) only because one infers (p or p) from one's prior belief that p. Similar points hold for the inference from (p and p) to p.

that she was female and left-handed. Here, Watson could infer and thereby gain first-time justified belief that the murderer was female by inference from the conjunction, the murderer was female and left-handed. Thus, an account appealing to warrant transmission can explain how a conjunction can sometimes be evidence for its conjuncts. It can do so because whether an argument exhibits a failure of warrant transmission is relative to background information (e.g. Jackson 1984, Davies 2009).

It seems, then, that appeal to the idea of warrant transmission seems to offer a unified account of the cases in which an evidence proposition that raises the probability of some hypothesis nonetheless isn't evidence for that hypothesis. Adding a warrant transmission condition to the probability-raising conception of evidential support not only reflects the role of evidence in discovery, but can accommodate the plausible idea that evidence is useful to us because it is more accessible than that for which it is evidence (Roush 2005, Kelly 2008b[15]). For instance, we may consult fossils now accessible to us in order to know what the earth was like in the far past. If evidence is more accessible to us than what it is evidence for, then our evidence for a proposition does not include that proposition itself. (Notice that the idea that evidence is typically more accessible to us than what it is evidence for need not be motivated by the implausible claim that we are infallible about our evidence.)

Much remains to be done to further defend and flesh out the idea of adding a warrant transmission condition to an account of evidential support. While there is not enough space to do that here, I want to point out a few issues for further examination. First, we need to develop a detailed account of the failure of the transmission of warrant. We introduced the notion by appeal to simple circular arguments in which one cannot acquire justified belief in the conclusion for the first time by inference from justified belief in the premises. However, it is controversial under what conditions an argument exhibits a failure of transmission of warrant (e.g. see Jackson 1984, Wright 2000, and Davies 2009, inter alia). Further, different accounts classify different arguments differently (for discussion, see Sgaravatti 2013). It is useful to note that while some of the most influential accounts do not include simple circular arguments of the form p therefore p, some better accounts do include simple circular arguments (e.g. Sgaravatti's doxastic formulation). Second, as we have already seen, whether an argument exhibits a failure of transmission of warrant is sometimes relative to background information. So, we need to ensure that our account of warrant transmission accounts for the role of background information. Third, we need to decide how to implement the idea of adding a warrant transmission condition to an account of evidential support. We could add a warrant transmission condition to a probabilistic account of evidential support, whether a probability-raising account or a threshold account. Alternatively, we could explore the idea of

[15] Kelly (2008b) states, 'In general, we rely on evidence in cases in which access to the truth would otherwise be problematic' (section 3).

understanding evidential support via warrant transmission while dropping a probabilistic condition on the relation between evidence and hypothesis. This latter suggestion connects in an interesting way with the recent work on evidential support by Pryor. One way to understand the argument of this paper is as appealing to certain cases of the failure of warrant transmission, namely the inference from p to p, to show that the probability-raising account does not provide sufficient conditions for evidential support. Interestingly, Pryor has been appealing to warrant transmission to cast doubt on the necessity direction of the probability-raising account of evidential support (Pryor 2004, 2013).

Pryor (2013) points out that a very large range of different views in epistemology share an assumption that he labels 'credulism'. According to this assumption, a subject can have justification to believe p without having antecedent justification to believe not-q, even though having justification to believe q would undermine her justification to believe that p. Philosophers may disagree about which pairs of propositions instantiate this structure. However, wherever the structure is instantiated, it seems that one could acquire justification to believe not-q on the basis of one's justification to believe that p. For instance, suppose that, as many philosophers think, one's experience provides justification to believe that one has hands even without one's having antecedent justification to believe that one is not a BIV. On this view, one could acquire first-time justification to believe that one is not a BIV by the following argument (e.g. Pryor 2004):

10. I am having an experience as of a hand.
11. So, I have a hand.
12. So, I am not a BIV.

On this view, the justification provided by 10) to 11) can help provide first-time justification to believe 12). However, accepting this puts pressure on the necessity direction of the probability-raising account of evidential support. For, as has been widely noted, having an experience as of a hand raises the probability that I am a BIV (e.g. White 2006). After all, if I were a BIV, I would have the experience as of being handed. So, whereas I've been concerned to use considerations of warrant transmission to deny the sufficiency direction of the probability-raising account of evidential support, Pryor has used considerations of warrant transmission to deny the necessity direction of the probability-raising account of evidential support.

One could then see the arguments of this paper and Pryor's as working together to suggest that the notion of evidential support should be explicated by appeal to the notion of transmission of warrant, rather than an increase in probability. While this thought is intriguing, its examination and possible development remain for a future occasion. My main conclusion is that it is not sufficient for some evidence to support a hypothesis that it raises its probability, and that adding a warrant transmission condition to a probability-raising account of evidential support can help deal with this problem. This conclusion is independent of both Pryor's suggestion that some evidence may support a hypothesis even without raising its probability, and the more radical suggestion that we should understand evidential support via warrant transmission rather than probabilistically.

8. AN OBJECTION: SELF-EVIDENCE

I have suggested a warrant transmission condition on a proposition's being evidence for a hypothesis, which rules it out that a proposition may be evidence for itself. However, it may be said, there are cases in which we plausibly justifiably believe and know that p, but would be hard pressed to specify any proposition other than p as evidence. How can my account deal with such cases?

For instance, suppose someone requests my evidence for my claim that I'm feeling cold now. It certainly seems difficult for me to cite a proposition that is both evidence for my claim and also distinct from that claim. However, we need not take such cases to show that we must sometimes allow a proposition to be evidence for itself. To avoid a potential infinite regress of justification, we should anyway allow that a subject may have immediate justification to believe a proposition, where that immediate justification does not depend on her having justification to believe some further proposition. Connecting immediate justification with evidence, we could either allow that a subject may have immediate justification to believe a proposition even without having evidence for it, where it is understood that a proposition is evidence for a subject only if she stands in a doxastic relation to it. Alternatively, we might extend the notion of evidence to include non-doxastic states such as experiences, and allow that having an experience with the content that p may constitute evidence for the proposition that p. To illustrate, suppose that a subject has immediate justification to believe that she has hands. It might be suggested that she has immediate justification to believe that she has hands, even while lacking any evidence that she has hands. For instance, an externalist might argue that she has justification to believe that she has hands in virtue of the fact that the process by which she formed this belief is reliable, even though she lacks any evidence that she has hands. Alternatively, it might be suggested that the subject has justification to believe that she has hands in virtue of her experience as of having hands, allowing evidence to consist both in propositions and non-proposition-like states, such as experiences. Both options make good the idea that she has justification to believe that she has hands even while denying that a proposition is evidence for itself.

9. CONCLUSION

I have been examining whether the popular probability-raising account of evidential support can avoid the problematic consequence that any proposition that is evidence for some hypothesis is evidence for itself. I first considered and rejected the suggestion that the probability-raising account could be defended by endorsing some particular invariantist account of what it is for a proposition to be part of a subject's evidence. I then examined whether a solution could be found by appealing either to contextualism about evidence or contextualism about evidential support. However, we saw that appeal to contextualism fails to help defend the probability-raising conception. In the light of this discussion, I've suggested that the best solution is an invariantist modification of the condition for an evidence proposition to be evidence for a hypothesis, a modification that appeals to the notion of warrant transmission.

REFERENCES

Achinstein, P. 2001. *The Book of Evidence*. Oxford: Oxford University Press.

Bach, K. 2002. 'Seemingly semantic intuitions.' In J. Campbell, M. O'Rourke, and D. Shier (eds.), *Meaning and Truth*, New York: Seven Bridges Press, 21–33.

Brown, J. 2006. 'Contextualism and warranted assertability manoeuvres.' *Philosophical Studies* 130: 407–35.

Brown, J. 2013. 'Infallibilism, evidence and pragmatics.' *Analysis* DOI: 10.1093/analys/ant071.

Davies, M. 2009. 'Two purposes of arguing and two epistemic projects.' In I. Ravenscroft (ed.), *Minds, Ethics and Conditionals: Themes from the Philosophy of Frank Jackson*, Oxford: Oxford University Press, 337–83.

DeRose, K. 1995. 'Solving the sceptical problem.' *Philosophical Review* 704: 1–52.

DeRose, K. 1998. 'Contextualism: An explanation and defence.' In J. Greco and E. Sosa (eds.), *Blackwell Guide to Epistemology*, Oxford: Blackwell, 187–205.

Goldman, A. 2009. 'Williamson on knowledge and evidence.' In P. Greenough and D. Pritchard (eds.), *Williamson on Knowledge*, Oxford: Oxford University Press, 73–91.

Jackson, F. 1984. 'Petitio and the purpose of arguing.' *Pacific Philosophical Quarterly* 65: 26–36.

Kelly, T. 2008a. 'Evidence: Fundamental concepts and the phenomenal conception.' *Philosophy Compass* 3/5: 933–55.

Kelly, T. 2008b. 'Evidence'. *The Stanford Encyclopedia of Philosophy* (Fall 2008 Edition), Edward N. Zalta (ed.), <http://plato.stanford.edu/archives/fall2008/entries/evidence/>.

Lewis, D. 1996. 'Elusive knowledge.' *Australasian Journal of Philosophy* 74: 549–67.

Neta, R. 2003. 'Contextualism and the problem of the external world.' *Philosophy and Phenomenological Research* 66, 1: 1–31.

Pryor, J. 2004. 'What's wrong with Moore's argument?' *Philosophical Issues* 14: 349–78.

Pryor, J. 2013. 'Problems with credulism.' In C. Tucker (ed.), *Seemings and Justification: New Essays on Dogmatism and Phenomenal Conservatism*, Oxford: Oxford University Press, 89–131.

Roush, S. 2005. *Tracking Truth*. Oxford: Oxford University Press.
Schaffer, J. 2004. 'From contextualism to contrastivism.' *Philosophical Studies* 119: 73–103.
Schaffer, J. 2005. 'Contrastive knowledge.' In T. Gendler and J. Hawthorne (eds.), *Oxford Studies in Epistemology*, 1, Oxford, Oxford University Press, 235–72.
Sgaravatti, D. 2013. 'Petitio Principii: A bad form of reasoning.' *Mind*, DOI: 10.1093/mind/fzt086.
Sinnott-Armstrong, W. 2008. 'Moderate classy Pyrrhonean moral skepticism.' *Philosophical Quarterly* 58, 232: 448–56.
Stanley, J. 2005. 'Semantics in context.' In G. Preyer and G. Peter (eds.), *Contextualism in Philosophy*, Oxford: Oxford University Press, 221–54.
White, R. 2006. 'Problems for dogmatism.' *Philosophical Studies* 131: 525–57.
Williamson, T. 2000. *Knowledge and its Limits*. Oxford: Oxford University Press.
Wright, C. 2000. 'Cogency and question begging: Some reflections on McKinsey's paradox and Putnam's proof.' *Philosophical Issues* 10: 140–63.

3. Accuracy, Coherence, and Evidence

Kenny Easwaran and Branden Fitelson

1. SETTING THE STAGE

This essay is about formal, synchronic, epistemic coherence requirements. We begin by explaining how we will be using each of these (five) key terms.

Formal epistemic coherence requirements involve properties of judgment sets that are logical (and, in principle, determinable a priori). These are to be distinguished from other less formal and more substantive notions of coherence that one encounters in the epistemological literature. For instance, so-called "coherentists" like BonJour (1985) use the term in a less formal sense, which implies (e.g.) that coherence is truth-conducive. While there will be conceptual connections between the accuracy of a doxastic state and its coherence (in the sense we have in mind), these connections will be quite weak (certainly too weak to merit the conventional honorific "truth-conducive"). All of the varieties of coherence to be discussed will be intimately related to deductive consistency. Consequently, whether a set of judgments is coherent will be determined by (i.e. will supervene on) logical properties of the set of propositions that are the objects of the judgments in question.

Synchronic epistemic coherence requirements apply to the doxastic states of agents at individual times. These are to be distinguished from diachronic coherence requirements (e.g. conditionalization, reflection, etc.), which apply to sequences of doxastic states across times. Presently, we will be concerned only with the former.[1]

This material has been presented in various places over the past several years, and the list of people with whom we've had useful relevant conversations is too long to enumerate here. But, in addition to three anonymous referees who provided very useful written comments on the penultimate version of this paper, we must single out (in alphabetical order) the following people who have been especially generous with valuable feedback regarding this project: Rachael Briggs, Fabrizio Cariani, Jim Joyce, Ole Hjortland, Hannes Leitgeb, Ben Levinstein, Richard Pettigrew, Florian Steinberger, and Jonathan Weisberg. Branden Fitelson would also like to thank the Alexander von Humboldt (AvH) Foundation for their generous support during his summer (2011, 2012) visits to the Munich Center for Mathematical Philosophy at Ludwig-Maximilians-Universität München, where he presented much of this material in seminars.

1 See Titelbaum (2013) for an excellent recent survey of the contemporary literature on (Bayesian) diachronic epistemic coherence requirements. Some, e.g. Moss (2013) and Hedden (2013), have argued that there are no diachronic epistemic rational requirements (i.e. that there are only synchronic epistemic rational requirements). We take no stand on this issue here. But, we will assume that there are (some) synchronic epistemic rational requirements of the sort we aim to explicate below (see n. 7).

Epistemic requirements are to be distinguished from, e.g., pragmatic requirements. Starting with Ramsey (1926), the most well-known arguments for probabilism as a formal, synchronic, coherence requirement for credences have depended on the pragmatic connection of belief to action. For instance, "Dutch Book" arguments and "Representation Theorem" arguments (Hájek 2008) aim to show that an agent with non-probabilistic credences (at a given time t) must (thereby) exhibit some sort of "pragmatic defect" (at t).[2] Following Joyce (1998, 2009), we will be focusing on non-pragmatic (viz. epistemic) defects implied by the synchronic incoherence (in a sense to be explicated below) of an agent's doxastic state. To be more precise, we will be concerned with two aspects of doxastic states that we take to be distinctively epistemic: (a) how accurate a doxastic state is, and (b) how much evidential support a doxastic state has. We will call these (a) alethic and (b) evidential aspects of doxastic states, respectively.[3]

Coherence requirements are global and wide-scope. Coherence is a global property of a judgment set in the sense that it depends on properties of an entire set in a way that is not (in general) reducible to properties of individual members of the set. Coherence requirements are wide-scope in Broome's (2007) sense. They are expressible using "shoulds" (or "oughts") that take wide-scope over some logical combination(s) of judgments. As a result, coherence requirements do not (in general[4]) require specific attitudes toward specific individual propositions. Instead, coherence requirements require the avoidance of certain combinations of judgments. We use the term "coherence"—rather than "consistency"—because (a) the latter is typically associated with classical deductive consistency (which, as we'll see shortly, we do not accept as a necessary requirement of epistemic rationality), and (b) the former is used by probabilists when they discuss analogous requirements for degrees of belief (viz. probabilism as a coherence requirement for credences). Because our general approach (which was inspired by Joycean arguments for probabilism) can be applied to many types of judgment—including both full

[2] We realize that "depragmatized" versions of these arguments have been presented (Christensen 1996). But, even these versions of the arguments trade essentially on the pragmatic role of doxastic attitudes (in "sanctioning" bets, etc.). In contrast, we will only be appealing to epistemic connections of belief to truth and evidence. That is, our arguments will not explicitly rely upon any connections between belief and action.

[3] The alethic/evidential distinction is central to the pre-Ramseyan debate between James (1896) and Clifford (1877). Roughly speaking, "alethic" considerations are "Jamesian," and "evidential" considerations are "Cliffordian." We will be assuming for the purposes of this article that alethic and evidential aspects exhaust the distinctively epistemic properties of doxastic states. But, our framework could be generalized to accommodate additional dimensions of epistemic evaluation (should there be such).

[4] There are two notable exceptions to this rule. It will follow from our approach that (a) rational agents should never believe individual propositions ($\perp$) that are logically self-contradictory, and (b) that rational agents should never disbelieve individual propositions ($\top$) that are logically true.

belief and partial belief[5]—we prefer to maintain a common parlance for the salient requirements in all of these settings.

Finally, and most importantly, when we use the term "requirements," we are talking about necessary requirements of ideal epistemic rationality.[6] The hallmark of a necessary requirement of epistemic rationality *N* is that if a doxastic state *S* violates *N*, then *S* is (thereby) epistemically irrational. However, just because a doxastic state *S* satisfies a necessary requirement *N*, this does not imply that *S* is (thereby) rational. For instance, just because a doxastic state *S* is coherent (i.e. just because *S* satisfies some formal, epistemic coherence requirement), this does not mean that *S* is (thereby) rational (as *S* may violate some other necessary requirement of epistemic rationality). Thus, coherence requirements in the present sense are (formal, synchronic) necessary conditions for the epistemic rationality of a doxastic state.[7] Our talk of the epistemic (ir)rationality of doxastic states is meant to be evaluative (rather than normative[8]) in nature. To be more precise, we will (for the most part) be concerned with the evaluation of doxastic states, relative to an idealized[9] standard of epistemic rationality. Sometimes we will speak

[5] In fact, the framework can be applied fruitfully to other types of judgment as well. See (Fitelson and McCarthy 2013) for an application to comparative confidence, which leads to a new foundation for comparative probability. For a survey of applications of the general framework, see (Fitelson 2014).

[6] Here, we adopt Titelbaum's (2013: ch. 2) locution "necessary requirement of (ideal) rationality" as well as (roughly) his usage of that locution (as applied to formal, synchronic requirements).

[7] For simplicity, we will assume that there exist some (synchronic, epistemic) rational requirements in the first place. We are well aware of the current debates about the very existence of rational requirements (e.g. coherence requirements). Specifically, we are cognizant of the salient debates between Kolodny (2007) and others, e.g. Broome (2007). Here, we will simply adopt the non-eliminativist stance of Broome et al., who accept the existence of (ineliminable) rational requirements (e.g. coherence requirements). We will not try to justify our non-eliminativist stance here, as this would take us too far afield. However, as we will explain below, even coherence eliminativists like Kolodny should be able to benefit from our approach and discussion (see n. 45). As such, we (ultimately) see our adoption of a non-eliminativist stance in the present context as a simplifying assumption.

[8] Normative principles support attributions of blame or praise of agents, and are (in some sense) action guiding. Evaluative principles support classifications of states (occupied by agents) as "defective" *vs.* "non-defective" ("bad" *vs.* "good"), relative to some evaluative standard (Smith 2005, §3).

[9] Deductive consistency and the other formal coherence requirements we'll be discussing are highly idealized rational epistemic requirements. They all presuppose a standard of ideal rationality that is insensitive to semantic and computational (and other) limitations of (actual) agents who occupy the doxastic states under evaluation. While this is, of course, a strong idealization (Harman 1986), it constitutes no significant loss of generality in the present context. This is because our aims here are rather modest. We aim (mainly) to do two things in this paper: (a) present the simplest, most idealized version of our framework and the (naïve) coherence requirements to which it gives rise, and (b) contrast these new requirements with the (equally simple and naïve) requirement of deductive consistency. Owing to the idealized/evaluative nature of our discussion, we will typically speak of the (ir)rationality of states, and not the (ir)rationality of agents who occupy them. Finally, we will sometimes speak simply of "rational requirements" or just "requirements." It is to be understood that these are shorthand for the full locution "necessary requirements of ideal epistemic rationality."

(loosely) of what agents "should" do—but this will (typically) be an evaluative sense of "should" (e.g. "should on pain of occupying a doxastic state that is not ideally epistemically rational"). If a different sense of "should" is intended, we will flag this by contrasting it with the idealized/evaluative "should" that features in our rational requirements. Now that the stage is set, it will be instructive to look at the most well-known "coherence requirement" in the intended sense.

2. DEDUCTIVE CONSISTENCY, THE TRUTH NORM, AND THE EVIDENTIAL NORM

The most well-known example of a formal, synchronic, epistemic coherence requirement for full belief is the (putative) requirement of deductive consistency.

> (CB) All agents *S* should (at any given time *t*) have sets of full beliefs (i.e. sets of full belief-contents) that are (classically) deductively consistent.

Many philosophers have assumed that (CB) is a necessary requirement of ideal epistemic rationality. That is, many philosophers have assumed that (CB) is true, if its "should" is interpreted as "should on pain of occupying a doxastic state that is not ideally epistemically rational." Interestingly, in our perusal of the literature, we haven't been able to find many (general) arguments in favor of the claim that (CB) is a rational requirement. One potential argument along these lines takes as its point of departure the (so-called) Truth Norm for full belief.[10]

> (TB) All agents *S* should (at any given time *t*) have full beliefs that are true.[11]

[10] We will use the term "norm" (as opposed to "requirement") to refer to local/narrow-scope epistemic constraints on belief. The Truth Norm (as well as the Evidential Norm, to be discussed below) is local in the sense that it constrains each individual belief—it requires that each proposition believed by an agent be true. This differs from the rational requirements we'll be focusing on here (viz. coherence requirements), which are global/wide-scope constraints on sets of beliefs. Moreover, the sense of "should" in norms will generally differ from the evaluative/global sense of "should" that we are associating with rational requirements (see n. 13).

[11] Our statement of (TB) is (intentionally) somewhat vague here. Various precisifications of (TB) have been discussed in the contemporary literature. See Thomson (2008), Wedgwood (2002), Shah (2003), Gibbard (2005), and Boghossian (2003) for some recent examples. The subtle distinctions between these various renditions of (TB) will not be crucial for our purposes. For us, (TB) plays the role of determining the correctness/accuracy conditions for belief (i.e. it determines the alethic ideal for belief states). In other words, the "should" in our (TB) is intended to mean something like "should on pain of occupying a doxastic state that is not entirely/perfectly correct/accurate." In this sense, the version of (TB) we have in mind here is perhaps most similar to Thomson's (2008: ch. 7).

Presumably, there is some sense of "should" for which (TB) comes out true, e.g. "should on pain of occupying a doxastic state that is not perfectly accurate" (see n. 11). But, we think most philosophers would not accept (TB) as a rational requirement.[12] Nonetheless, (TB) clearly implies (CB)—in the sense that all agents who satisfy (TB) must also satisfy (CB). So, if one violates (CB), then one must also violate (TB). Moreover, violations of (CB) are the sorts of things that one can (ideally, in principle) be in a position to detect a priori. Thus, one might try to argue that (CB) is a necessary requirement of ideal epistemic rationality, as follows. If one is (ideally, in principle) in a position to know a priori that one violates (TB), then one's doxastic state is not (ideally) epistemically rational. Therefore, (CB) is a rational requirement. While this (TB)-based argument for (CB) may have some prima facie plausibility, we'll argue that (CB) itself seems to be in tension with another plausible epistemic norm, which we call the Evidential Norm for full belief.

(EB) All agents *S* should (at any given time *t*) have full beliefs that are supported by the total evidence.

For now, we're being intentionally vague about what "supported" and "the total evidence" mean in (EB), but we'll precisify these locutions in due course.[13]

Versions of (EB) have been endorsed by various "evidentialists" (Clifford 1877; Conee and Feldman 2004). Interestingly, the variants of (EB) we have in mind conflict with (CB) in some ("paradoxical") contexts. For instance, consider the following example, which is a "global" version of the Preface Paradox.

> **Preface Paradox.** Let **B** be the set containing all of *S*'s justified first-order beliefs. Assuming *S* is a suitably interesting inquirer, this set **B** will be a very rich and complex set of judgments. And, because *S* is fallible, it is reasonable to believe that some of *S*'s first-order evidence will (inevitably) be misleading. As a result, it seems reasonable to believe that

[12] Some philosophers maintain that justification/warrant is factive (Littlejohn 2012; Merricks 1995). In light of the Gettier problem, factivity seems plausible as a constraint on the type of justification required for knowledge (Zagzebski 1994; Dretske 2013). However, factivity is implausible as a constraint on (the type of justification required for) rational belief. As such, we assume that "is supported by the total evidence" (i.e. "is justified/warranted") is not factive. This assumption is kosher here, since it cross-cuts the present debate regarding (CB). For instance, Pollock's defense of (CB) as a coherence requirement does not trade on the factivity of evidential support/warrant (see n. 15).

[13] The evidential norm (EB) is [like (TB)] a local/narrow-scope principle. It constrains each individual belief, so as to require that it be supported by the evidence. We will not take a stand on the precise content of (EB) here, since we will (ultimately) only need to make use of certain (weak) consequences of (EB). However, the "should" of (EB) is not to be confused with the "should" of (TB). It may be useful (heuristically) to read the "should" of (EB) as "should on pain of falling short of the Cliffordian ideal" and the "should" of (TB) as "should on pain of falling short of the Jamesian ideal" (see nn. 3 and 10).

some beliefs in **B** are false. Indeed, we think *S* herself could be justified in believing this very second-order claim. But, of course, adding this second-order belief to **B** renders *S*'s overall doxastic (full belief) state deductively inconsistent.

We take it that, in (some) such preface cases, an agent's doxastic state may satisfy (EB) while violating (CB). Moreover, we think that (some) such states need not be (ideally) epistemically irrational. That is, we think our Preface Paradox (and other similar examples) establish the following key claim:

(†) (EB) does not entail (CB). [i.e. the Evidential Norm does not entail that deductive consistency is a requirement of ideal epistemic rationality.]

We do not have space here to provide a thorough defense of (†).[14] Foley (1992) sketches the following, general "master argument" in support of (†).

If the avoidance of recognizable inconsistency were an absolute prerequisite of rational belief, we could not rationally believe each member of a set of propositions and also rationally believe of this set that at least one of its members is false. But this in turn pressures us to be unduly cautious. It pressures us to believe only those propositions that are certain or at least close to certain for us, since otherwise we are likely to have reasons to believe that at least one of these propositions is false. At first glance, the requirement that we avoid recognizable inconsistency seems little enough to ask in the name of rationality. It asks only that we avoid certain error. It turns out, however, that this is far too much to ask.

We think Foley is onto something important here. As we'll see, Foley's argument dovetails nicely with our approach to grounding coherence requirements for belief.

So far, we've been assuming that agents facing Prefaces (and similar paradoxes of deductive consistency) may be opinionated regarding the (inconsistent) sets of propositions in question (i.e. that the agents in question either believe or disbelieve each proposition in the set). In the next section, we consider the possibility that the appropriate response to the Preface Paradox (and other similar paradoxes) is to suspend judgment on (some or all) propositions implicated in the inconsistency.

[14] Presently, we are content to take (†) as a *datum*. However, definitively establishing (†) requires only the presentation of one example (preface or otherwise) in which (CB) is violated, (EB) is satisfied, and the doxastic state in question is not (ideally) epistemically irrational. We think our Preface Paradoxes suffice. Be that as it may, we think Christensen (2004), Foley (1992), and Klein (1985) have given compelling reasons to accept (†). And, we'll briefly parry some recent philosophical resistance to (†) below. One might even want to strengthen (†) so as to imply that satisfying (EB) sometimes requires the violation of (CB). Indeed, this stronger claim is arguably established by our Preface Paradox cases. In any event, we will, in the interest of simplicity, stick with our weaker rendition of (†).

3. SUSPENSION OF JUDGMENT TO THE RESCUE?

Some authors maintain that opinionation is to blame for the discomfort of the Preface Paradox (and should be abandoned in response to it). We are not moved by this line of response to the Preface Paradox. We will now briefly critique two types of "suspension strategies" that we have encountered.

It would seem that John Pollock (1983) was the pioneer of the "suspension strategy." According to Pollock, whenever one recognizes that one's beliefs are inconsistent, this leads to the "collective defeat" of (some or all of) the judgments comprising the inconsistent set. That is, the evidential support that one has for (some or all of) the beliefs in an inconsistent set is defeated as a result of the recognition of said inconsistency. If Pollock were right about this (in full generality), it would follow that if the total evidence supports each of one's beliefs, then one's belief set must be deductively consistent. In other words, Pollock's general theory of evidential support (or "warrant"[15]) must entail that (†) is false. Unfortunately, however, Pollock does not offer much in the way of a general argument against (†). His general remarks tend to be along the following lines (Pollock 1990, p. 231).[16]

> The set of warranted propositions must be deductively consistent.... If a contradiction could be derived from it, then reasoning from some warranted propositions would lead to the denial (and hence defeat) of other warranted propositions, in which case they would not be warranted.

The basic idea here seems to be that, if one (knowingly) has an inconsistent set of (justified) beliefs, then one can "deduce a contradiction" from this set, and then "use this contradiction" to perform a "*reductio*" of (some of) one's (justified) beliefs.[17] Needless to say, anyone who is already convinced that (†) is true will find this general argument against (†) unconvincing. Presumably, anyone who finds themselves in the midst of a situation that they take to be a counterexample to (EB) ⇒ (CB) should be reluctant to perform "*reductios*"

[15] Pollock uses the term "warranted" rather than "supported by the total evidence." But, for the purposes of our discussion of Pollock's views, we will assume that these are equivalent. This is kosher, since, for Pollock, "*S* is warranted in believing *p*" means "*S* could become justified in believing *p* through (ideal) reasoning proceeding exclusively from the propositions he is objectively justified in believing" (Pollock 1990: p. 87). Our agents, like Pollock's, are "idealized reasoners," so we may stipulate (for the purposes of our discussion of Pollock's views) that when we say "supported by the total evidence," we just mean whatever Pollock means by "warranted." Some (Merricks 1995) have argued that Pollock's notion of warrant is factive (see n. 12). This seems wrong to us (in the present context). If warrant (in the relevant sense) were factive, then Pollock wouldn't need such complicated responses to the paradoxes of consistency—they would be trivially ruled out, a fortiori. This is why, for present purposes, we interpret Pollock as claiming only that (EB) entails the consistency of (warranted) belief sets [(CB)], but not necessarily the truth of each (warranted) belief [(TB)].

[16] The ellipsis in our quotation contains the following parenthetical remark: "It is assumed here that an epistemic basis must be consistent." That is, Pollock gives no argument(s) for the claim that "epistemic bases" (which, for Pollock, are sets of "input propositions" of agents) must be consistent.

[17] Ryan (1991; 1996) gives a similar argument against (†). And, Nelkin (2000) endorses Ryan's argument, as applied to defusing the Lottery Paradox as a counterexample to ¬(†) [i.e. (EB) ⇒ (CB)].

of the sort Pollock seems to have in mind, since it appears that consistency is not required by their evidence. Here, Pollock seems to be assuming a closure condition (e.g. that "is supported by the total evidence" is closed under logical consequence/competent deduction) to provide a *reductio* of (†). It seems clear to us that those who accept (†) would/should reject closure conditions of this sort. We view (some) Preface cases as counterexamples to both consistency and closure of rational belief.[18]

While Pollock doesn't offer much of a general argument for ¬(†), he does address two apparent counterexamples to ¬(†): the Lottery Paradox and the Preface Paradox. Pollock (1983) first applied this "collective defeat" strategy to the Lottery Paradox. He later recognized (Pollock 1986) that the "collective defeat" strategy is far more difficult to (plausibly) apply in the case of the Preface Paradox. Indeed, we find it implausible on its face that the propositions of the (global) Preface "jointly defeat one another" in any probative sense. More generally, we find Pollock's treatment of the Preface Paradox quite puzzling and unpersuasive.[19] Be that as it may, it's difficult to see how this sort of "collective defeat" argument could serve to justify ¬(†) in full generality. What would it take for a theory of evidential support to entail ¬(†)—in full generality—via a Pollock-style "collective defeat" argument? We're not sure. But, we are confident that any explication of "supported by the total evidence" (or "warranted") that embraces a phenomenon of "collective defeat" that is robust enough to entail the falsity of (†) will also have some undesirable (even unacceptable) epistemological consequences.[20]

We often hear another line of response to the Preface that is similar to (but somewhat less ambitious than) Pollock's "collective defeat" approach. This line of response claims that there is something "heterogeneous" about the evidence in the Preface Paradox, and that this "evidential heterogeneity" somehow undermines the claim that one should believe all of the propositions that comprise the Preface Paradox. The idea seems to be[21] that the evidence one has for the first-order beliefs (in **B**) is a (radically) different kind

[18] See (Steinberger 2013) for a incisive analysis of the consequences of the Preface Paradox for various principles of deductive reasoning [i.e. "bridge principles" in the sense of (MacFarlane 2004)].

[19] We don't have the space here to analyze Pollock's (rather byzantine) approach to the Preface Paradox. Fortunately, Christensen (2004) has already done a very good job of explaining why "suspension strategies" like Pollock's can not, ultimately, furnish compelling responses to the Preface.

[20] For instance, it seems to us that any such approach will have to imply that "supported by the total evidence" is (generally) closed under logical consequence (or competent deduction), even under complicated entailments with many premises. See (Korb 1992) for discussion regarding (this and other) unpalatable consequences of Pollockian "collective defeat."

[21] We've actually not been able to find this exact line of response to the Preface anywhere in print, but we have heard this kind of line defended in various discussions and Q&As. The closest line of response we've seen in print is Leitgeb's (2013) approach, which appeals to the "heterogeneity" of the subject matter of the claims involved in the Preface. This doesn't exactly fall under our "evidential heterogeneity" rubric, but it is similar enough to be undermined by our Homogeneous Preface case.

of evidence than the evidence one has for the second-order belief (i.e. the belief that renders **B** inconsistent in the end). And, because these bodies of first-order and second-order evidence are so heterogeneous, there is no single body of evidence that supports both the first-order beliefs and the second-order belief in the Preface case. So, believing all the propositions of the Preface is not, in fact, the epistemically rational thing to do.[22] Hence, the apparent tension between (EB) and (CB) is merely apparent.

We think this line of response is unsuccessful, for three reasons. First, can't we just gather up the first-order and second-order evidential propositions, and put them all into one big collection of total Preface evidence? And, if we do so, why wouldn't the total Preface evidence support both the first-order beliefs and the second-order belief in the Preface case? Second, we only need one Preface case in which (EB) and (CB) do genuinely come into conflict in order to establish (†). And, it seems to us that there are "homogeneous" versions of the Preface which do not exhibit this (alleged) kind of "evidential heterogeneity." Here's one such example.

> **Homogeneous Preface Paradox**. John is an excellent empirical scientist. He has devoted his entire (long and esteemed) scientific career to gathering and assessing the evidence that is relevant to the following first-order, empirical hypothesis: (*H*) all scientific/empirical books of sufficient complexity contain at least one false claim. By the end of his career, John is ready to publish his masterpiece, which is an exhaustive, encyclopedic, fifteen-volume (scientific/empirical) book, which aims to summarize (all) the evidence that contemporary empirical science takes to be relevant to *H*. John sits down to write the Preface to his masterpiece. Rather than reflecting on his own fallibility, John simply reflects on the contents of (the main text of) his book, which constitutes very strong inductive evidence in favor of *H*. On this basis, John (inductively) infers *H*. But, John also believes each of the individual claims asserted in the main text of the book. Thus, because John believes (indeed, knows) that his masterpiece instantiates the antecedent of *H*, the (total) set of John's (rational/justified) beliefs is inconsistent.

In our Homogeneous Preface, there seems to be no "evidential heterogeneity" available to undermine the evidential support of John's ultimate doxastic state. Moreover, there seems to be no "collective defeat" looming here either. John is simply being a good empirical scientist (and a good inductive non-skeptic) here, by (rationally) inferring *H* from the total, *H*-relevant inductive scientific/empirical evidence. It is true that it was John himself who gathered (and analyzed, etc.) all of this inductive evidence and included it in one hugely

[22] Presumably, then, the rational thing to do is suspend judgment on some of the Preface propositions. But, which ones? As in the case of Pollock's "suspension strategy," it remains unclear (to us) precisely which propositions fail to be supported by the total evidence in the Preface Paradox (and why).

complex scientific/empirical book. But, we fail to see how this fact does anything to undermine the (ideal) epistemic rationality of John's (ultimate) doxastic state. So, we conclude that the "heterogeneity strategy" is not an adequate response to the Preface.[23] More generally, we think our Homogeneous Preface case undermines any strategy that maintains one should never believe all the propositions in any Preface.[24]

We maintain that (adequate) responses to the Preface Paradox need not require suspension of judgment on (any of) the Preface propositions. Consequently, we would like to see a (principled) response to the Preface Paradox (and other paradoxes of consistency) that allows for (full) opinionation with respect to the propositions in the Preface agenda. Indeed, we will provide just such a response (to all paradoxes of consistency) below.

Before presenting our framework (and response), we will compare and contrast our own view regarding the Preface Paradox (and other paradoxes of consistency) with the views recently expressed by a pair of philosophers who share our commitment to (†)—i.e. to the claim that Preface cases (and other similar cases) show that deductive consistency is not a necessary requirement of ideal epistemic rationality.

4. CHRISTENSEN AND KOLODNY ON COHERENCE REQUIREMENTS

We are not alone in our view that Prefaces (and other paradoxes of deductive consistency) suffice to establish (†).[25] For instance, David Christensen and Niko Kolodny agree with us about Prefaces and (†). But, Christensen and Kolodny react to the paradoxes of deductive consistency in a more radical way. They endorse

> ($\star$) There are no coherence requirements (in the relevant sense) for full belief.

[23] We said we rejected the "heterogeneous evidence" line of response to the Preface for three reasons. Our third reason is similar to the final worry we expressed above regarding Pollock's "collective defeat" strategy. We don't see how a "heterogeneity strategy" could serve to establish ¬(†) in full generality, without presupposing something very implausible about the general nature of evidential support, e.g. that evidential support is preserved by competent deduction (see n. 20).

[24] This includes Kaplan's (2013) line on the Preface, which appeals to the norms of "what we are willing to say in the context of inquiry." According to Kaplan, "what we are willing to say in the context of inquiry" is governed by a requirement of deductive cogency, which is stronger than (CB). Cogency implies (CB) plus closure (under competent deduction). John (the protagonist of our Homogeneous Preface Paradox) does not seem to us to be violating any norms of "what we are willing to say in the context of inquiry." It seems to us that nothing prevents John from being a perfectly rational scientific inquirer—even if he "says" every belief we ascribe to him in the Homogeneous Preface.

[25] Other authors besides Christensen (2004), Kolodny (2007), Foley (1992), and Klein (1985) have claimed that paradoxes of consistency place pressure on the claim that (EB) entails (CB). For instance, Kyburg (1970) maintains that the Lottery Paradox supports (†). We are focusing on Preface cases here, since we think they are, ultimately, more compelling than lottery cases (see n. 38).

That is to say, both Christensen and Kolodny endorse eliminativism regarding all (formal, synchronic, epistemic) coherence requirements for full belief. It is illuminating to compare and contrast the views of Christensen and Kolodny with our own views about paradoxes of consistency and proper responses to them.

Christensen (2004) accepts the following package of pertinent views.[26]

(C_1) Partial beliefs (viz. credences) are subject to a formal, synchronic, epistemic coherence requirement (of ideal rationality): probabilism.

($\star$) Full beliefs are not subject to any formal, synchronic, epistemic coherence requirements (of ideal rationality).

(C_2) Epistemic phenomena that appear to be adequately explainable only by appeal to coherence requirements for full belief (and facts about an agent's full beliefs) can be adequately explained entirely by appeal to probabilism (and facts about an agent's credences).

We agree with Christensen about (C_1). In fact, our framework for grounding coherence requirements for full belief is inspired by analogous arguments for probabilism as a coherence requirement for partial belief. We will return to this important parallel below. Christensen's (C_2) is part of an error theory regarding epistemological explanations that appear to involve coherence requirements for full belief as (essential) explanans. Some such error theory is needed—given ($\star$)—since epistemologists often seem to make essential use of such coherence-explanans.

Kolodny (2007), on the other hand, accepts the following pair:

(K_1) No attitudes (full belief, partial belief, or otherwise) are subject to any formal, synchronic, epistemic coherence requirements (of ideal rationality).

(K_2) Epistemic phenomena that appear to be adequately explainable only by appeal to coherence requirements for full belief (together with facts about an agent's full beliefs) can be adequately explained entirely by appeal to the Evidential Norm (EB), together with facts about an agent's full beliefs.

Kolodny's (K_1) is far more radical than anything Christensen accepts. Of course, (K_1) entails ($\star$), but it also entails universal eliminativism about

26 Strictly speaking, Christensen (2004) never explicitly endorses($\star$)or (C_2) in their full generality. He focuses on deductive consistency as a coherence-explanans, and he argues that it can be "eliminated" from such explanations, in favor of appeals only to probabilism (and facts about the agents credences). So, our ($\star$) and (C_2) may be stronger than the principles Christensen actually accepts. In recent personal communication, Christensen has voiced some sympathy with the (existence and explanatory power of) the coherence requirements for full belief developed here. Having said that, our "straw man Christensen" makes for a clearer and more illuminating contrast in the present context.

coherence requirements in epistemology. Kolodny doesn't think there are any (ineliminable) coherence requirements (or any ineliminable requirements of ideal rationality, for that matter), period. He doesn't even recognize probabilism as a coherence requirement for credences. As a result, Kolodny needs a different error theory to "explain away" the various epistemological explanations that seem to appeal essentially to coherence requirements for full belief. His error theory [(K_2)] uses the Evidential Norm for full belief (EB), along with facts about the agent's full beliefs, to explain away such appeals to "coherence requirements." So, Kolodny's error theory differs from Christensen's in a crucial respect: Kolodny appeals to local/narrow-scope norms for full belief to explain away apparent uses of coherence requirements for full belief; whereas, Christensen appeals to global/wide-scope requirements of partial belief to explain away apparent uses of coherence requirements for full belief. This is (partly) because Kolodny is committed to the following general claim:

> (K_3) Full beliefs are an essential (and ineliminable) part of epistemology (i.e. the full belief concept is ineliminable from some epistemological explanations).

We agree with Kolodny about (K_3). We, too, think that full belief is a crucial (and ineliminable) epistemological concept. (Indeed, this is one of the reasons we are offering a new framework for grounding coherence requirements for full belief!) Christensen, on the other hand (at least on our reading, see n. 26), seems to be unsympathetic to (K_3).

One last epistemological principle will be useful for the purposes of comparing and contrasting our views with the views of Christensen and Kolodny.[27]

> (‡) If there are any coherence requirements for full belief, then deductive consistency [(CB)] is one of them. [i.e. If $\neg(\star)$, then (CB).]

Christensen and Kolodny both accept (‡), albeit in a trivial way. They both reject the antecedent of (‡) [i.e. they both accept ($\star$)]. We, on the other hand, aim to provide a principled way of rejecting (‡). That is to say, we aim to ground new coherence requirements for full belief, which are distinct from deductive consistency. We think this is the proper response to the paradoxes of consistency [and (†)].

In the next section, we will present our formal framework for grounding coherence requirements for (opinionated) full belief. But, first, we propose a desideratum for such coherence requirements, inspired by the considerations adduced so far.

[27] The "If ..., then ..." in (‡) is a material conditional. That is, (‡) asserts: either ($\star$) or (CB).

($\mathcal{D}$) Coherence requirements for (opinionated) full belief should never come into conflict with either alethic or evidential norms for (opinionated) full belief. Furthermore, coherence requirements for (opinionated) full belief should be entailed by both the Truth Norm (TB) and the Evidential Norm (EB).

In light of (†), deductive consistency [(CB)] violates desideratum ($\mathcal{D}$). If a coherence requirement satisfies desideratum ($\mathcal{D}$), we will say that it is conflict-proof. Next, we explain how to ground conflict-proof coherence requirements for (opinionated) full belief.

5. OUR (NAÏVE) FRAMEWORK AND (SOME OF) ITS COHERENCE REQUIREMENTS

As it happens, our preferred alternative(s) to (CB) were not initially motivated by thinking about paradoxes of consistency. They were inspired by some recent arguments for probabilism as a (synchronic, epistemic) coherence requirement for credences. James Joyce (1998; 2008) has offered arguments for probabilism that are rooted in considerations of accuracy (i.e. in alethic considerations). We won't get into the details of Joyce's arguments here.[28] Instead, we present a general framework for grounding coherence requirements for sets of judgments of various types, including both credences and full beliefs. Our unified framework constitutes a generalization of Joyce's argument for probabilism. Moreover, when our approach is applied to full belief, it yields coherence requirements that are superior to (CB), in light of Preface cases (and other similar paradoxes of consistency).

Applying our framework to judgment sets $\mathbf{J}$ of type $\mathcal{J}$ only requires completing three steps. The three steps are as follows:

Step 1. Say what it means for a set $\mathbf{J}$ of type $\mathcal{J}$ to be perfectly accurate (at a possible world w). We use the term "vindicated" to describe the perfectly accurate set of judgments of type $\mathcal{J}$, at w, and we use $\mathring{\mathbf{J}}_w$ to denote this vindicated set.[29]

Step 2. Define a measure of distance between judgment sets, $d(\mathbf{J},\mathbf{J}')$. We use d to gauge a set $\mathbf{J}$'s distance from vindication at w [viz. $d(\mathbf{J},\mathring{\mathbf{J}}_w)$].

Step 3. Adopt a fundamental epistemic principle, which uses $d(\mathbf{J},\mathring{\mathbf{J}}_w)$ to ground a (synchronic, epistemic) coherence requirement for judgment sets $\mathbf{J}$ of type $\mathcal{J}$.

[28] There are some important disanalogies between Joyce's argument for probabilism and our analogous arguments regarding coherence requirements for full belief. Happily, the worry (articulated in Easwaran and Fitelson 2012) that Joyce's argument for probabilism may violate the credal analogue of ($\mathcal{D}$) does not apply to our present arguments (see n. 42).

[29] As a heuristic, you can think of $\mathring{\mathbf{J}}_w$ as the set of judgments of type $\mathcal{J}$ that an omniscient agent (i.e. an agent who is omniscient about the facts at world w) would have.

This is all very abstract. To make things more concrete, let's look at the simplest application of our framework—to the case of (opinionated) full belief. Let:

$$B(p) =_{df} S \text{ believes that } p$$
$$D(p) =_{df} S \text{ disbelieves that } p.$$

Our agents will be forming (opinionated) judgments on some salient agenda $\mathcal{A}$, which is a (possibly proper) subset of some finite boolean algebra of propositions. That is, for each $p \in \mathcal{A}$, S either believes p or S disbelieves p, and not both.[30] In this way, an agent can be represented by her "belief set" **B**, which is just the set of her beliefs (B) and disbeliefs (D) over some salient agenda $\mathcal{A}$. Similarly, we think of propositions as sets of (classical) possible worlds, so that a proposition is true at any world that it contains, and false at any world it doesn't contain.[31] With our (naïve) setup in place, we're ready for the three steps.

Step 1 is straightforward. It is clear what it means for a set **B** of this type to be perfectly accurate/vindicated at a world w. The vindicated set $\mathring{\mathbf{B}}_w$ is given by:

$\mathring{\mathbf{B}}_w$ contains $B(p)$ $[D(p)]$ just in case p is true [false] at w.

This is clearly the best explication of $\mathring{\mathbf{B}}_w$, since $B(p)$ $[D(p)]$ is accurate just in case p is true [false]. Given the accuracy conditions for B/D, Step 1 is uncontroversial.

[30] Our assumption of opinionation, relative to a salient agenda $\mathcal{A}$, results in no significant loss of generality for present purposes. As we have explained above, we do not think suspension of belief (on the Preface agenda—there are many propositions outside this agenda on which it may be reasonable to suspend) is an evidentially plausible way of responding to the Preface Paradox. Consequently, one of our present aims is to provide a response to paradoxes of consistency that allows for full opinionation (on the salient agendas). Moreover, there are other applications of the present framework for which opinionation is required. Briggs et al. (2014) show how to apply the present framework to the paradoxes of judgment aggregation, which presuppose opinionation on the salient agendas. Finally, we want to present the simplest and clearest version of our framework here. The naïve framework we present here can be generalized in various ways. Specifically, generalizing the present framework to allow for suspension of judgment (on the salient agendas) is, of course, desirable (Sturgeon 2008; Friedman 2013). See (Easwaran 2013) for a generalization of the present framework which allows for suspension of judgment on the salient agendas (see n. 39). And, see (Fitelson 2014) for several other interesting generalizations of the present framework.

[31] It is implicit in our (highly idealized) framework that agents satisfy a weak sort of logical omniscience, in the sense that if two propositions are logically equivalent, then they may be treated as the same proposition in all models of the present framework. As such, we're assuming that agents cannot have distinct attitudes toward logically equivalent (classical, possible-worlds) propositions. We have already explained why such idealizations are okay in the present context (see n. 9). However, it is important to note that we are not assuming agents satisfy a stronger sort of "omniscience"—an agent may believe some propositions while disbelieving some other proposition entailed by them (i.e. our logical omniscience presupposition does not imply any closure conditions on belief). In other words, our agents are aware of all logical relations, but their judgment sets may not be closed under them.

Step 2 is less straightforward, because there are a great many ways one could measure "distance between opinionated sets of beliefs/disbeliefs." For simplicity, we adopt perhaps the most naïve distance measure, which is given by:

$d(\mathbf{B}, \mathbf{B}') =_{df}$ the number of judgments on which **B** and **B**′ disagree.[32]

In particular, if you want to know how far your judgment set **B** is from vindication at w [i.e. if you want to know the value of $d(\mathbf{B}, \mathring{\mathbf{B}}_w)$] just count the number of mistakes you have made at w. To be sure, this is a very naïve measure of distance from vindication. As it turns out, however, we (ultimately) won't need to rely on such a strong (or naïve) assumption about $d(\mathbf{B}, \mathring{\mathbf{B}}_w)$. In the end, we'll only need a much weaker assumption about $d(\mathbf{B}, \mathring{\mathbf{B}}_w)$. But, for now, let's run with our naïve, "counting of mistakes at w" definition of $d(\mathbf{B}, \mathring{\mathbf{B}}_w)$. We'll return to this issue later.

Step 3 is the philosophically most important step. Before we get to our favored fundamental epistemic principle(s), we will digress briefly to discuss a stronger fundamental epistemic principle that one might find (*prima facie*) plausible. Given our naïve setup, it turns out that there is *a* choice of fundamental epistemic principle that yields deductive consistency [(CB)] as a coherence requirement for opinionated full belief. Specifically, consider the following principle:

> **Possible Vindication** (PV). There exists some possible world w at which all of the judgments in **B** are accurate. Or, to put this more formally, in terms of our distance measure d: $(\exists w)[d(\mathbf{B}, \mathring{\mathbf{B}}_w) = 0]$.

Given our naïve setup, it is easy to show that (PV) is equivalent to (CB).[33] As such, a defender of (CB) would presumably find (PV) attractive as a fundamental epistemic principle. However, as we have seen in previous sections, Preface cases (and other paradoxes of consistency) have led many philosophers (including us) to reject (CB) as a rational requirement. This motivates the adoption of fundamental principles that are weaker than (PV). Interestingly, as we mentioned above, our rejection of (PV) was not (initially) motivated by Prefaces and the like. Rather, our adoption of fundamental principles weaker than (PV) was motivated (initially) by analogy with Joyce's argument(s) for probabilism as a coherence requirement for credences.

In the case of credences, the analogue of (PV) is clearly too strong. The vindicated set of credences (i.e. the credences an omniscient agent would have) are

[32] This is called the Hamming distance between the binary vectors **B** and **B**′ (Deza and Deza 2009).

[33] Here, we're assuming a slight generalization of the standard notion of consistency. Standardly, consistency applies only to beliefs (not disbeliefs), and it requires that there be a possible world in which all the agent's beliefs are true. More generally, we may define consistency as the existence of a possible world in which all the agent's judgments (both beliefs and disbeliefs) are accurate. Given this more general notion of consistency, (PV) and (CB) are equivalent in the present framework.

such that they assign maximal confidence to all truths and minimal confidence to all falsehoods (Joyce, 1998). As a result, in the credal case, (PV) would require that all of one's credences be extremal. One doesn't need Preface cases (or any other subtle or paradoxical cases) to see that this would be an unreasonably strong (rational) requirement. It is for this reason that Joyce (and all others who argue in this way for probabilism) back away from the analogue of (PV) to strictly weaker epistemic principles—specifically, to accuracy-dominance avoidance principles, which are credal analogues of the following fundamental epistemic principle.

> **Weak Accuracy-Dominance Avoidance** (WADA). **B** is not weakly[34] dominated in distance from vindication. Or, to put this more formally (in terms of d), there does not exist an alternative belief set $\mathbf{B}'$ such that:
>
> (i) $(\forall w)[d(\mathbf{B}',\mathring{\mathbf{B}}_w) \leq d(\mathbf{B},\mathring{\mathbf{B}}_w)]$, and
> (ii) $(\exists w)[d(\mathbf{B}',\mathring{\mathbf{B}}_w) < d(\mathbf{B},\mathring{\mathbf{B}}_w)]$.

(WADA) is a very natural principle to adopt, if one is not going to insist that—as a requirement of rationality—it must be possible for an agent to achieve perfect accuracy in her doxastic state. In the credal case, the analogous requirement was clearly too strong to count as a rational requirement. In the case of full belief, one needs to think about Preface cases (and the like) to see why (PV) is too strong. Retreating from (PV) to (WADA) is analogous to what one does in decision theory, when one backs off a principle of maximizing (actual) utility to some less demanding requirement of rationality (e.g. dominance avoidance, maximization of expected utility, minimax, etc.).[35] Of course, there is a sense in which "the best action" is the one that maximizes actual utility; but, surely, maximization of actual utility is not a rational requirement. Similarly, there is clearly a sense in which "the best doxastic state" is the perfectly accurate [(TB)], or possibly perfectly accurate [(CB)/(PV)], doxastic state. But, in light of the paradoxes of consistency, (TB) and (CB) turn out not to be rational requirements either. One of the main problems with the existing literature on the paradoxes of consistency is that no principled alternative(s)

[34] Strictly speaking, Joyce et al. opt for the apparently weaker principle of avoiding strict dominance. However, in the credal case (assuming continuous, strictly proper scoring rules), there is no difference between weak and strict dominance (Schervish et al. 2009). In this sense, there is no serious disanalogy. Having said that, it is worth noting that, in the case of full belief, there is a significant difference between weak dominance and strict dominance. This difference will be discussed in some detail in §6. below. In the meantime, whenever we say "dominated" what we mean is weakly dominated in the sense of (WADA).

[35] The analogy to decision theory could be made even tighter. We could say that being accuracy-dominated reveals that you are in a position to recognize a priori that another option is guaranteed to do better at achieving the "epistemic aim" of getting as close to the truth as possible. This decision-theoretic stance dovetails nicely with the sentiments expressed by Foley (1992). See §8. for further discussion of (and elaboration on) this epistemic decision-theoretic stance.

to deductive consistency have been offered as coherence requirements for full belief. Such alternatives are just what our Joyce-style arguments provide.

If a belief set **B** satisfies (WADA), then we say **B** is non-dominated. This leads to the following, new coherence requirement for (opinionated) full belief:

(NDB) All (opinionated) agents S should (at any given time t) have sets of full beliefs (and disbeliefs) that are non-dominated.

Interestingly, (NDB) is strictly weaker than (CB). Moreover, (NDB) is weaker than (CB) in an appropriate way, in light of our Preface Paradoxes (and other similar paradoxes of consistency). Our first two theorems (each with an accompanying definition) help to explain why.

The first theorem states a necessary and sufficient condition for (i.e. a characterization of) non-dominance: we call it *Negative* because it identifies certain objects, the non-existence of which is necessary and sufficient for non-dominance. The second theorem states a sufficient condition for non-dominance: we call it *Positive* because it states that in order to show that a certain belief set **B** is non-dominated, it's enough to construct a certain type of object.

Definition 1 (Witnessing Sets). **S** *is a* ***witnessing set*** *iff (a) at every world w, at least half of the judgments*[36] *in* **S** *are inaccurate; and, (b) at some world more than half of the judgments in* **S** *are inaccurate.*

Theorem 1 (Negative). **B** *is non-dominated iff* **B** *contains no witnessing set.*
[*We will use "(NWS)" to abbreviate the claim that "no subset of* **B** *is a witnessing set." Thus, Theorem 1 can be stated equivalently as:* **B** *is non-dominated iff (NWS).*]

It is an immediate corollary of this first theorem that if **B** is deductively consistent [i.e if **B** satisfies (PV)], then **B** is non-dominated. After all, if **B** is deductively consistent, then there is a world w such that no judgments in **B** are inaccurate at w (n. 33). However, while deductive consistency guarantees non-dominance, the converse is not the case, i.e. non-dominance does not ensure deductive consistency. This will be most perspicuous as a consequence of our second theorem.

Definition 2. *A probability function* Pr **represents** *a belief set* **B** *iff for every $p \in \mathcal{A}$*

(i) **B** *contains $B(p)$ iff* $\Pr(p) > 1/2$, *and*
(ii) **B** *contains $D(p)$ iff* $\Pr(p) < 1/2$.

[36] Here, we rely on naïve counting. This is unproblematic, since all of our algebras are finite. Moreover, the coherence norm we'll propose in the end (see §7.) will not be based on counting and (as a result) will be applicable to both finite and infinite belief sets. All theorems are proved in the Appendix.

Theorem 2 (Positive). **B** *is non-dominated if there exists a probability function* Pr *that represents* **B**.[37]

To appreciate the significance of Theorem 2, it helps to think about a standard lottery case.[38] Consider a fair lottery with $n \geq 3$ tickets, exactly one of which is the winner. For each $j \leqslant n$, let p_j be the proposition that the jth ticket is not the winning ticket; let q be the proposition that some ticket is the winner; and, let these $n+1$ propositions exhaust the agenda $\mathcal{A}$. (Note that the agenda leaves out conjunctions of these propositions.) Finally, let LOTTERY be the following opinionated belief set on $\mathcal{A}$:

$$\{B(p_j) \mid 1 \leqslant j \leqslant n\} \cup \{B(q)\}.$$

In light of Theorem 2, LOTTERY is non-dominated. The probability function that assigns each ticket equal probability of winning represents LOTTERY. However, LOTTERY is not deductively consistent. Hence, (NDB) is strictly weaker than (CB).[39]

Not only is (NDB) weaker than (CB), it is weaker than (CB) in a desirable way. More precisely, in accordance with desideratum ($\mathcal{D}$), we will now demonstrate that (NDB) is entailed by both alethic considerations [(TB)/(CB)] and evidential considerations [(EB)]. While there is considerable disagreement about the precise content of the Evidential Norm for full belief (EB), there is widespread agreement (at least, among evidentialists) that the following is a necessary condition for satisfying (EB).

> **Necessary Condition for Satisfying (EB).** **B** satisfies (EB), i.e. all judgments in **B** are supported by the total evidence, ***only if***:
>
> ($\mathcal{R}$) There exists some probability function that probabilifies (i.e. assigns probability greater than 1/2 to) each belief in **B** and disprobabilifies (i.e. assigns probability less than 1/2 to) each disbelief in **B**.

Most evidentialists agree that probabilification—relative to some probability function—is a minimal necessary condition for justification. Admittedly, there

37 The question: "Does a belief set **B** have a representing probability function?" is decidable (Fitelson 2008). So is the question "Does a belief set **B** have a witnessing set?" This suffices to ensure that coherence properties (in our sense) of (finite) belief sets are formal in the intended sense of §1, above.

38 We are not endorsing the belief set LOTTERY in this example as epistemically rational. Indeed, we think that the Lottery Paradox is not as compelling—as a counterexample to (EB) $\Rightarrow$ (CB)—as the Preface Paradox is. On this score, we agree with Pollock (1990) and Nelkin (2000). We are just using this lottery example to make a formal point about the logical relationship between (CB) and (NDB).

39 It is worth noting that belief sets like LOTTERY can remain non-dominated even if we allow alternative judgment sets to suspend judgment on some or all of the propositions in the salient agenda. This is yet another reason why our restriction to opinionated judgment sets (over appropriate agendas of propositions) results in no significant loss of generality (see n. 30). See (Easwaran 2013) for further discussion.

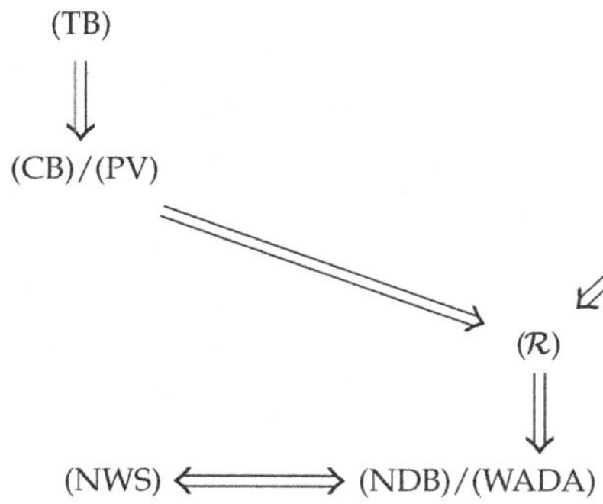

Figure 3.1 Logical relations between requirements and norms (so far)

is plenty of disagreement about which probability function is implicated in $(\mathcal{R})$.[40] But, because our Theorem 2 only requires the existence of *some* probability function that probabilifies *S*'s beliefs and dis-probabilifies *S*'s disbeliefs, it is sufficient to ensure (on most evidentialist views) that (EB) entails $(\mathcal{R})$. Assuming we're correct in our assessment that Prefaces (and other similar paradoxes of consistency) imply (†), this is precisely the entailment that fails for (CB), and the reason why (CB) fails to satisfy desideratum $(\mathcal{D})$ [i.e. why (CB) fails to be conflict-proof], while (NDB) does satisfy it.[41] Thus, by grounding coherence for full beliefs in the same way Joyce grounds probabilism for credences, we are naturally led to coherence requirements for (opinionated) full belief that are plausible alternatives to (CB).[42] This gives us a principled way to accept (†) while rejecting (‡), and it paves the way for a novel and compelling response to the Preface (and other similar paradoxes of consistency). Figure 3.1 depicts the logical relations between the epistemic requirements and norms we have discussed so far.

In the next two sections, we will elaborate on the family of coherence requirements generated by our framework.

40 Internalists like Fumerton (1995) require that the function $Pr(\cdot)$ that undergirds (EB) should be "internally accessible" to the agent (in various ways). Externalists like Williamson (2000) allow for "inaccessible" evidential probabilities. And, subjective Bayesians like Joyce (2005) say that $Pr(\cdot)$ should reflect the agent's subjective degrees of belief (viz. credences). Despite this disagreement, most evidentialists agree that (EB) entails $(\mathcal{R})$, which is all we need for present purposes.

41 Another way to see why there can't be preface-style counterexamples to (NDB) is to recognize that such cases would have to involve not only the (reasonable) belief that *some* of one's beliefs are false, but the much stronger (and unreasonable/irrational) belief that *most* of one's beliefs are false.

42 We have given a general, theoretical argument to the effect that the Evidential Norm for full belief [(EB)] entails the coherence requirement(s) for full belief that we favor. We know of no analogous general argument for credences. In (Easwaran and Fitelson 2012), we raised the possibility of counterexamples to the analogous theoretical claim: $(\mathcal{E})$ the evidential norm for credences (independently) entails probabilism. Joyce (2013) and Pettigrew (2013a) take steps toward general arguments for $(\mathcal{E})$.

6. A FAMILY OF NEW COHERENCE REQUIREMENTS FOR FULL BELIEF

The analysis above revealed two coherence requirements that are strictly weaker than deductive consistency: ($\mathcal{R}$) and (NDB). There is, in fact, a large family of such requirements. This family includes requirements that are even weaker than (NDB), as well as requirements that are stronger than ($\mathcal{R}$). Regarding the former, the most interesting requirement that is weaker than (NDB) is generated by replacing weak accuracy-dominance avoidance with strict accuracy-dominance avoidance, i.e. by adopting (SADA), rather than (WADA), as the fundamental epistemic principle.

> **Strict Accuracy-Dominance Avoidance** (SADA). **B** is not strictly dominated in distance from vindication. Or, to put this more formally (in terms of d), there does not exist an alternative belief set **B**′ such that:
>
> $$(\forall w)[d(\mathbf{B}',\mathring{\mathbf{B}}_w) < d(\mathbf{B},\mathring{\mathbf{B}}_w)].$$

It is obvious that (WADA) entails (SADA). That the converse entailment does not hold can be shown by producing an example of a doxastic state that is weakly, but not strictly, dominated in d-distance from vindication. We present such an example in the Appendix. What about requirements "in between" (CB) and ($\mathcal{R}$)?

One way to bring out requirements that are weaker than (CB) but stronger than ($\mathcal{R}$) is to think of (CB) and ($\mathcal{R}$) as "limiting cases" of the following parametric family.

> **Parametric Family of Probabilistic Requirements Between ($\mathcal{R}$) and (CB)**
>
> ($\mathcal{R}_r$) There exists a probability function Pr such that, for every $p \in \mathcal{A}$:
>
> (i) **B** contains $B(p)$ iff $\Pr(p) > r$, and
>
> (ii) **B** contains $D(p)$ iff $\Pr(p) < 1 - r$,
>
> where $r \in [1/2, 1)$.

What we have been calling ($\mathcal{R}$) is (obviously) equivalent to member ($\mathcal{R}_{1/2}$) of the above family. And, as the value of r approaches 1, the corresponding requirement ($\mathcal{R}_r$) approaches (CB) in logical strength. This gives rise to a continuum of coherence requirements that are "in between" (CB) and ($\mathcal{R}$) in terms of their logical strength. (CB) is equivalent to the following, extremal probabilistic requirement.

> **Extremal Probabilistic Equivalent to (CB)**
>
> ($\mathrm{CB}_{\Pr}$) There exists a probability function Pr such that, for every $p \in \mathcal{A}$:
>
> (i) **B** contains $B(p)$ iff $\Pr(p) = 1$, and
>
> (ii) **B** contains $D(p)$ iff $\Pr(p) = 0$.

To see that (CB_{Pr}) is equivalent to (CB), note that a belief set **B** is consistent (i.e. possibly perfectly accurate) just in case there is a truth-value assignment function that assigns $\top$ to all p such that $B(p) \in \mathbf{B}$ and $\bot$ to all p such that $D(p) \in \mathbf{B}$. But, this is equivalent to the existence of an indicator function that assigns 1 to all the believed propositions in **B** and 0 to all the disbelieved propositions in **B**. And, such indicator functions just are probability functions of the sort required by (CB_{Pr}).

In the next section, we'll look more closely at our family of new coherence requirements, with an eye toward narrowing the field.

7. A CLOSER LOOK AT OUR FAMILY OF NEW COHERENCE REQUIREMENTS

First, we note that there is a clear sense in which (NDB) seems to be too weak. (NDB) doesn't even rule out belief sets that contain contradictory pairs of beliefs. For instance, the belief set $\{B(P), B(\neg P)\}$ on the simple agenda $\{P, \neg P\}$ is not weakly dominated in d-distance from vindication. This can be seen in Table 3.1.

In Table 3.1, a "+" denotes an accurate judgment (at a world) and a "−" denotes an inaccurate judgment (at a world). As you can see, the belief set $\mathbf{B} =_{df} \{B(P), B(\neg P)\}$ contains one accurate judgment and one inaccurate judgment in each of the two salient possible worlds, i.e. $d(\mathbf{B}, \mathring{\mathbf{B}}_{w1}) = 1$ and $d(\mathbf{B}, \mathring{\mathbf{B}}_{w2}) = 1$. None of the other three possible (opinionated) belief sets on $\mathcal{A}$ weakly accuracy-dominates **B**. Specifically, let $\mathbf{B}' =_{df} \{B(P), D(\neg P)\}$, $\mathbf{B}'' =_{df} \{D(P), B(\neg P)\}$ and $\mathbf{B}''' =_{df} \{D(P), D(\neg P)\}$. Then:

$$1 = d(\mathbf{B}, \mathring{\mathbf{B}}_{w1}) < d(\mathbf{B}', \mathring{\mathbf{B}}_{w1}) = 2,$$
$$1 = d(\mathbf{B}, \mathring{\mathbf{B}}_{w2}) < d(\mathbf{B}'', \mathring{\mathbf{B}}_{w2}) = 2,$$
$$1 = d(\mathbf{B}, \mathring{\mathbf{B}}_{w1}) = d(\mathbf{B}''', \mathring{\mathbf{B}}_{w1}), \text{ and}$$
$$1 = d(\mathbf{B}, \mathring{\mathbf{B}}_{w2}) = d(\mathbf{B}''', \mathring{\mathbf{B}}_{w2}).$$

Therefore, none of $\mathbf{B}'$, $\mathbf{B}''$ or $\mathbf{B}'''$ weakly accuracy-dominates **B**, which implies that **B** satisfies (NDB). But, intuitively, **B** should count as incoherent. After all, **B** violates ($\mathcal{R}$), which implies that **B** cannot be supported by the total evidence—whatever the total evidence is. This suggests that (NDB) is too weak to serve as "the" (strongest, universally binding) coherence requirement

			$\mathbf{B}$		$\mathbf{B}'$		$\mathbf{B}''$		$\mathbf{B}'''$	
	P	$\neg P$	$B(P)$	$B(\neg P)$	$B(P)$	$D(\neg P)$	$D(P)$	$B(\neg P)$	$D(P)$	$D(\neg P)$
w_1	**F**	**T**	−	+	−	−	+	+	+	−
w_2	**T**	**F**	+	−	+	+	−	−	−	+

Table 3.1 *Why (NDB) seems too weak*

for (opinionated) full belief. Indeed, we think a similar argument could be given to show that no requirement that is (strictly) weaker than ($\mathcal{R}$) can be "the" coherence requirement for full belief. Dominance requirements like (NDB) have other shortcomings, besides being too weak.

Dominance avoidance conditions like (WADA) and (SADA) are defined in terms of the naïve "mistake-counting" measure of distance from vindication $d(\mathbf{B},\mathring{\mathbf{B}}_w)$. Such simple counting measures work fine for finite belief sets, but there seems to be no clear way to apply such naïve distance measures to infinite belief sets. On the other hand, probabilistic requirements like ($\mathcal{R}$) can be applied (in a uniform way) to both finite and infinite belief sets.

There is another problem (NDB) inherits from (WADA)'s reliance on the naïve, mistake-counting measure of distance from vindication $d(\mathbf{B},\mathring{\mathbf{B}}_w)$. This measure seems to require that each proposition in the agenda receive "equal weight" in the calculation of **B**'s distance from vindication. One might (for various reasons) want to be able to assign different "weights" to different propositions when calculating the overall distance from vindication of a doxastic state.[43] An examination of the proof of Theorem 2 (see the Appendix) reveals that if a belief set **B** satisfies ($\mathcal{R}$), then **B** will minimize expected distance from vindication, relative to its representing probability function Pr. The proof of this result requires only that the measure of distance from vindication be additive—i.e. that each judgment in **B** receive an "inaccuracy score" and that these "inaccuracy scores" are added up across the members of **B**. In other words, if we adopt ($\mathcal{R}$) as our (ultimate) coherence requirement, then—so long as our measure of distance from vindication is additive—coherent belief sets will be guaranteed to be non-dominated. So, another advantage of adopting ($\mathcal{R}$)—as opposed to (NDB)—as "the" coherence requirement for full belief is that it allows us to use any additive distance measure we like, while preserving the (Joycean) connection between coherence and the avoidance of accuracy-dominance.

What about requirements stronger than ($\mathcal{R}$)? For instance, could some requirement ($\mathcal{R}_r$)—with $r > 1/2$—be a better candidate than ($\mathcal{R}$) for "the" coherence requirement for (opinionated) full belief? We suspect this will depend on the context in which a doxastic state is being evaluated. Recall that the key premise in our argument that ($\mathcal{R}$) satisfies desideratum ($\mathcal{D}$) [i.e. that ($\mathcal{R}$) is conflict-proof] was the assumption that the Evidential Norm (EB) entails ($\mathcal{R}$). It is uncontroversial (among probabilist-evidentialists) that this entailment holds in all contexts. What happens to this entailment when we replace ($\mathcal{R}$) with ($\mathcal{R}_r$) and $r > 1/2$? When $r > 1/2$, it will no longer be uncontroversial—even among probabilist-evidentialists—that (EB) entails ($\mathcal{R}_r$) in all contexts. For instance, consider a Lockean, who thinks that one ought to believe P just in case $\Pr(P) > r$, where the value of the threshold r

[43] Joyce's (1998, 2009) argument(s) for probabilism allow(s) for different weights to be assigned to different propositions in the calculation of distance from vindication (of a credence function).

(and perhaps the evidential probability function Pr) depends on context. If $r > 1/2$, then such a probabilist-evidentialist will only accept the entailment (EB) $\Rightarrow$ ($\mathcal{R}_r$) in some contexts. As a result, when $r > 1/2$, the entailment (EB) $\Rightarrow$ ($\mathcal{R}_r$) will not hold uncontroversially, in a context-independent way. However, $r = 1/2$ seems too low to generate a strong enough coherence requirement in most (if not all) contexts, i.e. the requirement ($\mathcal{R}$) = ($\mathcal{R}_{1/2}$) seems too weak in most (if not all) contexts.

To see this, consider the class of minimal inconsistent sets of propositions of size n: $\mathbb{B}_n$. That is, each member of $\mathbb{B}_n$ is an inconsistent set of size n containing no inconsistent proper subset. For instance, each member of $\mathbb{B}_2$ will consist of a contradictory pair of propositions. We've already seen that believing both elements of a member of $\mathbb{B}_2$ is ruled out as incoherent by ($\mathcal{R}$) = ($\mathcal{R}_{1/2}$), but not by (NDB). However, believing each element of a member of $\mathbb{B}_3$ (e.g. believing each of the propositions in $\{P, Q, \neg(P \& Q)\}$) will *not* be ruled out by ($\mathcal{R}_{1/2}$). In order to rule out inconsistent belief sets of size three, we would need to raise the threshold r to $2/3$. In other words, ($\mathcal{R}_{2/3}$) is the weakest requirement in the ($\mathcal{R}_r$) family that rules out believing each member of a three-element minimal inconsistent set (i.e. each member of some set in $\mathbb{B}_3$). In general, we have the following theorem.[44]

Theorem 3. *For all $n \geq 2$ and for each set of propositions $\mathbf{P} \in \mathbb{B}_n$, if $r \geq \frac{n-1}{n}$ then ($\mathcal{R}_r$) rules out believing every member of $\mathbf{P}$, while if $r < \frac{n-1}{n}$, then ($\mathcal{R}_r$) doesn't rule out believing every member of $\mathbf{P}$.*

In Preface (or lottery) cases, n is typically quite large. As a result, in order to rule out such large inconsistent belief sets as incoherent, ($\mathcal{R}_r$) would require a large threshold $r = \frac{n-1}{n}$. For example, ruling out inconsistent belief sets of size 5 via ($\mathcal{R}_r$) requires a threshold of $r = 0.8$, and ruling out inconsistent belief sets of size 10 via ($\mathcal{R}_r$) requires a threshold of $r = 0.9$. We think this goes some way toward explaining why smaller inconsistent belief sets seem "less coherent" than larger inconsistent belief sets.[45] Having said that, we don't think there is a precise "universal threshold" r such that ($\mathcal{R}_r$) is "the" (strongest universally

[44] Hawthorne and Bovens (1999) prove some very similar formal results; and Christensen (2004) and Sturgeon (2008) appeal to such formal results in their discussions of the paradoxes of consistency. However, all of these authors presuppose that the probabilities involved in their arguments are the credences of the agent in question. Our probability functions need not be credence functions (see n. 40). Indeed, we need not even assume here that agents have degrees of belief. This is important for us, since we do not want to presuppose any sort of reduction (or elimination) of full belief to (or in favor of) partial belief (or any other quantitative concept). See (Fitelson 2014) for further discussion.

[45] More generally, it seems that the ($\mathcal{R}_r$) can do a lot of explanatory work. For instance, we mentioned above (n. 7) that even Kolodny—a coherence eliminativist—should be able to benefit from our framework and analysis. We think Kolodny can achieve a more compelling explanatory error theory by taking, e.g. ($\mathcal{R}$), rather than (CB), as his target. There is a much tighter conceptual connection between (EB)—which is Kolodny's central explanatory epistemic principle—and ($\mathcal{R}$). For this reason, we believe that a shift from (CB) to ($\mathcal{R}$) would make the explanatory aims of Kolodny's error theory easier to achieve. Finally, we suspect that debates about the existence of coherence requirements would become more interesting if we

binding) coherence requirement. There are precise values of r that yield clear-cut cases of universally binding coherence requirements ($\mathcal{R}_r$), e.g. $r = 1/2$. And, there are precise values of r which yield coherence requirements ($\mathcal{R}_r$) that we take to be clearly not universally binding, e.g. $r = 1 - \epsilon$, for some minuscule ϵ.[46] What, exactly, happens in between? We'll have to leave that question for a future investigation.[47]

In the next section, we'll elaborate briefly on an illuminating decision-theoretic analogy that we mentioned in passing (e.g. n. 35). Then, we'll tie up a few remaining theoretical loose ends. Finally, we'll consider a troublesome example for our framework, inspired by Caie's (2013) analogous recent counterexample to Joyce's argument for probabilism.

8. A DECISION-THEORETIC ANALOGY

If we think of closeness to vindication as a kind of epistemic utility (Pettigrew, 2013b, Greaves, 2013), then we may think of ($\mathcal{R}$) as an expected epistemic utility maximization principle. On this reading, ($\mathcal{R}$) is tantamount to the requirement that an agent's belief set should maximize expected epistemic utility, relative to some evidential probability function. Expected utility maximization principles are stronger (i.e. less permissive) than dominance principles, which (again) explains why ($\mathcal{R}$) is stronger than (NDB).

We could push this decision-theoretic analogy even further. We could think of the decision-theoretic analogue of (TB) as a principle of *actually* maximizing utility (AMU), i.e. choose an act that maximizes utility in the *actual* world. We could think of the decision-theoretic analogue of (CB) as a principle of *possibly* maximizing utility (PMU), i.e. choose an act that maximizes utility in some *possible* world. And, as we have already discussed (see n. 35), (NDB) would be analogous to a (weak) dominance principle in epistemic decision theory. The general correspondence between the epistemic norms and requirements discussed above and the analogous decision-theoretic principles is summarized in Table 3.2.[48]

stopped arguing about whether (CB) is a coherence requirement and started arguing about whether ($\mathcal{R}$) or (NDB) [or the other ($\mathcal{R}_r$)] are coherence requirements.

[46] When we say that $r = 1 - \epsilon$, for some minuscule ϵ, leads to a coherence requirement ($\mathcal{R}_r$) that is not universally binding, we are not making a claim about any specific probability function (see n. 44). For instance, we're not assuming a Lockean thesis with a "universal" threshold of $r = 1 - \epsilon$. It is important to remember that ($\mathcal{R}_r$) asserts only the existence of some probability function that assigns a value greater than r to all beliefs in **B** and less than $1 - r$ to all disbeliefs in **B**. This is why ($\mathcal{R}_{1-\epsilon}$) is strictly logically weaker than (CB), and (therefore) no more controversial than (CB).

[47] Fully answering this question would require (among other things) a more substantive analysis of the relationship between the Evidential Norm (EB) and the requirements ($\mathcal{R}_r$), in cases where $r > 1/2$. That sort of analysis is beyond the scope of this paper, but will be taken up in (Fitelson 2014).

[48] The double line between (CB) and ($\mathcal{R}$) in Table 3.2 is intended to separate rational requirements like ($\mathcal{R}$) from principles like (CB) that are too demanding to count as universally binding rational requirements. We're not sure exactly how to draw this line, but we think that reflection

Epistemic Principle	Analogous Decision-Theoretic Principle
(TB)	(AMU) Do ϕ only if ϕ maximizes utility in the *actual* world.
(CB)	(PMU) Do ϕ only if ϕ maximizes u in *some possible* world.
($\mathcal{R}$)	(MEU) Do ϕ only if ϕ maximizes *expected* utility.[49]
(WADA)	(WDOM) Do ϕ only if ϕ is *not weakly dominated* in utility.
(SADA)	(SDOM) Do ϕ only if ϕ is *not strictly dominated* in utility.

Table 3.2 *A decision-theoretic analogy*

Like (CB), the principle of "possible maximization of utility" (PMU) is not a requirement of rationality. And, as in the case of (CB), in order to see the failure of (PMU), one needs to consider "paradoxical" cases. See (Parfit 1988; Kolodny and MacFarlane 2010) for discussions of a "paradoxical" decision problem (the so-called "Miner Puzzle") in which the rational act is one that does not maximize utility in any possible world. From this decision-theoretic perspective, it is not surprising that deductive consistency (CB) turns out to be too demanding to be a universally binding rational requirement (Foley, 1992).[50]

9. TYING UP A FEW THEORETICAL LOOSE ENDS

As we have seen, the accuracy-dominance avoidance requirement (NDB) is equivalent to a purely combinatorial condition (NWS) defined in terms of witnessing sets. Similar purely combinatorial conditions exist for some of our other coherence requirements as well. Consider these variations on the concept of a witnessing set.

Definition 3 (Witnessing$_1$ Sets). **S** *is a* ***witnessing$_1$ set*** *iff at every world w, more than half of the judgments in* **S** *are inaccurate.*

on how the analogous line can be drawn on the decision-theoretic side of the analogy sheds some light on this question (Fitelson, 2014).

49 Strictly speaking, in order to make the analogy tight, we need to precisify (MEU) in the following way:

(MEU) Do ϕ only if ϕ maximizes expected utility, relative to *some* probability function.

This is weaker than the standard interpretation of (MEU), which involves maximizing expected utility relative to a *specific* probability function (viz. the agent's credence function in the decision situation).

50 For a nice survey of some of the recent fruits of this (epistemic) decision-theoretic stance, see (Pettigrew 2013b). And, see (Greaves 2013; Berker 2013) for some meta-epistemological worries about taking this sort of epistemic decision-theoretic stance. We don't think the worries expressed by Greaves and Berker are ultimately problematic for our present approach, but we don't have the space here to properly allay them.

Definition 4 (Witnessing$_2$ Sets). **S** *is a* ***witnessing$_2$ set*** *iff at every world w, at least half of the judgments in* **S** *are inaccurate.*

Corresponding to each of these types of witnessing sets is a requirement stating that no subset of **B** should be a witnessing set of that type. To wit:

(NW$_1$S) No subset of **B** is a witnessing$_1$ set.
(NW$_2$S) No subset of **B** is a witnessing$_2$ set.

It turns out that (NW$_1$S) and (NW$_2$S) are intimately connected with (SADA) and ($\mathcal{R}$), respectively. These connections are established by the following two theorems:

Theorem 4. **B** *is non-strictly-dominated iff* **B** *contains no witnessing$_1$ set.* [*In other words, the following equivalence holds: (SADA) $\Longleftrightarrow$ (NW$_1$S) .*]

Theorem 5. **B** *is probabilistically representable (in the sense of Definition 2) only if*[51] **B** *contains no witnessing$_2$ set* [*i.e.* ($\mathcal{R}$) $\Rightarrow$ (NW$_2$S)].

This brings us to our final theorem. Consider the following condition, which requires that there be no contradictory pairs of judgments in a belief set:

(NCP) **B** does not contain any contradictory pairs of judgments. That is, there is no proposition p such that either $\{B(p), B(\neg p)\} \subseteq \mathbf{B}$ or $\{D(p), D(\neg p)\} \subseteq \mathbf{B}$.

Theorem 6. **B** *is probabilistically representable (in the sense of Definition 2) only if*[52] **B** *satisfies both (NDB) and (NCP)* [*i.e.* ($\mathcal{R}$) $\Rightarrow$ (NDB & NCP)].

That ties up the remaining theoretical loose ends. Figure 3.2 depicts the known (n. 51) logical relationships between all the requirements and norms discussed above. In the next section, we discuss a worrisome example inspired by Caie's (2013) recent counterexample to Joyce's argument for probabilism.

10. A WORRISOME EXAMPLE FOR OUR FRAMEWORK

Michael Caie (2013) has recently given a problematic example for (perhaps even a counterexample to) Joyce's argument for probabilism. We don't have

[51] Johannes Marti has discovered a counterexample to the converse of this theorem. Marti's example is too complex to be included here (it involves 12 possible worlds), but it will be described in detail in (Fitelson 2014).

[52] Interestingly, the converse is false [i.e. ($\mathcal{R}$) $\not\Leftarrow$ (NDB & NCP)]. See the Appendix for a counterexample.

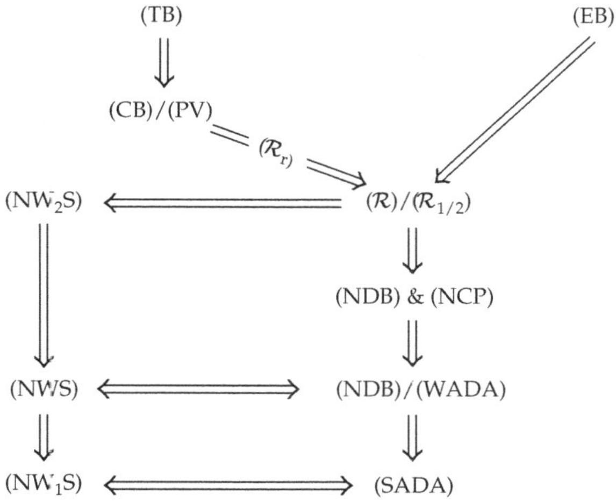

Figure 3.2 Logical relations between (all) requirements and norms

			B		**B′**		**B″**		**B‴**	
	P	$\neg P$	$B(P)$	$B(\neg P)$	$B(P)$	$D(\neg P)$	$D(P)$	$B(\neg P)$	$D(P)$	$D(\neg P)$
w_1	**F**	**T**	−	+	−	−	×	×	×	×
w_2	**T**	**F**	×	×	×	×	−	−	−	+

Table 3.3 *Caie-style example in our framework*

space here to fully analyze Caie's example. But, there is an obvious analogue of Caie's example in our framework for full belief. Consider the following (self-referential) claim:

(P) S does not believe that P.

That is, P says of itself that it is not believed by S. Consider the agenda $\mathcal{A} =_{df} \{P, \neg P\}$. There seems to be a sound argument for the (worrisome) claim that there are no coherent opinionated belief sets for S on $\mathcal{A}$. This can be seen via Table 3.3.

The "×"s in Table 3.3 indicate that these entries in the table are ruled out as logically impossible (given the definition of P). As such it appears that **B** and **B‴** strictly accuracy-dominate their (live) alternatives (i.e. it appears that **B** strictly dominates **B′** and **B‴** strictly dominates **B″**). As a result, all of the consistent opinionated belief sets on $\mathcal{A}$ would seem to be ruled out by (SADA). As for **B** and **B‴**, these belief sets consist of contradictory pairs of propositions. We argued earlier that ($\mathcal{R}$) entails (SADA) and rules out any belief set containing contradictory pairs, which seems to mean that ($\mathcal{R}$) rules out all (opinionated) belief sets on $\mathcal{A}$. However, one or both of these entailments might fail in the present case. Our earlier arguments assumed (as standard in probability theory) that none of the propositions exhibit dependence between doxastic states and the truth. Thinking about how to apply ($\mathcal{R}$) in this case requires reconsidering the relationship between a belief set and a probability function.

A probability function assigns numbers to various worlds. When evaluating a belief set with regards to various probability functions, we assumed that worlds specify the truth values of the propositions an agent might believe or disbelieve, while facts about what an agent believes in each world are specified by the belief set, and that every world and every belief set could go together. But with these propositions, there are problems in assessing each belief set with respect to a probability function that assigns positive probability to both worlds, since some belief sets and worlds make opposite specifications of the truth value of a single proposition. For instance, the proposition (P) is specified as false by $\mathbf{B}$ and $\mathbf{B}'$, but as true by w_2, which is why those cells of the table are ruled out.

There are two alternatives that one might pursue. On the first alternative, one considers "restricted worlds" that only specify truth values of propositions that don't concern the beliefs of the agent. On this alternative, there is only one restricted world, and thus only one relevant probability function (which assigns that restricted world probability 1). But this probability function doesn't say anything about the probability of (P), since (P) is a proposition that concerns the beliefs of the agent, and thus is specified by the belief set, and not the restricted worlds and the probability function. So ($\mathcal{R}$) doesn't even apply. But on this alternative, (WADA) and (SADA) still apply, and they rule out $\mathbf{B}'$ and $\mathbf{B}''$, while allowing $\mathbf{B}$ and $\mathbf{B}'''$, even though they involve contradictory pairs.

On the second alternative, one considers full worlds that specify truth values for all propositions, and evaluates each belief set with respect to every world, even though some combinations of belief set and world can't both be actual. On this alternative, Table 3.3 should be replaced by Table 3.1. In this new table (which includes impossible pairs), (WADA) and (SADA) no longer rule out $\mathbf{B}'$ and $\mathbf{B}''$. However ($\mathcal{R}$) does rule out $\mathbf{B}$ and $\mathbf{B}'''$. Thus, if one accepts all three of these principles whenever they apply, this second alternative gives the opposite coherence requirement to the first alternative.

This second alternative is endorsed by Briggs (2009, pp. 78–83) for talking about actual coherence requirements. She says that an agent with a belief set that is ruled out by the first alternative is "guaranteed to be wrong about something, even though his or her beliefs are perfectly coherent" (p. 79). The "guarantee" of wrongness only comes about because we allow the specification of the belief set to also specify parts of the world, instead of allowing the two to vary independently.

One might worry that the present context is sufficiently different from the types of cases that Briggs considered that one should go for the first alternative instead. But determining the right interpretation of ($\mathcal{R}$) in the current case is beyond the scope of this paper. Caie's examples show that some of our general results may need to be modified in cases where some of the relevant propositions either refer to or depend on the agent's beliefs.[53]

11. CONCLUSION

We have sketched a general framework for constructing coherence requirements for various types of judgment, by starting with a notion of vindication and a distance relation among judgment sets, and supplementing them with a fundamental epistemic principle. Many philosophers have perhaps implicitly assumed Possible Vindication [(PV)] as a fundamental epistemic principle, and thus derived deductive consistency [(CB)] as a coherence norm for full belief. We think that (PV) is too strong, and—following Joyce (1998, 2009)—we retreat to weak accuracy dominance avoidance (WADA). The resulting coherence requirement (NDB) is thus derived from considerations of accuracy, and therefore can't come into conflict with them. Moreover, on plausible understandings of evidential norms for belief [which require at least ($\mathcal{R}$)], this coherence requirement can't come in conflict with them either [depending on how we understand what is going on in examples involving state–act dependence (Caie 2012; 2013)]. As a result, our proposed coherence requirement meets desideratum ($\mathcal{D}$)—it doesn't conflict with either alethic or evidential norms. Consequently, our new requirement allows (some) Preface cases (with appropriate evidential structure) to conform to requirements of ideal epistemic rationality.

Thus, we think (NDB) is a coherence requirement that should replace (CB) in the relevant discussions. However, we suspect that something stronger, like ($\mathcal{R}$) [or perhaps even one of the ($\mathcal{R}_r$)], could provide a more robust coherence requirement. One question for future research is whether any such requirement can be derived within our framework, either by changing the notion of vindication, the distance measure, or the fundamental epistemic principle. Investigating these alternatives in the framework will be interesting in other ways as well. Is there some coherence requirement in this framework that deals well with infinite sets of judgments? What happens when we start with a notion of vindication that is evidential in nature, rather than alethic? Are there epistemic norms other than the alethic and evidential ones, and if so, do our coherence requirements conflict with them?

One final point: in accordance with desideratum ($\mathcal{D}$), our proposed coherence requirements are entailed by both evidential and alethic norms for

[53] Caie's self-referential example is a special case of a more general class of examples that involve dependencies (causal *or* semantic) between an agent's beliefs and their truth-values. For further discussion of this broader class of examples involving state-act dependencies, see (Carr 2013; Greaves 2013).

full belief (given appropriate fundamental epistemic principles). If one accepts both the Jamesian and Cliffordian claims that there are alethic and evidential norms on belief, then one might suspect that this is in fact the role of coherence requirements. They may be the most general (rational, global/wide-scope) requirements on belief that are entailed by all of the basic epistemic norms.

APPENDIX

Proof of Theorem 1

> **B** is non-dominated iff ($\Leftrightarrow$) **B** contains no witnessing set.

($\Rightarrow$) We prove the contrapositive. Suppose that $\mathbf{S} \subseteq \mathbf{B}$ is a witnessing set. Let $\mathbf{B}'$ agree with **B** on all judgments outside **S** and disagree with **B** on all judgments in **S**. By the definition of a witnessing set, $\mathbf{B}'$ must weakly dominate **B** in distance from vindication [$d(\mathbf{B}, \mathring{\mathbf{B}}_w)$]. Thus, **B** is dominated.
($\Leftarrow$) We prove the contrapositive. Suppose that **B** is dominated, i.e. that there is some $\mathbf{B}'$ that weakly dominates **B** in distance from vindication [$d(\mathbf{B}, \mathring{\mathbf{B}}_w)$]. Let $\mathbf{S} \subseteq \mathbf{B}$ be the set of judgments on which **B** and $\mathbf{B}'$ disagree. Then, **S** is a witnessing set.

Proof of Theorem 2

> **B** is non-dominated if ($\Leftarrow$) there is a probability function Pr that represents **B**.

Let Pr be a probability function that represents **B** in the sense of Definition 2. Consider the expected distance, as calculated by Pr, of a belief set **B**—the sum over all worlds w of $\Pr(w) \cdot d(\mathbf{B}, \mathring{\mathbf{B}}_w)$. Since $d(\mathbf{B}, \mathring{\mathbf{B}}_w)$ is a sum of components for each proposition (1 if **B** disagrees with w on the proposition and 0 if they agree), and since expectations are linear, the expected distance from vindication is the sum of the expectation of these components. The expectation of the component for disbelieving p is $\Pr(p)$ while the expectation of the component for believing p is $1 - \Pr(p)$. Thus, if $\Pr(p) > 1/2$ then believing p is the attitude that uniquely minimizes the expectation, while if $\Pr(p) < 1/2$ then disbelieving p is the attitude that uniquely minimizes the expectation. Thus, since Pr represents **B**, this means that **B** has strictly lower expected distance from vindication than any other belief set with respect to Pr (i.e. that **B** uniquely minimizes expected distance from vindication, or maximizes expected closeness to vindication, relative to Pr). Suppose, for *reductio*, that some $\mathbf{B}'$ (weakly) dominates **B**. Then, $\mathbf{B}'$ must be no farther from vindication than **B** in any world, and thus $\mathbf{B}'$ must have expected distance from vindication no greater than that of **B**. But **B** has strictly lower expected distance from vindication than any other belief set. Contradiction. Therefore, no $\mathbf{B}'$ can dominate **B**, and so **B** must be non-dominated.

$\mathcal{B}$	$\mathbf{B}_1$	$\mathbf{B}_2$
$\neg X \,\&\, \neg Y$	D	D
$X \,\&\, \neg Y$	D	D
$X \,\&\, Y$	D	D
$\neg X \,\&\, Y$	D	D
$\neg Y$	D	D
$X \equiv Y$	D	D
$\neg X$	B	B
X	B	D
$\neg(X \equiv Y)$	D	D
Y	D	D
$X \vee \neg Y$	B	B
$\neg X \vee \neg Y$	B	B
$\neg X \vee Y$	B	B
$X \vee Y$	D	B
$X \vee \neg X$	B	B
$X \,\&\, \neg X$	D	D

Table 3.4 *Examples showing (SADA) $\not\Rightarrow$ (WADA) and (NDB) $\not\Rightarrow$ ($\mathcal{R}$)*

Proofs that (SADA) $\not\Rightarrow$ (WADA) and (NDB) $\not\Rightarrow$ ($\mathcal{R}$).

Consider a sentential language $\mathscr{L}$ with two atomic sentences X and Y. The Boolean algebra $\mathcal{B}$ generated by $\mathscr{L}$ contains sixteen propositions (corresponding to the subsets of the set of four state descriptions of $\mathscr{L}$, i.e. the set of four salient possible worlds). Table 3.4 depicts $\mathcal{B}$, and two opinionated belief sets ($\mathbf{B}_1$ and $\mathbf{B}_2$) on $\mathcal{B}$.

We have the following four salient facts regarding $\mathbf{B}_1$ and $\mathbf{B}_2$:

1. $\mathbf{B}_1$ is weakly dominated (in distance from vindication) by belief set $\mathbf{B}_2$. [This can be verified via simple counting.] Thus, $\mathbf{B}_1$ violates (NDB)/(WADA).
2. $\mathbf{B}_1$ is not strictly dominated (in distance from vindication) by any belief set over $\mathcal{B}$. [This can be verified by performing an exhaustive search on the set of all possible belief sets over $\mathcal{B}$.[54]] Thus, $\mathbf{B}_1$ satisfies (SADA).
3. $\mathbf{B}_2$ is not weakly dominated (in distance from vindication) by any belief set over $\mathcal{B}$. [This can be verified by performing an exhaustive search on

[54] We have created a *Mathematica* notebook that verifies claims (1)–(4) of this Appendix. This notebook can be downloaded from <http://fitelson.org/ace.nb>.

the set of all possible belief sets over $\mathcal{B}$ (see n. 54).] Thus, $\mathbf{B}_2$ satisfies (WADA).

4. $\mathbf{B}_2$ is not represented (in the sense of Definition 2) by any probability function on $\mathcal{B}$. [This can be verified via Fitelson's (2008) decision procedure for probability calculus `PrSAT` (see n. 54).[55]] Thus, $\mathbf{B}_2$ violates ($\mathcal{R}$).

Proof of Theorem 3

> For all $n \geq 2$ and for each set of propositions $\mathbf{P} \in \mathbb{B}_n$, if $r \geq \frac{n-1}{n}$ then ($\mathcal{R}_r$) rules out believing every member of **P**, while if $r < \frac{n-1}{n}$, then ($\mathcal{R}_r$) doesn't rule out believing every member of **P**.

Let **P** be a member of $\mathbb{B}_n$, i.e. **P** consists of n propositions, there is no world in which all of these n propositions are true, but for each proper subset $\mathbf{P}' \subset \mathbf{P}$ there is a world in which all members of $\mathbf{P}'$ are true.

Let $\phi_1, \ldots, \phi_n$ be the n propositions in **P**. Let each w_i be a world in which ϕ_i is false, but all other members of **P** are true. Let Pr be the probability distribution that assigns probability $1/n$ to each world w_i and 0 to all other worlds. If $r < \frac{n-1}{n}$, then Pr shows that the belief set $\mathbf{B_P} := \{B(\phi_1), \ldots, B(\phi_n)\}$, which includes the belief that ϕ_i for each $\phi_i \in \mathbf{P}$, satisfies ($\mathcal{R}_r$). This establishes the second half of the theorem.

For the first half, we will proceed by contradiction. Thus, assume that **P** is a member of $\mathbb{B}_n$ such that the belief set $\mathbf{B_P} := \{B(\phi_1), \ldots, B(\phi_n)\}$, which includes the belief that ϕ_i for each $\phi_i \in \mathbf{P}$, is *not* ruled out by $\left(\mathcal{R}_{\frac{n-1}{n}}\right)$. Then there must be some Pr such that for each i, $\Pr(\phi_i) > \frac{n-1}{n}$. This means that for each i, $\Pr(\neg\phi_i) < 1/n$. Since the disjunction of finitely many propositions is at most as probable as the sum of their individual probabilities, this means that $\Pr(\neg\phi_1 \vee \ldots \vee \neg\phi_n) < 1$. But since **P** is inconsistent, $\neg\phi_1 \vee \ldots \vee \neg\phi_n$ is a tautology, and therefore must have probability 1. This is a contradiction, so $\mathbf{B_P}$ must be ruled out by $\left(\mathcal{R}_{\frac{n-1}{n}}\right)$.

Proof of Theorem 4

> **B** is not strictly dominated iff ($\Leftrightarrow$) **B** contains no witnessing$_1$ set.
> That is: (SADA) $\Leftrightarrow$ (NW$_1$S).

($\Rightarrow$) We'll prove the contrapositive. Suppose that $\mathbf{S} \subseteq \mathbf{B}$ is a witnessing$_1$ set. Let $\mathbf{B}'$ agree with **B** on all judgments outside **S** and disagree with **B** on all judgments in **S**. By the definition of a witnessing$_1$ set, $\mathbf{B}'$ must

[55] In fact, this can be verified easily by hand, since the set $\mathbf{B}_2$ contains two contradictory pairs of judgments: $\{D(Y), D(\neg Y)\}$ and $\{D(X \equiv Y), D(\neg(X \equiv Y))\}$. Moreover, we've already seen an example of this phenomenon in §7, when we showed (NDB) does not rule out contradictory pairs.

strictly dominate **B** in distance from vindication [$d(\mathbf{B},\mathring{\mathbf{B}}_w)$]. Thus, **B** is strictly dominated.
($\Leftarrow$) We prove the contrapositive. Suppose **B** is strictly dominated, i.e. that there is some **B**′ that strictly dominates **B** in distance from vindication [$d(\mathbf{B},\mathring{\mathbf{B}}_w)$]. Let $\mathbf{S} \subseteq \mathbf{B}$ be the set of judgments on which **B** and **B**′ disagree. Then, **S** is a witnessing$_1$ set.

Proof of Theorem 5

> **B** is probabilistically representable (in the sense of Definition 2) only if **B** contains no witnessing$_2$ set. That is, $(\mathcal{R}) \Longrightarrow$ (NW$_2$S).

In our proof of Theorem 2, we established that if Pr represents **B**, then **B** has strictly lower expected distance from vindication than any other belief set with respect to Pr. Assume, for *reductio*, that $\mathbf{S} \subseteq \mathbf{B}$ is a witnessing$_2$ set for **B**. Let **B**′ agree with **B** on all judgments outside **S** and disagree with **B** on all judgments in **S**. Then, by the definition of a witnessing$_2$ set, **B**′ must be no farther from vindication than **B** in any world. But this contradicts the fact that **B** has strictly lower expected distance from vindication than **B**′ with respect to Pr. So the witnessing$_2$ set must not exist.

Proof of Theorem 6

> **B** is probabilistically representable (in the sense of Definition 2) only if **B** satisfies both (NDB) and (NCP). That is, $(\mathcal{R}) \Longrightarrow$ (NDB & NCP).

Theorem 2 implies $(\mathcal{R}) \Longrightarrow$ (NDB). And, it is obvious that $(\mathcal{R}) \Longrightarrow$ (NCP), since no probability function can probabilify both members of a contradictory pair and no probability function can dis-probabilify both members of a contradictory pair.

Counterexample to the Converse of Theorem 6

> $(\mathcal{R}) \not\Longleftarrow$ (NDB & NCP).

Let there be six possible worlds, $w_1, w_2, w_3, w_4, w_5, w_6$. Consider the agenda $\mathcal{A}$ consisting of the following four propositions (i.e. $\mathcal{A} =_{df} \{p_1, p_2, p_3, p_4\}$).

$$p_1 = \{w_1, w_2, w_3\}$$
$$p_2 = \{w_1, w_4, w_5\}$$
$$p_3 = \{w_2, w_4, w_6\}$$
$$p_4 = \{w_3, w_5, w_6\}$$

Let $\mathbf{B} =_{df} \{B(p_1), B(p_2), B(p_3), B(p_4)\}$. **B** is itself a witnessing$_2$ set, since, in every possible world, exactly two beliefs (i.e. exactly half of the beliefs) in **B** are accurate. So by Theorem 5, **B** is not probabilistically representable. However,

B satisfies (NDB). To see this, note that every belief set on $\mathcal{A}$ has an expected distance from vindication of 2, relative to the uniform probability distribution. This implies that no belief set on $\mathcal{A}$ dominates any other belief set on $\mathcal{A}$. Finally, **B** satisfies (NCP), since every pair of beliefs in **B** is consistent.[56]

REFERENCES

Berker, S. (2013). Epistemic teleology and the separateness of propositions. *Philosophical Review* 122, 337–93.

Boghossian, P. (2003). The normativity of content. *Philosophical Issues* 13(1), 31–45.

BonJour, L. (1985). *The structure of empirical knowledge*. Cambridge University Press.

Briggs, R. (2009). Distorted reflection. *Philosophical Review* 118(1), 59–85.

Briggs, R., F. Cariani, K. Easwaran, and B. Fitelson (2014). Individual coherence and group coherence. In J. Lackey (ed.), *Essays in collective epistemology*. Oxford University Press.

Broome, J. (2007). Wide or narrow scope? *Mind* 116(462), 359–70.

Caie, M. (2012). Belief and indeterminacy. *Philosophical Review* 121(1), 1–54.

Caie, M. (2013). Rational probabilistic incoherence. *Philosophical Review* 122(4), 527–75.

Carr, J. (2013). What to expect when you're expecting. Manuscript.

Christensen, D. (1996). Dutch-book arguments depragmatized: Epistemic consistency for partial believers. *The Journal of Philosophy* 93(9), 450–79.

Christensen, D. (2004). *Putting logic in its place*. Oxford University Press.

Clifford, W. (1877). The ethics of belief. *Contemporary Review* 29, 289–309.

Conee, E. and R. Feldman (2004). *Evidentialism: Essays in epistemology*. Oxford University Press.

Deza, M. and E. Deza (2009). *Encyclopedia of distances*. Springer.

Dretske, F. (2013). Gettier and justified true belief: Fifty years on. *The Philosophers' Magazine* 61. URL: http://philosophypress.co.uk/?p=1171.

Easwaran, K. (2013). Dr. Truthlove, or How I learned to stop worrying and love Bayesian probabilities. Manuscript.

Easwaran, K. and B. Fitelson (2012). An "evidentialist" worry about Joyce's argument for probabilism. *Dialectica* 66(3), 425–33.

Fitelson, B. (2008). A decision procedure for probability calculus with applications. *The Review of Symbolic Logic* 1(01), 111–25.

[56] We can extend this counterexample to an example that isn't restricted to $\mathcal{A}$, but in fact is opinionated across the whole algebra $\mathcal{B}$ of propositions constructible out of the six possible worlds. Here's how to extend **B** to a belief set **B**′ over $\mathcal{B}$. For the propositions p that are true in four or more worlds, **B**′ contains $B(p)$. For the propositions q that are true in two or fewer worlds, **B**′ contains $D(q)$. For each pair of complementary propositions that are true in exactly three worlds, **B**′ includes belief in one and disbelief in the other (i.e. **B**′ may include any consistent pair of attitudes one likes toward each pair of complementary propositions that are true in exactly three worlds). By construction, this set **B**′ satisfies (NCP). Additionally, the belief set **B**′ is tied for minimal expected distance from vindication on $\mathcal{B}$, relative to the uniform probability distribution on $\mathcal{B}$. Therefore, **B**′ satisfies (NDB).

Fitelson, B. (2014). *Coherence*. Book manuscript (in progress).
Fitelson, B. and D. McCarthy (2013). Toward an epistemic foundation for comparative confidence. Manuscript.
Foley, R. (1992). *Working without a net*. Oxford University Press.
Friedman, J. (2013). Suspended judgment. *Philosophical Studies* 162(2), 165–81.
Fumerton, R. (1995). *Metaepistemology and skepticism*. Rowman & Littlefield.
Gibbard, A. (2005). Truth and correct belief. *Philosophical Issues* 15(1), 338–50.
Greaves, H. (2013). Epistemic decision theory. *Mind* 122(485): 915–52.
Hájek, A. (2008). Arguments for—or against—probabilism? *The British Journal for the Philosophy of Science* 59(4), 793–819.
Harman, G. (1986). *Change in view: Principles of reasoning*. MIT Press.
Hawthorne, J. and L. Bovens (1999). The preface, the lottery, and the logic of belief. *Mind* 108(430), 241–64.
Hedden, B. (2013). Time-slice rationality. Manuscript.
James, W. (1896). The will to believe. *The New World* 5, 327–47.
Joyce, J. M. (1998). A nonpragmatic vindication of probabilism. *Philosophy of Science* 65(4), 575–603.
Joyce, J. M. (2005). How probabilities reflect evidence. *Philosophical Perspectives* 19(1), 153–78.
Joyce, J. M. (2009). Accuracy and coherence: Prospects for an alethic epistemology of partial belief. In F. Huber and C. Schmidt-Petri (eds.), *Degrees of belief*. Springer.
Joyce, J. M. (2013). The role of evidence in an accuracy-centered epistemology. Manuscript.
Kaplan, M. (2013). Coming to terms with our human fallibility: Christensen on the preface. *Philosophy and Phenomenological Research* 87(1), 1–35.
Klein, P. (1985). The virtues of inconsistency. *The Monist* 68(1), 105–35.
Kolodny, N. (2007). How does coherence matter? *Proceedings of the Aristotelian Society* 107, 229–63.
Kolodny, N. and J. MacFarlane (2010). Ifs and oughts. *Journal of Philosophy* 107, 115–43.
Korb, K. B. (1992). The collapse of collective defeat: Lessons from the lottery paradox. In *Proceedings of the Biennial Meeting of the Philosophy of Science Association*, pp. 230–6.
Kyburg, H. E. (1970). Conjunctivitis. In M. Swain (ed.), *Induction, acceptance and rational belief*, pp. 55–82. Reidel.
Leitgeb, H. (2013). The stability theory of belief. Manuscript.
Littlejohn, C. (2012). *Justification and the truth-connection*. Cambridge University Press.
MacFarlane, J. (2004). In what sense (if any) is logic normative for thought? Manuscript.
Merricks, T. (1995). Warrant entails truth. *Philosophy and Phenomenological Research* 55(4), 841–55.
Moss, S. (2013). Time-slice epistemology and action under indeterminacy. Manuscript.

Nelkin, D. K. (2000). The lottery paradox, knowledge, and rationality. *The Philosophical Review* 109(3), 373–409.

Parfit, D. (1988). What we together do. Manuscript.

Pettigrew, R. (2013a). Accuracy and evidence. *Dialectica*, 67: 579–96.

Pettigrew, R. (2013b). Epistemic utility and norms for credences. *Philosophy Compass*, 8(10), 897–908.

Pollock, J. L. (1983). Epistemology and probability. *Synthese* 55(2), 231–52.

Pollock, J. L. (1986). The paradox of the preface. *Philosophy of Science* 53, 246–58.

Pollock, J. L. (1990). *Nomic probability and the foundations of induction*. Oxford University Press.

Ramsey, F. P. (1926). Truth and probability. In R. Braithwaite (ed.), *The foundations of mathematics and other logical essays* [1931], pp. 156–98. Routledge.

Ryan, S. (1991). The preface paradox. *Philosophical Studies* 64(3), 293–307.

Ryan, S. (1996). The epistemic virtues of consistency. *Synthese* 109(2), 121–41.

Schervish, M., T. Seidenfeld, and J. Kadane (2009). Proper scoring rules, dominated forecasts, and coherence. *Decision Analysis* 6(4), 202–21.

Shah, N. (2003). How truth governs belief. *The Philosophical Review* 112(4), 447–82.

Smith, M. (2005). Meta-ethics. In F. Jackson and M. Smith (eds.) *The Oxford Handbook of Contemporary Philosophy*, pp. 3–30, Oxford University Press.

Steinberger, F. (2013). Explosion and the normativity of logic. Manuscript.

Sturgeon, S. (2008). Reason and the grain of belief. *Noûs* 42(1), 139–65.

Thomson, J. J. (2008). *Normativity*. Open Court.

Titelbaum, M. G. (2013). *Quitting certainties: A Bayesian framework modeling degrees of belief*. Oxford University Press.

Wedgwood, R. (2002). The aim of belief. *Philosophical Perspectives* 16, 267–97.

Williamson, T. (2000). *Knowledge and its limits*. Oxford University Press.

Zagzebski, L. (1994). The inescapability of Gettier problems. *Philosophical Quarterly* 44(174), 65–73.

4. Fallibilism and Multiple Paths to Knowledge

Wesley H. Holliday

If knowledge required the elimination of all logically possible alternatives, there would be no knowledge (at least of contingent truths).

Alvin Goldman (1976, 775)

There are always, it seems, possibilities that our evidence is powerless to eliminate.... If knowledge... requires the elimination of all competing possibilities... then, clearly we seldom, if ever, satisfy the conditions for applying the concept.

Fred Dretske (1981, 365)

1. INTRODUCTION

Being a fallibilist isn't easy. A fallibilist about empirical knowledge, in Lewis's (1996) sense, holds that an agent can know a contingent empirical proposition P, even if she has not ruled out *every last way* that P could be false.[1] In this sense, it seems that most contemporary epistemologists are fallibilists, at least relative to some way of understanding what it is to "rule out" an alternative. And with good reason: if knowing a contingent empirical proposition P required ruling out *every last way* that P could be false, then we would have little if any empirical knowledge. Radical skepticism would reign. Yet fallibilism, despite its promise for defending the possibility of knowledge, also faces problems. To borrow an analogy sometimes applied to philosophical projects, trying to fill in the details of a fallibilist theory of knowledge is like trying to install an unstretched carpet: flatten a problematic lump in one place and a new one appears elsewhere. But then again, the alternative of radical

For helpful feedback on this paper, I wish to thank Justin Bledin, John Campbell, Peter Hawke, Thomas Icard, Ethan Jerzak, Krista Lawlor, John Perry, Michael Rieppel, Shane Steinert-Threlkeld, Justin Vlasits, Seth Yalcin, and two anonymous referees for *Oxford Studies in Epistemology*. For helpful conversations or correspondence on issues discussed in the paper, I wish to thank Johan van Benthem, Keith DeRose, John MacFarlane, Sherrilyn Roush, Crispin Wright, and Stephen Yablo.

[1] The term "fallibilism" means many different things to many different people. I explain in more detail what I mean by "fallibilism" in §2.1.

skepticism about knowledge is like having the rug pulled out from under your feet.

The primary goal of this paper is to argue that what I call the *standard alternatives picture*, assumed by many fallibilist theories, should be replaced by a new *multipath picture* of knowledge. In §2, I identify the problematic lumps in the standard picture: fallibilists working with this picture cannot maintain even the most uncontroversial (single-premise, logical) epistemic closure principles without having to make extreme assumptions about the ability of humans to know empirical truths without empirical investigation. In §3, I show how the multipath picture, motivated by independent arguments, saves fallibilism from this problem. The multipath picture is based on taking seriously the idea that there can be multiple paths to knowing some propositions about the world. An overlooked consequence of fallibilism is that these multiple paths to knowledge may involve ruling out different sets of alternatives, which should be represented in our picture of knowledge. In §4, I consider inductive knowledge and strong epistemic closure principles from this multipath perspective.

In what follows, I presuppose familiarity with the kinds of *skeptical hypotheses* that motivate fallibilism about knowledge (see, e.g., Dretske 1970, 1981, 2005). For lack of space, I cannot review the standard examples here. Instead, I leave it to the reader's imagination to fill in abstract discussions of skepticism, fallibilism, and epistemic closure with specific scenarios and propositions. Lewis (1996, 549) said it best: "Let your paranoid fantasies rip—CIA plots, hallucinogens in the tap water, conspiracies to deceive, old Nick himself—and soon you find that uneliminated possibilities of error are everywhere. Those possibilities of error are far-fetched, of course, but possibilities all the same. They bite into even our most everyday knowledge. We never have infallible knowledge."

1.1. Scenarios and Propositions

Let us begin with some preliminary points of terminology and notation used throughout.

We start with a set W of triples $\langle \mathtt{w}, \mathtt{a}, \mathtt{t} \rangle$ where $\mathtt{w}$ is a *way the world could (or could not) be including agent* $\mathtt{a}$ *at time* $\mathtt{t}$.[2] I use "w", "v", "u", etc., for members of W, which I will call *scenarios*. For each scenario w, let W_w be the subset of W containing those scenarios that are metaphysically possible relative to w. Everything in this paper is compatible with the view that $W_w = W_v = W$ for all w and v, so that no scenarios are metaphysically impossible relative to any others, and compatible with the rejection of this view. I leave these as parameter choices for the reader. However, for simplicity I assume that W does not include any "logically impossible" scenarios (see below).

[2] In possible-worlds parlance, W would be a set of "centered possible (or impossible) worlds" (see Lewis 1979 on centered worlds and King 2007 on impossible worlds), but this need not be a context-independent "intended standard model of super-reality" (Stalnaker 1986: 122). As Stalnaker remarks, "The formalism of possible worlds semantics assumes that

Following standard set-theoretic notation, I use "$\in$" for the membership relation, "$\notin$" to deny the membership relation, "$\subseteq$" for the subset relation, "$\nsubseteq$" to deny the subset relation, and "$\subsetneq$" for the *strict* subset relation ($A \subseteq B$ but $B \nsubseteq A$); for any sets A and B, $A - B = \{w \in A \mid w \notin B\}$ is the complement of B in A, $A \cup B$ is the union of A and B, and $A \cap B$ is their intersection; given a set X *of sets*, $\bigcup X$ (resp. $\bigcap X$) is the union (resp. intersection) of all members of X; and given an indexed family $\{A_i\}_{i \in I}$ of sets, $\bigcup_{i \in I} A_i$ (resp. $\bigcap_{i \in I} A_i$) is the union (resp. intersection) of all the A_i sets.

My topic is knowledge of *propositions*. I use "P", "Q", "S", etc., for propositions and "$\mathcal{P}$" for the set of all propositions under consideration. I assume that propositions are *true* or *false* at scenarios in W and that propositions can have truth-functional structure: if P is a proposition, so is the negation of P, denoted by "$\neg P$"; if P and Q are propositions, so is the disjunction of P and Q, denoted by "$P \vee Q$"; and so on for other truth-functions.[3] If P does not have the structure of a truth-function applied to one or more propositions, call it *TF-atomic*.[4] As usual, an assignment of truth values to TF-atomic propositions determines a truth value for every proposition; and Q is a *TF-consequence* (resp. *TF-equivalent*) of P iff any such assignment makes Q true if (resp. iff) it makes P true.

For any proposition P, define $\boldsymbol{P} = \{w \in W \mid P \text{ is true at } w\}$, the set of scenarios at which P is true.[5] Given a classical understanding of negation, disjunction, conjunction, etc., and the ban on logically impossible scenarios, we have $\boldsymbol{\neg P} = W - \boldsymbol{P}$, $\boldsymbol{P \vee Q} = \boldsymbol{P} \cup \boldsymbol{Q}$, $\boldsymbol{P \wedge Q} = \boldsymbol{P} \cap \boldsymbol{Q}$, etc. Let us also define $\boldsymbol{P}_w = \boldsymbol{P} \cap W_w$, the set of scenarios metaphysically possible relative to w at which P is true. Relative to w, P is metaphysically *necessary* (resp. *possible*) iff $\boldsymbol{P}_w = W_w$ (resp. $\boldsymbol{P}_w \neq \emptyset$), P is metaphysically *contingent* iff $\emptyset \neq \boldsymbol{P}_w \neq W_w$, and P *metaphysically entails* Q (resp. is *metaphysically equivalent* to Q) iff $\boldsymbol{P}_w \subseteq \boldsymbol{Q}$

possible states of the world are disjoint alternatives, and that everything that can be said *within a given context* can be said by distinguishing between these alternatives.... Nothing in the formalism of possible worlds semantics, or in the intuitive conception of a way things might be, or a possible state of the world, excludes an interpretation in which possible worlds are alternative states of some limited subject matter. Possible worlds must be complete, relative to the distinctions that can be made within the given interpretation, but they might be quite partial relative to another interpretation, or relative to an external intuitive commentary on the interpretation" (118–19). Compare Lewis (1996, 552): "we needn't decide whether they must always be maximally specific possibilities, or whether they need only be specific enough for the purpose at hand. A possibility will be specific enough if it cannot be split into subcases in such a way that anything we have said about possibilities, or anything we are going to say before we are done, applies to some subcases and not to others. For instance, it should never happen that proposition P holds in some but not all sub-cases; or that some but not all sub-cases are eliminated by S's evidence." For simplicity, I will not relativize the set W to contexts, but these remarks should be kept in mind. The framework developed here can also be generalized to include what Perry (1986) calls *partial ways* the world could be (see Holliday 2014b).

[3] See King 2011 for a survey of views of structured propositions.

[4] I deliberately use the term "TF-atomic" instead of "atomic." A proposition that has a complex structure may count as TF-atomic, because it does not have the structure of a truth-function applied to one or more propositions.

[5] As explained in §2.4, one may take $\boldsymbol{P} = \{w \in W \mid P \text{ is true at } w \textit{ considered as actual}\}$.

(resp. $\boldsymbol{P}_w = \boldsymbol{Q}_w$). According to some non-structured proposition views (Stalnaker 1981, Lewis 1986), if for all scenarios w based on the way our world is, $\boldsymbol{P}_w = \boldsymbol{Q}_w$, then $P = Q$; but for propositions *qua* objects of knowledge, I do not make this strong assumption for standard reasons and for a reason specific to fallibilism, discussed in §4.2.

Finally, I use "C", "C'", etc. for *contexts* of knowledge attribution or assessment. Nothing in what follows depends on what contexts *are*, beyond the assumption that contexts play a certain "functional role" (namely by being something to which the functions in §2.1 are relativized). Following DeRose (2009, 187), I say that an agent in a scenario w does or does not "count as knowing proposition P in context C" or "relative to C." Yet I intend all that follows to be consistent with invariantism as well as contextualism and relativism; invariantists can assume that there is only one constant context C.

2. THE STANDARD ALTERNATIVES PICTURE

In this section, I introduce a *standard alternatives picture* of knowledge, show how a family of fallibilist theories fit into this picture, and then argue that the picture is fundamentally flawed.

2.1. *Relevancy Set and Uneliminated Set*

The starting point of the standard alternatives picture is the idea that for each proposition to be known, there is "a set of situations each member of which contrasts with what is [to be] known... and must be evidentially excluded if one is to know" (Dretske 1981, 373). Dretske proposes that we "call the set of possible alternatives that a person must be in an evidential position to exclude (when he knows that P) the *Relevancy Set*" (371). Similarly, let us call the set of alternatives for P that the person has *not* excluded the *Uneliminated Set*. According to this picture, there are two *functions* r and u, each of which takes as input a proposition P, scenario w, and possibly a context C, and returns a set of *alternatives*, which I take to be *scenarios* (for reasons explained later):

- $\mathsf{r}_C(P, w)$ = the set of ("relevant") alternatives such that the agent in scenario w counts as knowing proposition P relative to context C only if she has eliminated these alternatives;
- $\mathsf{u}_C(P, w)$ = the set of ("uneliminated") alternatives that the agent in scenario w has *not eliminated* as alternatives for P relative to context C.

The reasons for relativizing these sets to a scenario and possibly a context are well known. First, since objective features of an agent's situation in a scenario w may affect what alternatives are relevant in w and therefore what it takes to know P in w (see Dretske 1981, 377 and Derose 2009, 30f. on "subject factors"), we allow that $\mathsf{r}_C(P, w)$ may differ from $\mathsf{r}_C(P, v)$ for a distinct scenario v in which the agent's situation is different. Second, if we allow—unlike Dretske—that features of the conversational context C of those attributing knowledge to the agent (or the context of assessment of a knowledge attribution, in the sense

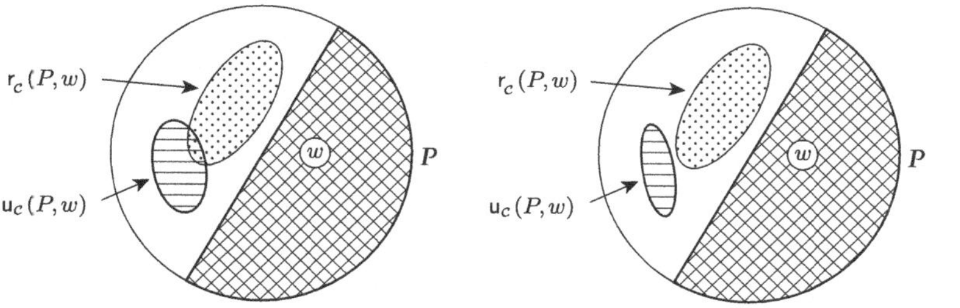

Figure 4.1 (Knows) violated on left vs. satisfied on right

of MacFarlane 2005) can also affect what it takes to count as knowing P in w relative to C (see DeRose 2009, 30f. on "attributor factors"), then we should allow that $\mathsf{r}_C(P,w)$ may differ from $\mathsf{r}_{C'}(P,w)$ for a distinct context C'. Similarly, if we allow that what counts as eliminating an alternative may vary with context (see DeRose 2009, 30n29) or depend on the agent's situation, then our u function should take in a context and scenario as well.

According to the standard alternatives picture,[6] an agent in scenario w counts as knowing P relative to context C if and only if (or at least only if) (the agent believes P and) the following holds:

$$\mathsf{r}_C(P,w) \cap \mathsf{u}_C(P,w) = \emptyset. \qquad \text{(Knows)}$$

Fig. 4.1 shows the (Knows) condition violated vs. satisfied. Each of the large circles represents the set W of scenarios under consideration. The crosshatched region is the set $\boldsymbol{P}$ of scenarios in which the proposition P is true, including scenario w. The Relevancy Set and Uneliminated Set for P in w relative to context C are shown in the ellipses with dots and horizontal lines, respectively, in the blank $\neg P$-zone. If these sets overlap, as on the left, then the agent in w does not know P relative to C; if they do not overlap, as on the right, then the agent in w knows P relative to C.[7]

In §2.3, I will show that a family of fallibilist theories fit into this picture as special cases, distinguished in part by the structural constraints they impose on the r and u functions. Some theories with more moving parts have another pair of functions r'_C and u'_C, also requiring $\mathsf{r}'_C(P,w) \cap \mathsf{u}'_C(P,w) = \emptyset$ for knowledge (see §2.3), but I will concentrate on theories with one pair of functions.

In virtue of what is an alternative in $\mathsf{r}_C(P,w)$ or $\mathsf{u}_C(P,w)$? For r, one can give "thick" or "thin" accounts of what it takes for a scenario to be in $\mathsf{r}_C(P,w)$, depending on whether the account is independent of epistemic notions like *knowledge*. In §2.3, we will see some thick accounts with such independence,

[6] By calling this picture "standard," I am not claiming that *all* contemporary views of knowledge fit into it.

[7] The sizes of the various regions in the diagram are not intended to reflect the sizes of the corresponding sets, and the locations of the regions are not intended to reflect the "distance" of scenarios from w.

but we already have a good thin account in the first bullet point above. Of course, this pushes us to the question about elimination and $\mathsf{u}_C(P,w)$.[8] But let us first consider the decision about what our "alternatives" are: *scenarios* or *propositions* or something else?

I take alternatives to be *scenarios*. What really matters is that the set of all $\neg P$-alternatives in a context should form a nontrivial *partition* of the set of $\neg P$-scenarios, so the alternatives are disjoint.[9] (Recall the quotes from Stalnaker and Lewis in footnote 2.) We could call the cells in such a partition "Alternatives," and let $\mathsf{r}^A_C(P,w)$ and $\mathsf{u}^A_C(P,w)$ be sets of Alternatives. But since I think of elimination in terms of scenarios, I take $\mathsf{r}_C(P,w)$ and $\mathsf{u}_C(P,w)$ to be sets of scenarios.[10] This approach fits with what I consider the best-developed of previous fallibilist theories, discussed in §2.3. It also has other advantages, especially over taking the set of $\neg P$-alternatives to be the set of *all propositions incompatible with P*, which violates the disjointness of alternatives in a context.

For example, Vogel's (1999, 163) argument that *probability* cannot provide a sufficient condition for relevance of alternatives depends on assuming the proposition-based view of alternatives (see footnote 52 in §3.5). Moreover, the puzzling question (see Stine 1976, 258) of whether $\neg P$ is a relevant alternative to P—and if so, what it takes to "eliminate" $\neg P$ other than knowing P—suggests that the level of propositions might not be the best level at which to locate alternatives. It seems that one can give a more substantive account of what it is for a *scenario* to be (un)eliminated, since one may refer to the experiences or beliefs of the agent in that scenario, compared to those of the agent in another scenario. By contrast, accounts of what it is for a proposition to be eliminated seem not to take us very far from the idea of *knowing* the negation of the proposition.

According to Lewis (1996), "a possibility [v] is *uneliminated* iff the subject's perceptual experience and memory in [v] exactly match his perceptual experience and memory in actuality" (553). I will postpone discussion of whether such match is necessary.[11] All of the theories I consider seem to agree on at

[8] Dretske (1981) gives thin accounts of both r and u in terms of knowledge: "let us call the set of possible alternatives that a person must be in an evidential position to exclude (when he knows that P) the *Relevancy Set* (*RS*). In saying that he must be in a position to exclude these possibilities I mean that his evidence or justification for thinking these alternatives are *not* the case must be good enough to say he *knows* they are not the case" (371). Lawlor (2013) gives a thicker account of what makes an alternative one that must be eliminated for knowledge, i.e. an alternative in $\mathsf{r}_C(P,w)$: it is "an alternative to p that a reasonable person would want ruled out by reasons or evidence before judging that S knows p" (152), where the notion of a *reasonable person* is given a substantive independent characterization (Lawlor 2013, §5.1).

[9] Or more generally, the set of alternatives for P in a given context should form a nontrivial partition of W.

[10] Of course, this is the partition view where each cell contains only one scenario. Another option would be to use Alternatives, but only from partitions with the property that if one of the scenarios in an Alternative is in $\mathsf{u}_C(P,w)$, then all of the scenarios in that Alternative are in $\mathsf{u}_C(P,w)$, following the quote from Lewis in footnote 2. Then we could define u^A from u: an Alternative is in $\mathsf{u}^A_C(P,w)$ iff all scenarios in that Alternative are in $\mathsf{u}_C(P,w)$.

[11] Cf. Goldman's (1976, 779–84) detailed discussion of the notion of a *perceptual equivalent* of a state of affairs, which does not require exact match of perceptual appearances (781).

least this much: for $v \in \mathrm{W} - \boldsymbol{P}$, it is *sufficient* for $v \in \mathsf{u}_{\mathcal{C}}(P, w)$ that v and w are subjectively indistinguishable, appear the same way, etc., to the agent, given her total experience and memory, where this requires that the agent's ("narrow") beliefs are the same in v and w. Many theorists would also agree that v and w are subjectively indistinguishable to the agent if she is in the same physical state in both, so this would provide another sufficient condition for $v \in \mathsf{u}_{\mathcal{C}}(P, w)$.

Given these sufficient conditions, it follows that for many contingent propositions P about the world external to the agent, $\mathsf{u}_{\mathcal{C}}(P, w) \cap (\mathrm{W}_w - \boldsymbol{P}) \neq \emptyset$. For given a scenario w, perhaps in which the agent believes P, there is another possible scenario v in which the agent is in the same physical state, or at any rate a scenario that is subjectively indistinguishable, but in which P is false, so $v \in \mathsf{u}_{\mathcal{C}}(P, w) \cap (\mathrm{W}_w - \boldsymbol{P})$.[12] This is a reflection of the separation between mind and world.[13]

Given that $\mathsf{u}_{\mathcal{C}}(P, w) \cap (\mathrm{W}_w - \boldsymbol{P}) \neq \emptyset$ for so many empirical propositions P, radical skepticism about empirical knowledge follows from the "infallibilist" assumption that knowing a proposition P requires ruling out *all* possible $\neg P$-scenarios, which in terms of Fig. 4.1 requires that the dotted region covers the *entire* $\neg P$-zone (at least within W_w, if W_w is a smaller circle than W):

infallibilism—for all w, P, and $\mathcal{C}$: $\mathrm{W}_w - \boldsymbol{P} \subseteq \mathsf{r}_{\mathcal{C}}(P, w)$.[14]

[12] I am not claiming (what certain kinds of externalists about perception would deny) that given a scenario w in which the agent believes P, there is always another possible scenario v in which the agent has the same *type of experience* or the *same evidence*, but in which P is false; for I am not assuming that subjective indistinguishability entails the same type of experience or evidence. I am also not claiming that if w and v are subjectively indistinguishable, then $\mathsf{u}_{\mathcal{C}}(P, w) = \mathsf{u}_{\mathcal{C}}(P, v)$, i.e. that the agent in w has eliminated exactly the same alternatives as the agent in v.

[13] Examples abound in the literature on skepticism, but let us consider another. In the actual scenario w, Jones, who lives in the U.S., receives a postcard from Smith, who is visiting the U.K. The postcard is signed by Smith in his unique handwriting, stamped and dated by U.K. postal officials, and so on. Jones recognizes all of this, and he correctly takes Smith to be a perfectly reliable reporter of his vacation whereabouts. According to everyone but radical skeptics, on the basis of receiving such a postcard, Jones can know that *Smith visited the U.K. some days ago* (P). Yet everyone must also admit that there are possible scenarios v in which everything appears the same to Jones (during his whole life up until now) as in w, but the postcard was not sent by Smith, and Smith never visited the U.K., so $v \in \mathsf{u}_{\mathcal{C}}(P, w) \cap (\mathrm{W}_w - \boldsymbol{P})$. Some of these scenarios are ones in which skeptical hypotheses incompatible with P obtain: in some of them, *the postcard was forged by a team of deceivers* (SH_1); in others, *all the world and Jones's memories were created five seconds before he received the postcard* (SH_2); and so on. Of course, such deceptive possibilities arise for a tremendous number of other propositions that Jones believes about the external world.

[14] It is sometimes suggested that one has an "infallibilist" conception of knowledge if one accepts the following principle: if an agent knows that P, then her evidential probability for P is 1. According to the present conception of infallibilism and fallibilism, that suggestion is incorrect. As defined below, a fallibilist may hold that (i) an agent knows that P, so (ii) the agent's evidential probability for P is 1, even though (iii) there may be some scenarios that are subjectively indistinguishable from the agent's actual scenario—and in that sense are uneliminated—in which P is false. The fallibilism is in the conjunction of (i) and (iii). Dretske (1981; 1971) is such a fallibilist who holds that (i) implies (ii). (Note that such a view is not inconsistent with Dretske's denial of closure, because he does not hold that probability 1 is

It follows from infallibilism and $\mathsf{u}_C(P,w) \cap (\mathrm{W}_w - \boldsymbol{P}) \neq \emptyset$ that $\mathsf{r}_C(P,w) \cap \mathsf{u}_C(P,w) \neq \emptyset$, so by (Knows), P is not known. Thus, in order to avoid radical skepticism, one must *at least* deny infallibilism:

> fallibilism—for some w, P, and C: $\mathrm{W}_w - \boldsymbol{P} \not\subseteq \mathsf{r}_C(P,w)$.

This is an extremely weak version of fallibilism: in effect, fallibilism about at least one possible case of knowledge. A stronger, but still extremely weak, version of fallibilism says that there is some proposition Q that is true in all of the relevant alternatives to P but not in all possible $\neg P$-scenarios:

> e-fallibilism—for some w, P, Q, and C: $\mathsf{r}_C(P,w) \subseteq \boldsymbol{Q}$ and $\mathrm{W}_w - \boldsymbol{P} \not\subseteq \boldsymbol{Q}$.

Here "e-fallibilism" stands for *expressible fallibilism*, since it says that we can *express* with Q something that the relevant alternatives have in common with each other (and perhaps some other scenarios) but not with all possible $\neg P$-scenarios. It would be a strange version of fallibilism that denied there was even one such proposition P for which we could express our fallibilism in this way. Note that if for every set of scenarios there is a corresponding proposition true in exactly those scenarios, then fallibilism is equivalent to e-fallibilism, taking Q to be the proposition corresponding to $\mathsf{r}_C(P,w)$. Also note that e-fallibilism does not even require that the proposition Q be incompatible with P, i.e. $\boldsymbol{Q}_w \subseteq \mathrm{W} - \boldsymbol{P}$. For that, one could assume what I will call *expressible contrast fallibilism*:

> ec-fallibilism—for some w, P, Q, and C: $\mathsf{r}_C(P,w) \subseteq \boldsymbol{Q}$ and $\boldsymbol{Q}_w \subsetneq \mathrm{W}_w - \boldsymbol{P}$.

The reason for considering such weak principles will become apparent later. I will argue that being even a weak fallibilist is tricky, although not for the reasons that some philosophers think.[15]

sufficient for knowledge. Some propositions will have probability 1, although they are not known, because they are entailed by other propositions with probability 1 that are known. See Dretske 2006.)

[15] Not, for example, for worries about *concessive knowledge attributions* of the form "I know that P, but it's possible that $\neg P$" or "I know that P, but it might be that $\neg P$" (Rysiew 2001, Stanley 2005). I see no reason why a fallibilist in one of the senses stated above should be committed to the felicity of such claims. Fallibilists hold that an agent can know P even if $\mathsf{u}_C(P,w) \cap (\mathrm{W} - \boldsymbol{P}) \neq \emptyset$, but what does this have to do with the semantics/pragmatics of claims with the epistemic modals "possible" and "might"? According to Yalcin (2011, 309), an utterance of "it might be that $\neg P$" expresses (roughly) that there is a $\neg P$-scenario v compatible with what the agent believes, which does not follow from there being a $\neg P$-scenario $v \in \mathsf{u}_C(P,w)$. Indeed, it is compatible with there being a $\neg P$-scenario $v \in \mathsf{u}_C(P,w)$ that the agent in w believes P with the utmost certainty. It is noteworthy in this connection that Dretske (1981), a strong fallibilist, remarks: "it does seem reasonable to insist that if S knows that P, he does not believe that he might be wrong. In other words, if the bird-watcher really believes that the bird he sees might be a grebe, then he does not know it is a Gadwall" (378n8) (cf. footnote 14). Of course, fallibilists are committed to there being contexts in which it would be true to say (if it can be said without changing the context) "the agent knows P, but the agent has not eliminated all $\neg P$-scenarios." But here "eliminated" is a theoretical term, so we should not conclude that pre-theoretic intuitions about natural language pose any problem for fallibilism here.

In addition to satisfying the above fallibilist conditions, all of the theories to be considered in the standard alternatives framework satisfy two further kinds of conditions. First, following Dretske's characterization of the Relevancy Set for a proposition P as "a set of situations each member of which contrasts with what is [to be] known," i.e. a set of $\neg P$-scenarios, we have

contrast/enough—$\mathsf{r}_C(P,w) \subseteq \mathrm{W} - \boldsymbol{P}$,

which says that the alternatives one must eliminate to know P are $\neg P$-scenarios. (From now on I will leave the universal quantification over w, P, and C implicit.) A stronger version is

M-contrast/enough—$\mathsf{r}_C(P,w) \subseteq \mathrm{W}_w - \boldsymbol{P}$,

which says that the alternatives are all $\neg P$-scenarios based on ways the world metaphysically *could be* (so an agent's ignorance cannot be witnessed by "impossible worlds"). Second, following Lewis's (1996) *Rule of Actuality*, that "actuality is always a relevant alternative" (554), we have

r-RofA—$w \notin \boldsymbol{P}$ implies $w \in \mathsf{r}_C(P,w)$,

which says that whenever w is a $\neg P$-scenario, it is a relevant alternative that one must eliminate in order to know P in w. However, it is immediate from the sufficient condition for $v \in \mathsf{u}_C(P,w)$ given above that an agent cannot eliminate her actual scenario:

u-RofA—$w \notin \boldsymbol{P}$ implies $w \in \mathsf{u}_C(P,w)$.

It follows from r-RofA and u-RofA together that if $w \notin \boldsymbol{P}$, then $w \in \mathsf{r}_C(P,w) \cap \mathsf{u}_C(P,w) \neq \emptyset$, so by (Knows), P is not known. Hence only truths can be known.

In this framework we can also state necessary and sufficient conditions for epistemic closure. Let **R** be some relation that a sequence of propositions can bear to another proposition. Here is a general schema for an *empirical* epistemic closure principle with respect to **R**: if an agent knows propositions $P_1, \ldots, P_n$, which together bear **R** to proposition Q, then, as MacFarlane (2014, 177) puts it, the agent "could come to know [Q] without further empirical investigation."[16] This requires that

$$\mathsf{r}_C(Q,w) \subseteq \bigcup_{1 \leq i \leq n} \mathsf{r}_C(P_i,w), \tag{1}$$

to guarantee that if the agent has eliminated enough scenarios to know $P_1, \ldots, P_n$, then she has eliminated enough to know Q. Note, though, that this guarantee assumes that if a scenario $v \in \mathsf{r}_C(P_i,w) \cap \mathsf{r}_C(Q,w)$ is eliminated as an

[16] Something more may be required to know Q, such as "putting two and two together" and inferring Q from $P_1, \ldots, P_n$, or simply *coming to believe Q as a result of the same experiences that make the agent believe $P_1, \ldots, P_n$*, but no more empirical investigation of the world is required to know Q than to know $P_1, \ldots, P_n$ (assuming the agent has already had sufficient experience to enable her to grasp the concepts required for understanding Q).

alternative for P_i, then v is also eliminated as an alternative for Q. In terms of *un*elimination:

$$\forall i \leq n : \mathsf{r}_C(P_i, w) \cap \mathsf{r}_C(Q, w) \cap \mathsf{u}_C(Q, w) \subseteq \mathsf{u}_C(P_i, w). \tag{2}$$

Together (1) and (2) imply that if (Knows) holds for $P_1, \ldots, P_n$, then it holds for Q.[17]

As for specific closure principles, **R** could be the relation that the sequence $P_1, \ldots, P_n$ bears to Q iff Q is a TF-consequence of $\{P_1, \ldots, P_n\}$, i.e. of $P_1 \wedge \cdots \wedge P_n$. If $n = 1$, I call this *single-premise closure under TF-consequence.* If n is allowed to be arbitrary, I call this *multi-premise closure under TF-consequence.* Sometimes I will not specify **R** explicitly, and I will write "$(\mathrm{K}P_1 \,\&\ldots\&\, \mathrm{K}P_n) \Rightarrow \mathrm{K}Q$" to abbreviate the principle that if an agent knows $P_1, \ldots, P_n$ relative to C, then she could know Q relative to C without further empirical investigation. For example, *closure under conjunction elimination*, $\mathrm{K}(P \wedge Q) \Rightarrow \mathrm{K}Q$, says that if an agent knows $P \wedge Q$ (P_1), then she could come to know Q without further empirical investigation; *closure under known material implication*, $(\mathrm{K}P \,\&\, K(P \to Q)) \Rightarrow \mathrm{K}Q$, says that if an agent knows P (P_1) and $P \to Q$ (P_2), then she could come to know Q without further empirical investigation; and so on. Note that closure under known material implication is a multi-premise closure principle.[18] In the next section, we will see a crucial pair of conditions that affect whether this principles holds.

2.2. *The RS and RO Parameters*

Fallibilists working with the standard alternatives picture face two questions. First, can one say whether a scenario v is simply "relevant" for the agent in a scenario w, independently of any proposition in question; or must one instead say that v is relevant in w *as an alternative for* a particular proposition Q, allowing that v may not be relevant in w as an alternative for a different proposition P? Second, can one say whether v is simply "eliminated" by the agent in w, independently of any proposition in question; or must one instead say that v is eliminated in w *as an alternative for* a particular Q, allowing that v may not be eliminated in w as an alternative for a different P?

Consider the first question. Dretske's (1981) idea was that *for each proposition, there is a Relevancy Set for that proposition*, motivating the following definition of $\mathsf{RS}_{\forall\exists}$ theories:

> $\mathsf{RS}_{\forall\exists}$ theories hold that for every context C, for every scenario w, and for every ($\forall$) proposition P, there is ($\exists$) a set of *relevant* (*in* w) $\neg P$*-scenarios*,

[17] *Proof*: if (Knows) does not hold for Q, then there is some $v \in \mathsf{r}_C(Q, w) \cap \mathsf{u}_C(Q, w)$. Since $v \in \mathsf{r}_C(Q, w)$, it follows by (1) that $v \in \mathsf{r}_C(P_i, w)$ for some $1 \leq i \leq n$; then since $v \in \mathsf{r}_C(P_i, w) \cap \mathsf{r}_C(Q, w) \cap \mathsf{u}_C(Q, w)$, it follows by (2) that $v \in \mathsf{u}_C(P_i, w)$. Thus, $\mathsf{r}_C(P_i, w) \cap \mathsf{u}_C(P_i, w) \neq \emptyset$, so (Knows) does not hold for P_i.

[18] One could treat any multi-premise closure principle as a single-premise principle by loading the other premises into **R**, e.g. taking **R** to be the relation that P bears to Q iff the agent knows $P \to Q$, but this trick is not helpful.

$\mathsf{r}_C(P,w) \subseteq \mathrm{W} - \boldsymbol{P}$, such that in order to know P relative to C the agent in w has to eliminate the scenarios in $\mathsf{r}_C(P,w)$.

By contrast, Heller (1999) considers (and rejects) a version of the relevant alternatives (RA) theory in which "there is a certain set of worlds selected as relevant, and S must be able to rule out the not-p worlds within that set" (197), which suggests the following definition of $\mathsf{RS}_{\exists\forall}$ theories:

> $\mathsf{RS}_{\exists\forall}$ theories hold that for every context C and scenario w, there is ($\exists$) a set of *relevant* (*in* w) *scenarios*, $\mathsf{R}_C(w)$, such that for every ($\forall$) proposition P, in order to know P relative to C the agent in w has to eliminate the $\neg P$-scenarios in $\mathsf{R}_C(w)$, i.e. the scenarios in $\mathsf{R}_C(w) \cap (\mathrm{W} - \boldsymbol{P})$.

As a simple logical point, every $\mathsf{RS}_{\exists\forall}$ theory is a $\mathsf{RS}_{\forall\exists}$ theory (take $\mathsf{r}_C(P,w) = \mathsf{R}_C(w) \cap (\mathrm{W} - \boldsymbol{P})$), but not necessarily vice versa. From now on, when I refer to $\mathsf{RS}_{\forall\exists}$ theories, I have in mind theories that are not also $\mathsf{RS}_{\exists\forall}$ theories. As I will explain below, this distinction is at the heart of the disagreement about epistemic closure that pits Dretske (1970) and Nozick (1981), who defend $\mathsf{RS}_{\forall\exists}$ theories, against Stine (1976) and Lewis (1996), who defend $\mathsf{RS}_{\exists\forall}$ theories.

To be precise, let us define the following condition on r, of which $\mathsf{RS}_{\forall\exists}$ is the denial:

> $\mathsf{RS}_{\exists\forall}$—*there is* ($\exists$) $\mathsf{R}_C(w) \subseteq \mathrm{W}$ such that *for all* ($\forall$) P:
> $\mathsf{R}_C(P,w) = \mathsf{r}_C(w) \cap (\mathrm{W} - \boldsymbol{P})$.

In a *contextualist* $\mathsf{RS}_{\exists\forall}$ theory, such as Lewis's (1996) RA theory, the set of relevant scenarios may change as context changes. Still, for any given context C, there is a set $\mathsf{R}_C(w)$ of relevant (in w) scenarios, which does not depend on a particular proposition in question. The $\mathsf{RS}_{\forall\exists}$ vs. $\mathsf{RS}_{\exists\forall}$ distinction is about how theories view the relevant alternatives with respect to a *fixed context*.

Let us now return to the second question above: can one say, independently of any proposition in question, that v is eliminated by the agent in w? According to Lewis's (1996) notion of elimination, the answer is "yes": whether there is exact match of experience and memory in v and w does not depend on any proposition in question. Hence for every scenario w, there is a fixed set of "uneliminated" scenarios $\mathsf{U}_C(w) \subseteq \mathrm{W}$, singled out independently of any proposition in question. However, as we shall see in §2.3, according to the notions of elimination implicit in *sensitivity* and *safety* theories of knowledge, the answer is "no"; it may be that v is eliminated as an alternative for a proposition P but not as an alternative for a proposition Q. Parallel to the definition of $\mathsf{RS}_{\exists\forall}$ above, we define the following RO (for "ruling out") condition on u, of which $\mathsf{RO}_{\forall\exists}$ is the denial:

> $\mathsf{RO}_{\exists\forall}$—*there is* ($\exists$) $\mathsf{U}_C(w) \subseteq \mathrm{W}$ such that *for all* ($\forall$) P:
> $\mathsf{U}_C(P,w) = \mathsf{u}_C(w) \cap (\mathrm{W} - \boldsymbol{P})$.

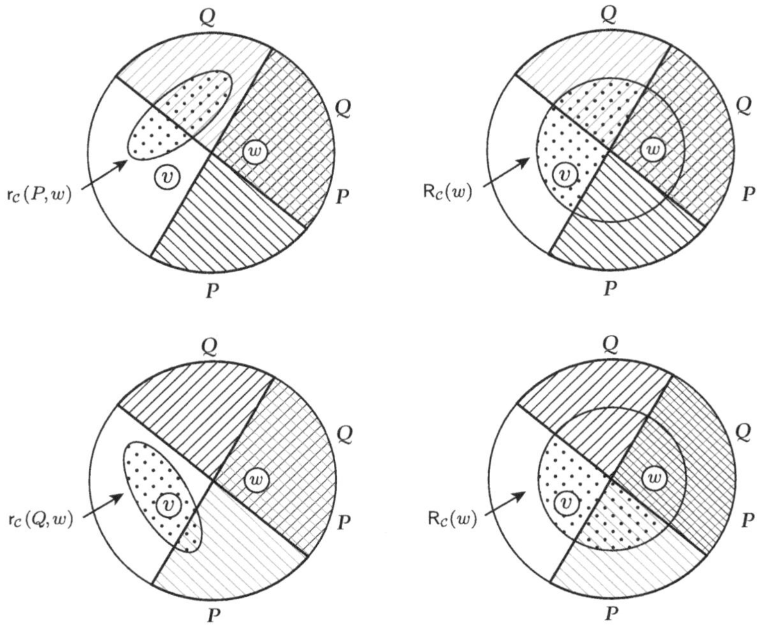

Figure 4.2 $\mathsf{RS}_{\forall\exists}$ (left) vs. $\mathsf{RS}_{\exists\forall}$ (right)

Fig. 4.2 shows the difference between $\mathsf{RS}_{\forall\exists}$ and $\mathsf{RS}_{\exists\forall}$. Observe that v is a $\neg P$-scenario and a $\neg Q$-scenario. On the $\mathsf{RS}_{\forall\exists}$ side (left), while v is a scenario that must be eliminated in order to know Q (where $\boldsymbol{Q}$ is the darker semicircle in the lower row), it is not a scenario that must be eliminated in order to know P (where $\boldsymbol{P}$ is the darker semicircle in the upper row). By contrast, on the $\mathsf{RS}_{\exists\forall}$ side (right), where the inner circles represent the fixed set $\mathsf{R}_c(w)$ of relevant scenarios, no such split-decision on v is possible; so v is a scenario that must be eliminated in order to know P and in order to know Q. The pictures for $\mathsf{RO}_{\forall\exists}$ vs. $\mathsf{RO}_{\exists\forall}$ would be the same if we were to substitute u for r and U for R. As I will explain in §2.3, the theories of Lewis (1996), Sosa (1999), DeRose (1995), Dretske (1981), Nozick (1981), and Heller (1999) have the parameter settings in Fig. 4.3.

I claimed above that the distinction between $\forall\exists$ and $\exists\forall$ parameter settings is at the heart of the disagreement about epistemic closure. Assuming $\mathsf{RS}_{\exists\forall}$ and $\mathsf{RO}_{\exists\forall}$, the (**Knows**) condition becomes

$$\begin{array}{ccccc} \mathsf{r}_c(P,w) & \cap & \mathsf{u}_c(P,w) & = \emptyset \\ \| & & \| & \\ \mathsf{R}_c(w)\cap(\mathrm{W}-\boldsymbol{P}) & \cap & \mathsf{U}_c(w)\cap(\mathrm{W}-\boldsymbol{P}) & = \emptyset, \end{array}$$

which is equivalent to

$$\mathsf{R}_c(w)\cap\mathsf{U}_c(w)\subseteq\boldsymbol{P}. \tag{3}$$

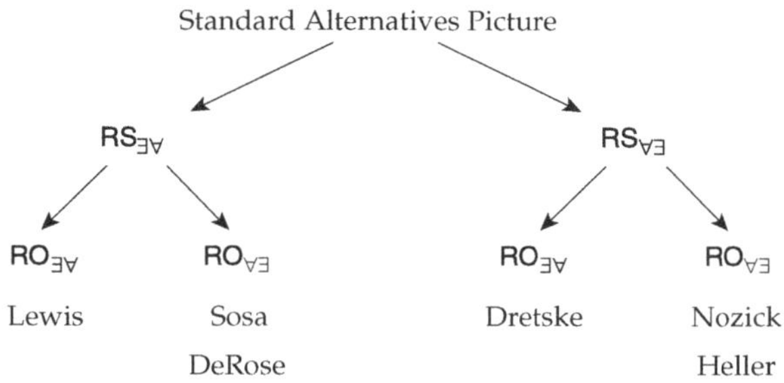

Figure 4.3 Theories classified by RS and RO parameter settings

Now it is easy to see why $\exists\forall$ settings are to closure under known material implication. If the agent knows P and $P \rightarrow Q$, then as instances of (3) we have $\mathsf{R}_c(w) \cap \mathsf{U}_c(w) \subseteq \boldsymbol{P}$ and $\mathsf{R}_c(w) \cap \mathsf{U}_c(w) \subseteq \boldsymbol{P} \rightarrow \boldsymbol{Q}$, which imply $\mathsf{R}_c(w) \cap \mathsf{U}_c(w) \subseteq \boldsymbol{Q}$, so the agent has done enough elimination of scenarios to know Q. Indeed, this is why closure under known implication holds on Lewis's (1996) theory. By contrast, if we do not assume $\mathsf{RS}_{\exists\forall}$ and $\mathsf{RO}_{\exists\forall}$, then as shown in Fig. 4.2, a $(\neg P \wedge \neg Q)$-scenario v that is relevant (or uneliminated) as an alternative for Q may not be relevant (or uneliminated) as an alternative for P, even if the agent knows the implication $P \rightarrow Q$, which opens up the possibility of a failure of closure under known implication (recall the end of §2.1). Indeed, this is why closure under known implication fails on Dretske's (1970) theory; an agent may know a mundane proposition P, because uneliminated skeptical scenarios v are not in $\mathsf{r}_c(P, w)$, and yet fail to know Q, the *denial* of the skeptical hypothesis, because those v are in $\mathsf{r}_c(Q, w)$.[19]

2.3. *Unification*

Let us see how some standard fallibilist theories are special cases of the standard alternatives picture. I will define the r and u functions according to each theory. With the exception of Lewis's (1996) theory, each theory requires belief for knowledge: if the agent in w does not believe P, then she does not know P; if the agent in w does believe P, then, as the reader can verify, the (Knows) condition $\mathsf{r}_c(P, w) \cap \mathsf{u}_c(P, w) = \emptyset$ coincides with the knowledge condition of the theory. For $\mathsf{RS}_{\exists\forall}$ theories, I will simply define $\mathsf{R}_c(w)$, from which r is derived by $\mathsf{r}_c(P, w) = \mathsf{R}_c(w) \cap (\mathrm{W} - \boldsymbol{P})$, and similarly for u. For example, for Lewis's (1996) RA theory, we have:

> $\mathsf{R}_c(w)$ = the set of scenarios that are not properly ignored in context C when attributing knowledge to the agent in scenario w;
>
> $\mathsf{U}_c(w)$ = the set of scenarios in which the agent's perceptual experience and memory exactly match that of the agent in w.

[19] Recall the postcard example in footnote 13, and take Q to be $\neg(SH_1 \vee \cdots \vee SH_n)$.

By contrast, for Dretske's (1981) RA theory (recall footnote 8) stated in terms of scenarios, we have:

> $\mathsf{r}_{\mathcal{C}}(P,w)$ = the set of $\neg P$-scenarios that the agent in w "must be in an evidential position to exclude" in order to know P (371);
>
> $\mathsf{U}_{\mathcal{C}}(w)$ = the set of scenarios that the agent in w is not in an evidential position to exclude.

For Heller's (1989, 1999) RA theory, we have the following definitions, "cashing out S's ability to rule out a not-p world in terms of her not believing p in that world" (1999, 198):

> $\mathsf{r}_{\mathcal{C}}(P,w)$ = the set of closest $\neg P$-scenarios according to an ordering[20] (dependent on the context $\mathcal{C}$) of scenarios "according to how realistic they are" (Heller, 1989, 25);
>
> $\mathsf{u}_{\mathcal{C}}(P,w)$ = the set of $\neg P$-scenarios where the agent[21] believes P.

Thus, for Heller $\mathsf{r}_{\mathcal{C}}(P,w) \cap \mathsf{u}_{\mathcal{C}}(P,w) = \emptyset$ says that the agent does not believe P in any of the closest $\neg P$-scenarios according to the ordering. For the similar *sensitivity* theories in the tradition of Nozick (1981) (without *adherence* and with counterfactuals understood following Lewis 1973[22]) we have:

> $\mathsf{r}_{\mathcal{C}}(P,w)$ = the set of closest $\neg P$-scenarios according to an ordering (possibly dependent on the context $\mathcal{C}$) of scenarios for evaluating counterfactuals at w;
>
> $\mathsf{u}_{\mathcal{C}}(P,w)$ = the set of $\neg P$-scenarios where the agent believes P (by the same method as in w).[23]

[20] Heller rejects the idea that $\mathsf{r}_{\mathcal{C}}(P,w)$ contains only the closest $\neg P$-scenarios according to a Lewisian similarity ordering $\leqslant_w^{\mathcal{C}}$ (see below in the text for this notation), arguing that any "close enough" $\neg P$-scenarios must be included as well. But since Heller (1999, 201f.) holds that the set of possible scenarios that are "close enough" to w, $CloseEnough_{\mathcal{C}}(w)$, is independent of any proposition in question, Heller's view is equivalent to the view that $\mathsf{r}_{\mathcal{C}}(P,w)$ is the set of *closest* $\neg P$-scenarios according to a more coarse-grained ordering $\preceq_w^{\mathcal{C}}$, of which $\leqslant_w^{\mathcal{C}}$ is a refinement: define $v \preceq_w^{\mathcal{C}} u$ iff $v \in CloseEnough_{\mathcal{C}}(w)$ or $v \leqslant_w^{\mathcal{C}} u$. Then assuming that whenever $u \in CloseEnough_{\mathcal{C}}(w)$ and $v \leqslant_w^{\mathcal{C}} u$, we have $v \in CloseEnough_{\mathcal{C}}(w)$, the claimed equivalence follows from the fact that for any set A of scenarios:

$$Closest_{\preceq_w^{\mathcal{C}}}(\mathrm{A}) = Closest_{\leqslant_w^{\mathcal{C}}}(\mathrm{A}) \cup (CloseEnough_{\mathcal{C}}(w) \cap \mathrm{A}).$$

[21] Where $w = \langle \mathtt{w}, \mathtt{a}, \mathtt{t} \rangle$, by "the agent" I mean $\mathtt{a}$ (recall §1.1). Those who reject "trans-world identity" may substitute "a counterpart of $\mathtt{a}$." Nothing here turns on this subtlety, so I will ignore it in what follows.

[22] Nozick (1981, 680n8) tentatively proposes alternative truth conditions for counterfactuals. However, he also indicates that sensitivity may be understood in terms of Lewis's semantics for counterfactuals. This has become the standard practice in the literature. For example, see Vogel 1987, Comesaña 2007, and Alspector-Kelly 2011.

[23] Here I follow Luper-Foy's (1984, 29) statement of the sensitivity condition with "methods," which differs slightly from Nozick's, which we could write down as well. For simplicity I omit "methods" for adherence below.

Theories that add an *adherence* condition use another pair r' and u' of functions such that $\mathsf{r}'_{\mathcal{C}}(P,w) = \mathsf{R}'_{\mathcal{C}}(w) \cap \boldsymbol{P}$ where:

> $\mathsf{R}'_{\mathcal{C}}(w)$ = the set of scenarios that are "close" or "nearby" to w (relative to $\mathcal{C}$);
>
> $\mathsf{u}'_{\mathcal{C}}(P,w)$ = the set of P-scenarios where the agent does not believe P.[24]

Thus, $\mathsf{r}'_{\mathcal{C}}(P,w) \cap \mathsf{u}'_{\mathcal{C}}(P,w) = \emptyset$ iff the agent believes P in all of the close P-scenarios.[25] Nozick's (1981) full tracking theory adds this requirement to the sensitivity requirement above.[26]

Finally, turning to *safety* theories in the tradition of Sosa (1999), we have:

> $\mathsf{R}_{\mathcal{C}}(w)$ = the set of scenarios that are "close" or "nearby" to w (relative to $\mathcal{C}$);
>
> $\mathsf{u}_{\mathcal{C}}(P,w)$ = the set of $\neg P$-scenarios where the agent believes P (on the same basis as in w).

Thus, $\mathsf{r}_{\mathcal{C}}(P,w) \cap \mathsf{u}_{\mathcal{C}}(P,w) = \emptyset$ iff there are no close scenarios where the agent falsely believes P (on the same basis on which she believes P in w). Parallel to the fact that Nozick's tracking theory requires sensitivity and adherence, DeRose's (1995) "double safety" theory requires safety and adherence.

One can now check that the above definitions imply the classifications in Fig. 4.3.

It is important to realize that while safety theories are $\mathsf{RS}_{\exists\forall}$ theories, which may lead one to think that they support full epistemic closure, they are also $\mathsf{RO}_{\forall\exists}$ theories, so it is not at all obvious that they support full epistemic closure.[27] From the fact that in all close scenarios where the agent believes $P \wedge Q$, $P \wedge Q$ is true (and in all close scenarios where $P \wedge Q$ is true, the agent believes $P \wedge Q$), it obviously does not follow that in all close scenarios where the agent believes P, P is true. So an agent can have a (double) safe belief that $P \wedge Q$, even though she has an unsafe belief that P. But an agent who knows $P \wedge Q$ knows P, so safety theorists have some explaining to do.[28]

24 One may not wish to call this a set of "uneliminated" scenarios, but there is nonetheless a structural analogy between $\mathsf{r}'_{\mathcal{C}}$ and $\mathsf{u}'_{\mathcal{C}}$ on the one hand and $\mathsf{r}_{\mathcal{C}}$ and $\mathsf{u}_{\mathcal{C}}$ on the other.

25 Nozick (1981, 680n8) suggests interpreting adherence counterfactuals $P \,\Box\!\!\rightarrow BP$ with true antecedents in such a way that the sphere over which $P \rightarrow BP$ needs to hold may differ for different propositions P. By contrast, I am interpreting adherence as a kind of $\exists\forall$ condition, in a sense that generalizes that of §2.2: there is a fixed set $\mathsf{R}'_{\mathcal{C}}(w)$ of scenarios such that for all propositions P, to know P one needs to meet an epistemic success condition in the P-worlds in $\mathsf{R}'_{\mathcal{C}}(w)$. A $\forall\exists$ interpretation of adherence that, e.g., allows the adherence sphere for $P \vee Q$ to go beyond that of P, would create yet another source of closure failure in Nozick's theory.

26 Nozick used the term "variation" for what I call "sensitivity" and used "sensitivity" to cover both variation and adherence; but the narrower use of "sensitivity" is now standard.

27 For those safety theorists who propose only necessary conditions for knowledge, see Remark 4.2 in Holliday Forthcoming on the relation between closure failures for necessary conditions for knowledge and closure failures for knowledge.

28 For discussion of closure failures for safety, see Murphy 2005, 2006, Alspector-Kelly 2011, and Holliday Forthcoming.

Now that we have definitions of r and u for each theory, we can investigate the properties of r and u implied by these definitions. For example, consider the theories according to which $\mathsf{r}_C(P,w)$ is the set of closest $\neg P$-scenarios according to some kind of ordering. We can extract a lot of information about r from this assumption. First, let us assume (cf. Lewis 1973, §2.3) that for each scenario w, there is a binary relation $\leqslant_w^C$ on W_w that is a *total preorder*,[29] *weakly centered* on w,[30] where we read "$v \leqslant_w^C u$" as "v is *at least as close* to w as u is." Let us also assume that $\leqslant_w^C$ is *well-founded*, which means that for every set $\mathrm{A} \subseteq \mathrm{W}$ of scenarios, if $\mathrm{A} \cap \mathrm{W}_w$ is nonempty, then

$$Closest_{\leqslant_w^C}(\mathrm{A}) = \{v \in \mathrm{A} \mid \forall u \in \mathrm{A} \cap \mathrm{W}_w : v \leqslant_w^C u\},$$

the set of closest scenarios to w among those in A, is also nonempty. This implies that for any proposition P, if P is possible relative to w ($\boldsymbol{P} \cap \mathrm{W}_w \neq \emptyset$), then there is a set of closest P-scenarios to w ($\mathrm{Closest}_{\leqslant_w^C}(\boldsymbol{P}) \neq \emptyset$), as epistemologists working with ordering-based theories typically assume. With this setup, we can completely characterize the properties of r for the ordering-based theories.

Theorem 1. Given a family $\{\leqslant_w^C\}_{w \in \mathrm{W}}$ of orderings as above for each context C, the function r defined by $\mathsf{r}_C(P,w) = Closest_{\leqslant_w^C}(\mathrm{W}_w - \boldsymbol{P})$ satisfies all of the following conditions:

M-equiv—if $\boldsymbol{P}_w = \boldsymbol{Q}_w$, then $\mathsf{r}_C(P,w) = \mathsf{r}_C(Q,w)$;
M-contrast/enough—$\mathsf{r}_C(P,w) \subseteq \mathrm{W}_w - \boldsymbol{P}$;
r-RofA—$w \notin \boldsymbol{P}$ implies $w \in \mathsf{r}_C(P,w)$;
noVK—$\boldsymbol{P}_w \neq \mathrm{W}_w$ implies $\mathsf{r}_C(P,w) \neq \emptyset$;
alpha—$\mathsf{r}_C(P \wedge Q,w) \subseteq \mathsf{r}_C(P,w) \cup \mathsf{r}_C(Q,w)$;
beta—if $\boldsymbol{P}_w \subseteq \boldsymbol{Q}$ and $\mathsf{r}_C(P,w) \cap \mathsf{r}_C(Q,w) \neq \emptyset$, then $\mathsf{r}_C(Q,w) \subseteq \mathsf{r}_C(P,w)$.

Conversely, given any function r satisfying these conditions, there is a family $\{\leqslant_w^C\}_{w \in \mathrm{W}}$ of orderings for each C such that for all P and w, $\mathsf{r}_C(P,w) = Closest_{\leqslant_w^C}(\mathrm{W}_w - \boldsymbol{P})$.[31]

I omit the proof of Theorem 1, since it is essentially a variation on a well-known result of Arrow (1959), but here formulated using analogues of Sen's (1971) α and β conditions applied to r.[32] With the possible exception of beta, all of the conditions should be self-explanatory. Most important for

[29] i.e. reflexive, transitive, and such that for all $u, v \in \mathrm{W}_w$, either $u \leqslant_w^C v$ or $v \leqslant_w^C u$.

[30] i.e. $w \in \mathrm{W}_w$ and for all $v \in \mathrm{W}_w$, $w \leqslant_w^C v$.

[31] For all $x, y \in \mathrm{W}_w$, define $x \leqslant_w^C y$ iff either (i) for all propositions P, $y \notin \mathsf{r}_C(P,w)$, or (ii) there is some proposition Q such that $x \in \mathsf{r}_C(Q,w)$ and $y \in \mathrm{W} - \boldsymbol{Q}$.

[32] I have written alpha in the equivalent form that Sen (1971, §9, n1) calls α^* and beta in the form given by Bordes (1976, §2). My conditions look different than theirs at first because my r function picks the "best" $\neg P$-scenarios, whereas the economist's *choice function* picks the best P-scenarios. Another minor difference is that the r function takes in a proposition, whereas a choice function takes in a set. A proof of Theorem 1 in the case where the input to r is a set is in Holliday 2012, §3.A, and the proof there can be easily adapted for the present setup.

our purposes in the next section will be the condition noVK for *no vacuous knowledge.*

2.4. The Problems of Vacuous Knowledge and Containment

All of the fallibilist theories developed so far in the standard alternatives picture have at least one of two serious problems, depending on whether they are $\mathsf{RS}_{\exists\forall}$ theories or $\mathsf{RS}_{\forall\exists}$ theories.

Assuming $\mathsf{RS}_{\exists\forall}$, fallibilism implies that the set $\mathsf{R}_c(w)$ of relevant/nearby scenarios is a *strict* subset of W_w. Thus, there can be *contingent* propositions Q ($\boldsymbol{Q}_w \neq \mathrm{W}_w$) true throughout $\mathsf{R}_c(w)$ ($\mathsf{R}_c(w) \subseteq \boldsymbol{Q}$), as shown on the left of Fig. 4.4 after Proposition 1 in §2.5, where $\boldsymbol{Q}$ is the region with diagonal lines and $\mathsf{R}_c(w)$ is the region with stars. But then $\mathsf{RS}_{\exists\forall}$ implies $\mathsf{r}_c(Q,w) = \mathsf{R}_c(w) \cap (\mathrm{W} - \boldsymbol{Q}) = \emptyset$; and if $\mathsf{r}_c(Q,w) = \emptyset$, then as long as the agent believes Q, she *knows* it, for *any* u function! No matter what (lack of) experience the agent has, and no matter what experience and beliefs the agent would have had under other circumstances, the agent supposedly knows the *contingent* proposition Q.[33]

For example, according to Lewis's $\mathsf{RS}_{\exists\forall}$ theory, even if the agent has never opened her eyes or ears, she knows any contingent Q that is true throughout the set $\mathsf{R}_c(w)$ of relevant scenarios; and according to the $\mathsf{RS}_{\exists\forall}$ safety theory, no matter how insensitive an agent's beliefs are to reality, she knows (or at least safely believes) any contingent Q that is true throughout the set $\mathsf{R}_c(w)$ of nearby scenarios, provided she believes it. Vogel (1999) recognizes this problem for some versions of the RA theory, observing that if we allow "for detailed empirical knowledge without evidence, then anyone who happens to arrive at the appropriate belief, no matter how, will enjoy that knowledge. This outcome is wrong; knowledge is dearer than that" (171f.). I call this problem the problem of *vacuous knowledge*, following Heller (1999), who also realizes that the $\mathsf{RS}_{\exists\forall}$ assumption is to blame.

However, Heller and I view the problem differently. For Heller, the problem seems to be that when a contingent Q is true throughout $\mathsf{R}_c(w)$, $\mathsf{RS}_{\exists\forall}$ theories do not place a requirement on the agent to eliminate any $\neg Q$-scenarios in order to know Q. In my view, the problem is that $\mathsf{RS}_{\exists\forall}$ theories do not place on the agent *any requirement to eliminate any scenarios* in order to know Q. This distinction will come up again in the Answer to the First Reply below and in §3.3 and §4.1.

It will not help here to claim that Kripke (1980) has given examples of a priori knowable contingent truths. For one thing, we can take $\boldsymbol{Q}$ to be the set of scenarios v such that Q is true at v *considered as actual*, so $\emptyset \neq \boldsymbol{Q}_w \neq \mathrm{W}_w$ means

[33] According to what Vogel (1999, 168) calls "Backsliding" RA theories, which blur the roles of the r and u functions, alternatives can become "irrelevant" when one has good evidence against them. A Backsliding RA theorist might claim that $\mathsf{r}_c(Q,w) = \emptyset$ holds only if the agent has done a lot of empirical investigation. But this is not how the r function works for any of the theories discussed here. See Vogel 1999 for arguments against Backsliding accounts.

that Q is *deeply* contingent (see Davies and Humberstone 1980). Then $\mathsf{RS}_{\exists\forall}$ theories allow knowledge of deeply contingent truths with no requirement of eliminating scenarios. But even if one thinks there are some special counter-examples to Evans's (1979) famous claim that "it would be intolerable for there to be a statement which is both knowable *a priori* and deeply contingent" (161), such examples are as much beside the point here as Kripke's.

The main point is this: $\mathsf{RS}_{\exists\forall}$ theories imply that *every proposition Q* with $\mathsf{R}_c(w) \subseteq \boldsymbol{Q}$ is knowable with no requirement of eliminating scenarios, and there is no guarantee that every such Q fits the mold of one of the recherché examples of (deeply) contingent but a priori knowable propositions. Instead, $\mathsf{RS}_{\exists\forall}$ theorists tell us that such Q may include the denials of skeptical hypotheses, not only what I call *self-side* skeptical hypothesis about how we are hooked up to the world (BIVs, etc.), but also *world-side* skeptical hypotheses about which objects there are and what they are like in particular locations in the external world (disguised mules, etc.). But if a theory implies that propositions about which objects there are and what they are like in particular locations in the external world are knowable with no requirement of eliminating scenarios—that's intolerable.[34]

As perhaps the first epistemologist to postulate the $\mathsf{RS}_{\exists\forall}$ condition, Stine (1976) seemed to embrace the vacuous knowledge consequence that I take to be damning; but since then epistemologists have recognized that there appears to be a serious problem that must be addressed.[35] I will now consider three replies to the vacuous knowledge objection to $\mathsf{RS}_{\exists\forall}$, answering each.

[34] I am not objecting to the view that one may be *entitled*, in the sense of Wright (2004), to accept contingent empirical propositions, such as the negations of skeptical hypotheses, without doing empirical work for them—provided the view does not add that one thereby *knows* the propositions; Wright (2004) is careful not to make this further claim. Relatedly, I am not objecting to the view that one may *justifiably take for granted*, in the sense of Sherman and Harman (2011), contingent empirical propositions, such as the negations of skeptical hypotheses, without doing empirical work for them; Sherman and Harman are explicit that one cannot come to know a proposition just by justifiably taking it for granted. Turning from justified *taking for granted* to justified *belief*, White (2006, §9) argues that we have a priori "default" justification for believing, or that we are "entitled" to believe, the negations of skeptical hypotheses; but he does not claim that we have a priori *knowledge* of the negations of skeptical hypotheses. Contrary to White, Schiffer (2004, 178) argues that "There is nothing in the concept of a priori justified belief to warrant the claim that we're a priori justified in disbelieving skeptical hypotheses." Although Schiffer proposes a revised concept of justification* according to which we are a priori justified* in disbelieving skeptical hypothesis, he also does not claim that we have a priori *knowledge* of the negations of skeptical hypotheses. Thanks to an anonymous referee for posing the question of how what I claim is intolerable relates to the views of Wright, White, and Schiffer.

[35] About Dretske's (1970) zebra case, Stine (1976, 258) writes: "[O]ne does know what one takes for granted in normal circumstances. I do know that it is not a mule painted to look like a zebra. I do not need evidence for such a proposition ... [I]f the negation of a proposition is not a relevant alternative, then I know it—obviously, without needing to provide evidence." Cohen (1988, 99) responds: "Here, I think Stine's strategy for preserving closure becomes strongly counter-intuitive. Even if it is true that some propositions can be known without evidence, surely this is not true of the proposition that S is not deceived by a cleverly disguised mule." The key point is to consider the *kinds* of propositions that $\mathsf{RS}_{\exists\forall}$ theories imply can be known with no requirement of eliminating scenarios.

First Reply—knowledge of deeply contingent empirical truths does require epistemic work, but this "epistemic work" may involve something less than *eliminating* scenarios. Vogel (1999, 159n12) considers and rejects something like this reply: the RA theory that assumes $\mathsf{RS}_{\exists\forall}$ "is committed to the thesis that one can know that an irrelevant alternative is false even though one can't rule it out.... The RA theorist might still require that you have *some* minimal evidence against irrelevant alternatives in order to know that they are false. However, holding onto this scruple will make it more difficult, if not impossible, for the RA theorist to resist skepticism."

Answer—in addition to the problem of skepticism noted by Vogel,[36] there is another problem. While having "minimal evidence" may not require eliminating $\neg P$-scenarios, where P is the proposition to be known, does it not require eliminating *some* scenarios, perhaps as alternatives to related propositions? (Cf. §4.1 on inductive knowledge.) If it does, then we must reject $\mathsf{RS}_{\exists\forall}$, since it allows agents to know deeply contingent truths with *no* requirement of eliminating scenarios.

Second Reply—the "double safety" theory is an $\mathsf{RS}_{\exists\forall}$ theory that avoids the problem of vacuous knowledge. For even if one's belief that Q is *vacuously safe*, in virtue of the fact that Q is true throughout the set $\mathsf{R}_c(w)$ of nearby scenarios, it is not vacuously *adherent*, since it is an epistemic achievement that in all of the nearby scenarios where Q is true, the adherent agent believes Q.[37]

Answer—adherence doesn't help. Kripke (2011) showed that if an agent's belief that P is sensitive, then normally her belief that *P and I believe that P* will be both sensitive and adherent. Kripke rightly concludes that adherence "is almost without force, a broken reed. What can be the point of a condition whose rigor can almost always be overcome by conjoining 'and I believe (via M) that p' ...?" (184). A similar point applies to the adherence part of double safety. Suppose an agent's belief that P is vacuously safe, since there are no $\neg P$-scenarios among the nearby scenarios. It follows by an argument similar to Kripke's that the agent's belief that *P and I believe that P* will normally be double-safe, even if her belief that P is not. So on the double safety theory, it is normally sufficient to know that *P and I believe that P* that one has a vacuously safe belief that P. Thus, double safety does not solve the problem of vacuous knowledge, but only relocates it.

Third Reply—allowing knowledge of deeply contingent empirical propositions with no requirement of eliminating scenarios may seem bad, but it's alright, because "[s]imply *mentioning* any particular case of this knowledge, aloud or even in silent thought, is a way to ... create a context in which it is

36 And by Cohen (1988, 111): "Radical skeptical hypotheses are immune to rejection on the basis of *any* evidence. There would appear to be no evidence that could count against the hypothesis that we are deceived by a Cartesian demon.... Radical skeptical hypotheses are designed to neutralize any evidence that could be adduced against them."

37 Heller (1999, 207) considers and rejects this reply. By contrast, DeRose (2000, 135) endorses a similar position.

no longer true to ascribe the knowledge in question to yourself or others" (Lewis 1996).

Answer—there are a number of problems with this reply, three of which I will discuss:

First, there is a *motivation* problem. When Stine (1976) first posited $\mathsf{RS}_{\exists\forall}$, the motivation was clear: defend closure from Dretske. But then when faced with the problem that $\mathsf{RS}_{\exists\forall}$ leads to vacuous knowledge, Lewis (1996) appeals to a super-shifty version of contextualism, according to which whenever you try to claim the vacuous knowledge that is rightly yours according to closure, context change invariably prevents you from truly claiming it (so Lewis concedes that "Dretske gets the phenomenon right" (564) after all). Sure, you can endorse a fixed-context closure principle in the abstract, but be careful not to instantiate it with any *specific* propositions and trigger an instant, irresistible change in context! But with closure made impotent in this way, was it worth getting into this vacuous knowledge mess to defend it? As Dretske (2005, 19) observes of super-shifty contextualism, "it is a way of preserving closure for the heavyweight implications while abandoning its usefulness in acquiring knowledge of them,"[38] or rather, while abandoning its usefulness in reasoning about agents' knowledge of them—a bad trade for the problem of vacuous knowledge. Moreover, if one wants to stick with super-shifty contextualism and fixed-context closure, one can do so without being committed to vacuous knowledge, using the *multipath picture* proposed below (see §4.2).

Second, there is a *mechanism* problem. Most contemporary contextualists do not think that sayings or thinkings invariably introduce relevant counterpossibilities as Lewis claims.[39] So it is unclear what general mechanism would prevent those who have vacuous knowledge from sometimes truly claiming that they do. If this is so, then Lewis's "unclaimable knowledge" reply collapses.

Third, there is a *missing-the-point* problem. What is problematic about vacuous knowledge is not just that agents could *truly claim* to have it—which they probably could according to post-Lewisian contextualism—but rather that they could *have it* at all. Cohen (2000, 105) correctly sees this: "it looks as if the [$\mathsf{RS}_{\exists\forall}$] contextualist is committed to the view that we have contingent *a priori* knowledge. And of course, these cases do not fit the structure of the reference-fixing cases called to our attention by Kripke. Of course, I am not entirely happy with this result." Cohen concludes that this is a "bullet" he is "prepared to bite" (106). But contextualists need not bite this bullet if they

[38] To put it this way is misleading, since a closure principle is not something that agents use in acquiring knowledge (except about other agents' knowledge). It is something that we use in *reasoning about agents' knowledge*.

[39] Cohen (1998, 303n24) suggests that Lewis's Rule of Attention may need to be defeasible; Ichikawa (2011, §4) disavows it; and Blome-Tillman (2009: 246–7) argues that it is too strong. DeRose (2009, Ch. 4) suggests that members of a conversation may resist context changes by sticking to their own "personally indicated epistemic standards."

opt for a contextualist version of the multipath picture of knowledge to be introduced in §3.

So much for $\mathsf{RS}_{\exists\forall}$ theories then. On to $\mathsf{RS}_{\forall\exists}$ theories. $\mathsf{RS}_{\forall\exists}$ theories that take $\mathsf{r}_c(P,w)$ to be the set of closest $\neg P$-scenarios according to some kind of ordering avoid the problem of vacuous knowledge. In fact, they satisfy the general **noVK** (no vacuous knowledge) principle in Theorem 1, which says that if P is (deeply) contingent, then knowing P requires eliminating some scenarios. This is one of Heller's (1999) main arguments for his $\mathsf{RS}_{\forall\exists}$ theory over $\mathsf{RS}_{\exists\forall}$ theories.

Unfortunately, the ordering-based $\mathsf{RS}_{\forall\exists}$ theories that avoid the problem of vacuous knowledge face what I call the *problem of containment*. While it may be a virtue that these theories invalidate controversial multi-premise closure principles like closure under known implication, it is not a virtue that they allow closure failures to spread far beyond those controverial principles, to uncontroversial single-premise closure principles. Nozick (1981, 228) was well aware that even such a weak closure principle as $\mathrm{K}(P \wedge Q) \Rightarrow \mathrm{K}P$ is invalid according to his theory. He resisted the idea that $\mathrm{K}P \Rightarrow \mathrm{K}(P \vee Q)$ is invalid, but his theory clearly invalidates it (see Holliday 2014a and the Appendix).

In Holliday 2014a, I systematically investigate this problem of containment for a family of what I call "subjunctivist-flavored" theories, including basic version of the RA, sensitivity/tracking, and safety theories. The main Closure Theorem gives an exact characterization of the closure properties of knowledge according to these theories. Surprisingly, it turns out that despite the differences within the family of subjunctivist-flavored theories, the valid epistemic closure principles are essentially the same for these different theories. The problem is that these theories allow egregious failures of single-premise closure, failures of principles as weak as $\mathrm{K}(P \wedge Q) \Rightarrow \mathrm{K}(P \vee Q)$ and $(\mathrm{K}P \,\&\, \mathrm{K}Q) \Rightarrow \mathrm{K}(P \vee Q)$.

The source of the problem with the ordering-based $\mathsf{RS}_{\forall\exists}$ theories is that they do not satisfy a necessary condition for single-premise closure under TF-consequence (recall (1) in §2.1):

TF-cover—if Q is a TF-consequence of P, then $\mathsf{r}_c(Q,w) \subseteq \mathsf{r}_c(P,w)$,

which says that the empirical work needed to know P *covers* the empirical work needed to know Q. One can easily check that if $\mathsf{r}_c(S,w)$ is always the set of closest $\neg S$-scenarios according to an ordering, then r does not satisfy **TF-cover**, which explains the failures of single-premise closure.

Is there any way to avoid these problems of containment and of vacuous knowledge?

2.5. *An Impossibility Result*

In the standard alternatives picture, it is impossible to avoid both problems from §2.4, even if we restrict our attention to a limited domain of propositions. Call a set Σ of propositions an *area* iff whenever $P \in \Sigma$ and Q is a TF-consequence of P, then $Q \in \Sigma$. Then we have the following result.

Proposition 1. For any scenario w, context C, and area Σ, the following principles are *inconsistent* in the standard alternatives picture:

> contrast/enough$_\Sigma$—$\forall P \in \Sigma$: $\mathsf{r}_C(P, w) \subseteq \mathrm{W} - \boldsymbol{P}$;
> e-fallibilism$_\Sigma$—$\exists P \in \Sigma\ \exists Q \in \mathcal{P}$: $\mathsf{r}_C(P, w) \subseteq \boldsymbol{Q}$ and $\mathrm{W}_w - \boldsymbol{P} \not\subseteq \boldsymbol{Q}$;
> noVK$_\Sigma$— $\forall P \in \Sigma$: $\boldsymbol{P}_w \neq \mathrm{W}_w$ implies $\mathsf{r}_C(P, w) \neq \emptyset$;
> TF-cover$_\Sigma$—$\forall P, Q \in \Sigma$: if Q is a TF-consequence of P, then $\mathsf{r}_C(Q, w) \subseteq \mathsf{r}_C(P, w)$.

Here is the essence of the proof: by e-fallibilism$_\Sigma$ there are propositions P and Q as on the right side of Fig. 4.4 (where $\boldsymbol{Q}$ may overlap with $\boldsymbol{P}$). Consider $P \vee Q$ and the set $\boldsymbol{P} \vee \boldsymbol{Q}$, which is the union of the two regions, $\boldsymbol{P}$ and $\boldsymbol{Q}$, with diagonal lines. Where should we draw $\mathsf{r}_C(P \vee Q, w)$? Since $P \vee Q$ is a TF-consequence of P, TF-cover$_\Sigma$ requires that $\mathsf{r}_C(P \vee Q, w)$ be a subset of $\mathsf{r}_C(P, w)$; but contrast/enough$_\Sigma$ requires that $\mathsf{r}_C(P \vee Q, w)$ be a subset of the *blank* region. The only way both can hold is if $\mathsf{r}_C(P \vee Q) = \emptyset$. But this contradicts noVK, given that $\boldsymbol{P} \vee \boldsymbol{Q}$ does not include all of W_w.

Proof. By e-fallibilism$_\Sigma$, there are propositions $P \in \Sigma$ and $Q \in \mathcal{P}$ such that

$$\mathsf{r}_C(P, w) \subseteq \boldsymbol{Q} \tag{4}$$

and

$$\mathrm{W}_w - \boldsymbol{P} \not\subseteq \boldsymbol{Q}. \tag{5}$$

Since $P \vee Q$ is a TF-consequence of P, $P \in \Sigma$ implies $P \vee Q \in \Sigma$, and TF-cover$_\Sigma$ implies

$$\mathsf{r}_C(P \vee Q, w) \subseteq \mathsf{r}_C(P, w), \tag{6}$$

which with (4) implies

$$\mathsf{r}_C(P \vee Q, w) \subseteq \boldsymbol{Q} \subseteq \boldsymbol{P} \cup \boldsymbol{Q}. \tag{7}$$

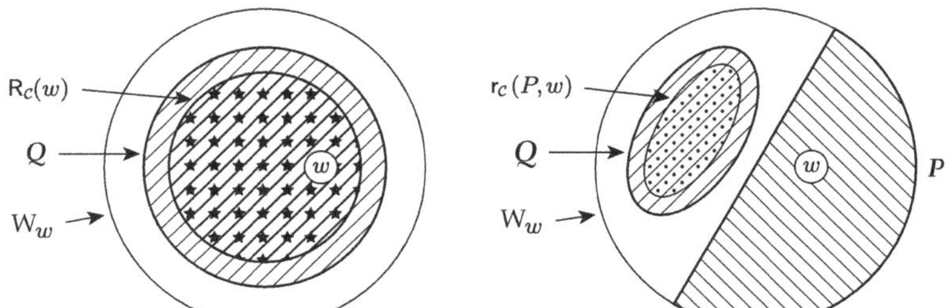

Figure 4.4 Vacuous knowledge given $\mathsf{RS}_{\exists\forall}$ (left) and a diagram for Proposition 1 (right)

However, contrast/enough$_\Sigma$ implies

$$\mathsf{r}_C(P \vee Q, w) \subseteq \mathrm{W} - (\boldsymbol{P} \vee \boldsymbol{Q}) = \mathrm{W} - (\boldsymbol{P} \cup \boldsymbol{Q}), \tag{8}$$

which with (7) implies

$$\mathsf{r}_C(P \vee Q, w) = \emptyset. \tag{9}$$

Finally, (5) implies

$$(\boldsymbol{P} \vee \boldsymbol{Q})_w \neq \mathrm{W}_w, \tag{10}$$

which with noVK$_\Sigma$ implies

$$\mathsf{r}_C(P \vee Q, w) \neq \emptyset, \tag{11}$$

which contradicts (9). □

Note that Proposition 1 does not use the full strength of TF-cover, but only its instance $KP \Rightarrow \mathrm{K}(P \vee Q)$. Also note that we could get the same result using $\mathrm{K}P \Rightarrow \mathrm{K}((P \vee Q) \wedge (P \vee \neg Q))$ and $\mathrm{K}(S \wedge R) \Rightarrow \mathrm{K}S$. In any case, I agree with Dretske (1970, 1009), Kripke (2011, 202), and Nozick (1981, 230n64) (not what his theory says, but what he says) that $\mathrm{K}P \Rightarrow \mathrm{K}(P \vee Q)$ should not fail.[40]

The only principle in Proposition 1 that I have not yet defended is contrast/enough. All of the theories discussed in §2.3 satisfy this principle, but can we escape Proposition 1 by giving up constrast/enough? Not in the standard alternatives picture (but see §3.3). The reason is that in the standard alternatives picture, giving up contrast/enough means claiming that there are

[40] I am assuming that the agent grasps the concepts needed to understand the new disjunct Q (cf. Williamson 2000, 283). Also recall the meaning of the notation $KP \Rightarrow KP'$ from §2.1 Although I agree with Dretske, Kripke, and Nozick in endorsing $\mathrm{K}P \Rightarrow \mathrm{K}(P \vee Q)$ so understood, not everyone does. According to Yablo's (2011; 2012; 2014) view of closure, knowing $P \vee Q$ relative to a context $\mathcal{C}$ may require more empirical investigation than knowing P relative to $\mathcal{C}$, since the *subject matter* of $P \vee Q$ is not in general included in that of P. Although I take Yablo's move to connect subject matter and epistemic closure to be deep and important (see footnote 58), I disagree with the specifics of his view of the connection (see Holliday 2014b). While I cannot do justice to his view here, I will briefly register points of disagreement. On one way of developing Yablo's view (based on "reductive truthmaking" and the relation of "content-parthood"), $K(P \wedge Q) \Rightarrow K(P \vee Q)$ does not hold—so an agent may know a conjunction and yet have more empirical work to do to know the disjunction of the conjuncts—and although $\mathrm{K}(P \wedge Q) \Rightarrow \mathrm{K}P$ holds when P and Q are TF-atomic, it does not hold in general for arbitrary P and Q—so an agent may know a conjunction without knowing the conjuncts—which is hard to swallow. (Note that the propositions in the consequents of those closure principles do not seem to "change the subject" relative to the propositions in the antecedents.) On another way of developing his view (based on "recursive truthmaking" and the relation of "inclusive entailment"), the principle $\mathrm{K}(P \wedge Q) \Rightarrow \mathrm{K}(P \vee \neg Q)$ does not hold, which is also hard to swallow. Indeed, so is the denial of $\mathrm{K}P \Rightarrow \mathrm{K}(P \vee Q)$, even though this principle—apparently unlike the previous ones—has something new appear in its consequent. One must be careful to distinguish two ideas: first, the less radical idea that in a context $\mathcal{C}$ relative to which an agent has done enough empirical work to count as knowing P, someone's raising the issue of Q by bringing up $P \vee Q$ may change the context to a $\mathcal{C}'$ relative to which the agent has not done enough empirical work to count as knowing $P \vee Q$ (or P); second, the more radical idea that knowing $P \vee Q$ relative to a context $\mathcal{C}$ may require more empirical work than knowing P relative to $\mathcal{C}$. The second idea, like the denial of the conjunctive closure principles, strikes me as unnecessary and undesirable (admittedly, we are in near-bedrock territory here).

some propositions P such that it is *necessary* in order to know P that one eliminate some P-scenarios. But if anything is sufficient for knowing P (as far as empirical work goes), it is eliminating *all* $\neg P$-scenarios, a kind of epistemic supererogation. Suppose I were to say, "I agree that you've ruled out every possible way in which P could be false, but that's not enough for you to know that P is true; you also have to rule out such-and-such ways in which P could be true." This seems absurd.[41]

I take the impossibility result in Proposition 1 to show that there is something seriously wrong with the standard alternatives picture. Remember that it is not enough to escape this result to argue that *there are* some cases in which knowing a deeply contingent empirical proposition imposes no requirement of empirically eliminating scenarios. Rather, to escape this impossibility result, one would have to argue that there is *no area* of propositions knowledge of which requires empirical investigation and with respect to which we are very weak fallibilists maintaining a very weak closure principle. This strikes me as an incredible claim. Until a credible argument for this claim appears, I conclude that fallibilists must seek a replacement for the standard alternatives picture.

3. THE MULTIPATH PICTURE OF KNOWLEDGE

In this section, I propose a new framework for fallibilism that solves the problems raised for the standard alternatives picture in §2. I call it the *multipath picture of knowledge*.

Recall the starting point of the standard alternatives picture: for each proposition to be known, there is "*a* [*single*] *set* of situations each member of which *contrasts* with what is [to be] known... and must be evidentially excluded if one is to know" [emphasis added] (Dretske, 1981, 373). Against these *single alternative set* and *contrast* assumptions, I will argue:

- In some cases, there is no set of situations all of which must be excluded if one is to know a proposition P; instead, there are **multiple sets** of situations (scenarios), such that if one is to know P, one must exclude *all of the situations in at least one of those sets.*

41 Some might think that Gettier cases show we should reject contrast/enough. For example: not having any idea what time it is, you check a clock that—unbeknownst to you—has been stopped for weeks on 5:43; as it happens, the time is now 5:43; but you do not come to know this from the stopped clock. Where F is the proposition that *the time is 5:43* and S is the proposition that *the clock has stopped*, one might think this is a case in which knowing F *requires* ruling out $(F \wedge S)$-possibilities, which would explain your ignorance of F (since you have not ruled those out) and violate contrast/enough. But this is a mistake. What explains your ignorance of F is that since you have only looked at a stopped clocked, you have not ruled out various relevant scenarios in which F is *false* and the time is something other than 5:43. If by some other means you had ruled out every scenario in which F is false, then it would be absurd to say "I agree that you've ruled out every scenario in which the time is something other than 5:43, but you still don't know the time is 5:43 unless you rule out such-and-such scenarios in which the time is 5:43."

- In some cases, it is sufficient (as far as empirical investigation goes) for an agent to know a proposition P that she *only eliminates non-contrasting scenarios in which P is* **true**.

A key observation will be that while the *single alternative set* and *contrast* assumptions may seem plausible for propositions that are "atomic" from a truth-functional or quantificational perspective (but see §4), fallibilists should reject these assumptions for logically complex propositions.

3.1. *Against the Single Alternative Set Assumption*

Suppose that an agent wants to know whether $P \vee Q$ is true, where P and Q are contingent empirical propositions. Further suppose that $P \vee Q$ is in fact true. Then there are at least three paths by which the agent could come to know it: she could start eliminating $\neg P$-scenarios, and if she comes to know P, then she is done (at least with empirical investigation); or she could start eliminating $\neg Q$-scenarios, and if she comes to know Q, then she is done (with empirical investigation); or she could come to know $P \vee Q$ without coming to know which disjunct is true, perhaps by eliminating all $(\neg P \wedge \neg Q)$-scenarios without eliminating any $(\neg P \wedge Q)$-scenarios or any $(P \wedge \neg Q)$-scenarios. This is hardly a novel observation. But it raises the question of why any fallibilist should think that for a proposition like $P \vee Q$, there is a *single* set of scenarios that must be evidentially excluded if one is to know $P \vee Q$. It seems instead that there may be at least *three* sets of scenarios such that if one is to know $P \vee Q$, one must evidentially exclude all of the scenarios in at least one of those three sets, corresponding to the three paths to knowledge of $P \vee Q$ described above.

If we were *infallibilists*, there would be no need for these multiple "alternative sets" for $P \vee Q$. According to infallibilism, coming to know P requires eliminating all $(\neg P \wedge \neg Q)$-scenarios; so does coming to know Q; and so does coming to know $P \vee Q$ without coming to know which disjunct is true. Moreover, as argued in §2.5, eliminating all contrasting scenarios should be *enough* to know a proposition. Thus, infallibilists need only consider one alternative set for $P \vee Q$: to know $P \vee Q$ it is necessary and sufficient (as far as empirical work goes) that one eliminate all $(\neg P \wedge \neg Q)$-scenarios.

But we are *fallibilists*. According to fallibilism, coming to know P might not require eliminating *all* $(\neg P \wedge \neg Q)$-scenarios, at least not for every Q. Indeed, it might not require eliminating *any* $(\neg P \wedge \neg Q)$-scenarios.[42] But then since it is enough to know $P \vee Q$ that one eliminate all $(\neg P \wedge \neg Q)$-scenarios, it is immediate that we need multiple alternative sets for $P \vee Q$, corresponding to the multiple paths to knowing $P \vee Q$ above: the scenarios that one must eliminate in order to know P may be different from those that one must eliminate in order to know Q, which may be different from those that one must eliminate in order to know $P \vee Q$ without knowing either disjunct.

[42] The claim that for every Q, knowing P requires eliminating some $(\neg P \wedge \neg Q)$-scenario is essentially equivalent to infallibilism; and if for every set of scenarios there is a proposition true in exactly those scenarios, then the claim is exactly equivalent to infallibilism.

3.2. *Multiple Alternative Sets*

What §3.1 shows is that we should replace the r function of the standard alternatives picture, which assigns to each triple of a context $\mathcal{C}$, proposition P, and scenario w, a single set $\mathsf{r}_{\mathcal{C}}(P,w) \subseteq \mathrm{W}$ of scenarios, with a new "multipath" $\mathbf{r}$ function that assigns to each such $\mathcal{C}$, P, and w, a set

$$\mathbf{r}_{\mathcal{C}}(P,w) = \{\mathrm{A}_1, \mathrm{A}_2, \ldots\}$$

of sets $\mathrm{A}_i \subseteq \mathrm{W}$ of scenarios. For example, for $P \vee Q$ we may have $\mathbf{r}(P \vee Q, w) = \{\mathrm{A}_1, \mathrm{A}_2, \mathrm{A}_3\}$, where A_1 is the set of scenarios to be eliminated in the path to knowing $P \vee Q$ via P; A_2 is the set of scenarios to be eliminated in the path to knowing $P \vee Q$ via Q; and A_3 is the set of scenarios to be eliminated in the path to knowing $P \vee Q$ without knowing either P or Q individually.

The foregoing points about disjunctive propositions apply to existential propositions as well. Assuming propositions can have quantificational structure as well as truth-functional structure, one could come to know $\exists x P(x)$ by coming to know $P(a)$, or by coming to $P(b)$, etc., or by coming to know $\exists x P(x)$ without coming to know $P(c)$ for any c. As a consequence of fallibilism, the alternative sets for these different paths to knowing $\exists x P(x)$ may be different. In this paper I concentrate on truth-functional structure, but a full treatment would include quantificational structure as well.

According to the *multipath picture of knowledge*, to know a proposition P, it is necessary and sufficient (as far as empirical elimination goes) that one eliminate all of the alternatives in *at least one* of the alternative sets for P (as on the right side of Fig. 4.5 with A_2):

$$\exists \mathrm{A} \in \mathbf{r}_{\mathcal{C}}(P,w) : \mathrm{A} \cap \mathsf{u}_{\mathcal{C}}(P,w) = \emptyset, \qquad \textbf{(Knows)}$$

where u is the same function as before.[43,44] As we shall see, this is just the generalization that fallibilists need in order to avoid the problems raised for them in the standard alternatives picture.

In explaining the multipath picture, I deliberately use the term "path to knowing" instead of "way of knowing." There are often multiple "ways of knowing" a proposition in the sense that one can come to know it by eliminating a *single set* of alternatives in a number of ways: by sight, sound, smell, etc. I reserve the idea of multiple "paths to knowing" for the case in which for

[43] For reasons of space, I cannot go into the theory of the u function here. For simplicity (but not in the final analysis), one may assume that for each w and $\mathcal{C}$ there is a set $\mathsf{U}_{\mathcal{C}}(w)$ of uneliminated scenarios such that for all P, $\mathsf{u}_{\mathcal{C}}(P,w) = \mathsf{U}_{\mathcal{C}}(w)$. Note that this is not the same as $\mathsf{RO}_{\exists\forall}$ from §2.2, since it does not imply an analogue of $\mathsf{contrast}$ for u. This will be important given the argument of §3.3.

[44] An $\mathbf{r}$ function may contain some eliminable redundancy in the following sense. Given $\mathbf{r}_{\mathcal{C}}(P,w)$, define

$$^{-}\mathbf{r}_{\mathcal{C}}(P,w) = \{\mathrm{A} \in \mathbf{r}_{\mathcal{C}}(P,w) \mid \text{there is no } \mathrm{A}' \in \mathbf{r}_{\mathcal{C}}(P,w) : \mathrm{A}' \subsetneq \mathrm{A}\}.$$

$^{-}\mathbf{r}_{\mathcal{C}}(P,w)$ may contain fewer alternative sets than $\mathbf{r}_{\mathcal{C}}(P,w)$, but **(Knows)** holds for $\mathbf{r}$ iff **(Knows)** holds for $^{-}\mathbf{r}$.

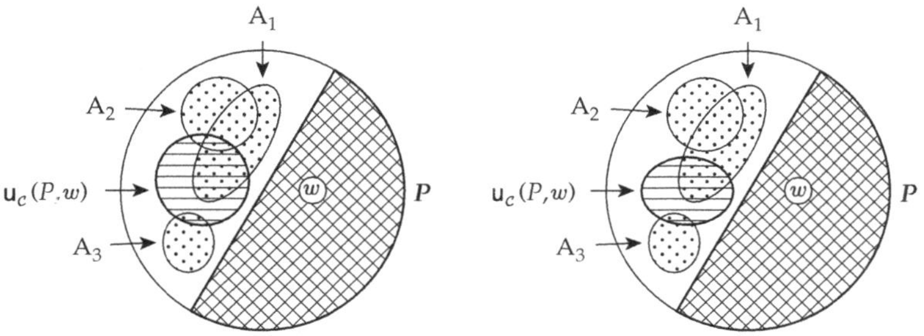

Figure 4.5 **(Knows)** violated on left vs. satisfied on right

a given proposition there are *multiple sets of alternatives* such that in order to know the proposition, it suffices to eliminate all of the alternatives in one of those sets, which one may often do in a number of *ways*.

The standard alternatives picture is equivalent to a special case of the multipath picture. Assuming

singlepath—$|\mathbf{r}_c(P,w)| = 1$,

according to which each proposition has only one alternative set, we can move back and forth between the singlepath r function and multipath **R** function as follows:

$$\mathbf{r}_c(P,w) = \{\mathsf{r}_c(P,w)\}; \tag{12}$$

$$\mathsf{r}_c(P,w) = \bigcup \mathbf{r}_c(P,w). \tag{13}$$

It follows from these equations and **singlepath** that $\mathsf{r}_c(P,w) \cap \mathsf{u}_c(P,w) = \emptyset$ iff there is some $\mathrm{A} \in \mathbf{r}_c(P,w)$ such that $\mathrm{A} \cap \mathsf{u}_c(P,w) = \emptyset$, so (Knows) would be equivalent to (**Knows**).

Having rejected **singlepath** with the argument from disjunctive and existential propositions, let us consider multipath generalizations of singlepath principles. First, the singlepath principle

r-RofA—$w \notin \boldsymbol{P}$ implies $w \in \mathsf{r}_c(P,w)$

from §2.1 generalizes to the multipath principle

r-RofA—$w \notin \boldsymbol{P}$ implies $w \in \bigcap \mathbf{r}_c(P,w)$,

which says that if P is false at w, then w is in every alternative set for P. As before, since w is always uneliminated for the agent in w, i.e. $w \in \mathsf{u}_c(P,w)$ by u-RofA, only truths can be known.

Second, the singlepath principle

ec-fallibilism—for some P and Q: $\mathsf{r}_c(P,w) \subseteq \boldsymbol{Q}$ and $\boldsymbol{Q}_w \subsetneq \mathrm{W}_w - \boldsymbol{P}$

from §2.1 generalizes to the multipath principle

ec-fallibilism—for some P, Q, and A: $A \in \mathbf{r}_C(P,w)$, $A \subseteq \boldsymbol{Q}$, and $\boldsymbol{Q}_w \subsetneq W_w - \boldsymbol{P}$,

according to which there are propositions P and Q and a path to knowing P that only requires eliminating Q-scenarios, rather than all $\neg P$-scenarios, giving us *expressible* fallibilism.

3.3. *Against the Contrast Assumption*

Next recall the contrast/enough assumption stated in the standard alternatives framework:

contrast/enough—$r_C(P,w) \subseteq W - \boldsymbol{P}$.

In §2.5, I argued that the standard alternatives framework requires contrast/enough, because it should always be *enough* to know a proposition P that one eliminates *all* $\neg P$-scenarios.

In the multipath alternatives framework, contrast/enough splits into two principles:

contrast—$\forall A \in \mathbf{r}_C(P,w)$: $A \subseteq W - \boldsymbol{P}$;
enough—$\exists A \in \mathbf{r}_C(P,w)$: $A \subseteq W - \boldsymbol{P}$.

As before, fallibilists should accept **enough**, which ensures that it is enough to know P that one eliminates all $\neg P$-scenarios. Yet fallibilists should reject **contrast** and even the weaker principle

semi-contrast—$\forall A \in \mathbf{r}_C(P,w)$: $A \neq \emptyset$ implies $A \cap (W - \boldsymbol{P}) \neq \emptyset$,

which says that every nonempty alternative set for P contains some $\neg P$-scenario. Instead, we should allow one of the dotted alternative sets in Fig. 4.5 to overlap with or even be inside the crosshatched $\boldsymbol{P}$-region.

The argument is simple. By **ec-fallibilism**,[45] there are propositions P and Q such that knowing P only requires eliminating Q-scenarios, where $\boldsymbol{Q}_w \subsetneq W_w - \boldsymbol{P}$ and hence $(\boldsymbol{P} \vee \boldsymbol{Q})_w \neq W_w$. But then since one path to knowing the contingent proposition $P \vee Q$ is via knowing P, and since knowing this P only requires ruling out Q-scenarios, which are of course $(P \vee Q)$-scenarios, it follows that there is a path to knowing $P \vee Q$ that only requires eliminating $(P \vee Q)$-scenarios.[46] This contradicts **contrast** and **semi-contrast**.

[45] Or even a weaker **e-fallibilism** generalizing e-fallibilism.

[46] Here is a trickier argument using existential propositions. Start with a standard fallibilist view according to which there are many propositions P that Jones can know by eliminating $\neg P$-scenarios, without having to eliminate the $\neg P$-scenarios that a radical skeptic would raise against him, such as subjectively indistinguishable scenarios in which *Jones is a brain in a vat* (*BIV*). One such proposition P that Jones can know, just by getting a good look at Smith's body, is that *Smith is not a BIV*. (Knowing that *someone else* is not a BIV is less of a problem!) Now where Q is the proposition that *someone is not a BIV*, surely if Jones knows P, then with a step of logic he can know Q. Finally, noticing that our initial assumption implies that Jones can know P by eliminating only scenarios in which Q is true (for they are scenarios in which *Jones*

What may have fooled some fallibilists into assuming **contrast** for all propositions is that it may seem plausible when applied to logically atomic propositions. However, when we turn to the study of epistemic closure, we must consider logically complex propositions, for which universal **contrast** is not plausible from a fallibilist perspective. See the Appendix for further discussion of the relation between **contrast** and complex propositions.

In the disjunction counterexample to **contrast** assuming **ec-fallibilism**, one reason it makes sense for an alternatives set A for $P \vee Q$ to overlap with $\boldsymbol{P \vee Q}$ (i.e. $\mathrm{A} \cap \boldsymbol{P \vee Q} \neq \emptyset$) is that A is also an alternative set for a *stronger* proposition, P, with which A does not overlap (i.e. $\mathrm{A} \cap \boldsymbol{P} = \emptyset$). One might think this is always the case when an alternative set for a proposition S overlaps with $\boldsymbol{S}$:

> **overlap**—$\forall \mathrm{A} \in \mathbf{r}_c(S, w)$: $\mathrm{A} \cap \boldsymbol{S} \neq \emptyset$ implies $\exists P \in \mathcal{P}$: $\boldsymbol{P}_w \subsetneq \boldsymbol{S}$, $\mathrm{A} \cap \boldsymbol{P} = \emptyset$, and $\mathrm{A} \in \mathbf{r}_c(P, w)$.

Nothing in my arguments turns on fallibilists accepting **overlap**, but it is noteworthy that **overlap** is consistent with all of the other principles I propose, as shown in the following section.

3.4. *Problem Solved*

Given the general arguments above, the multipath picture of knowledge should be attractive to all fallibilists. But I have yet to give one of the strongest arguments in its favor: it solves the problem represented by the impossibility result of Proposition 1.

First, observe that the singlepath principle

> noVK—$\boldsymbol{P}_w \neq \mathrm{W}_w$ implies $\mathsf{r}_c(P, w) \neq \emptyset$

from §§2.3–2.4 generalizes, following equation (12), to the multipath principle

> **noVK**—$\boldsymbol{P}_w \neq \mathrm{W}_w$ implies $\emptyset \notin \mathbf{r}_c(P, w)$,

which also says that if P is (deeply) contingent, then knowing P requires eliminating scenarios.[47]

Second, observe that the singlepath principle

> TF-cover—if Q is a TF-consequence of P, then $\mathsf{r}_c(Q, w) \subseteq \mathsf{r}_c(P, w)$

from §2.4 generalizes to the multipath principle

> **TF-cover**—if Q is a TF-consequence of P, then $\forall \mathrm{A} \in \mathbf{r}_c(P, w) \exists \mathrm{B} \in \mathbf{r}_c(Q, w)$: $\mathrm{B} \subseteq \mathrm{A}$,

is not a BIV and hence *someone* is not a BIV), it follows that Jones can know Q by eliminating only scenarios in which Q is true. Notably, this argument is compatible with **noVK** in §3.4.

[47] Note that if $\mathbf{r}_c(P, w) = \emptyset$, then P cannot be known, given the existential character of (**Knows**).

which says that the empirical work for knowing P via *any* path covers the empirical work for knowing Q via *some* path. This principle is necessary for single-premise closure under TF-consequence.[48]

It is now provable that all of the principles I have recommended for fallibilists are consistent in the multipath picture. Compare the negative Proposition 1 with the following positive result.

Proposition 2 (The Five Postulates). In the multipath picture, the following Five Postulates (for all w, P, and $\mathcal{C}$) are *jointly consistent* with **ec-fallibilism**:

r-RofA—$w \notin \boldsymbol{P}$ implies $w \in \bigcap \mathbf{r}_{\mathcal{C}}(P, w)$;

enough—$\exists \mathrm{A} \in \mathbf{r}_{\mathcal{C}}(P, w)$: $\mathrm{A} \subseteq \mathrm{W} - \boldsymbol{P}$;

noVK—$\boldsymbol{P}_w \neq \mathrm{W}_w$ implies $\emptyset \notin \mathbf{r}_{\mathcal{C}}(P, w)$;

TF-cover—if Q is a TF-consequence of P, then $\forall \mathrm{A} \in \mathbf{r}_{\mathcal{C}}(P, w)\ \exists \mathrm{B} \in \mathbf{r}_{\mathcal{C}}(Q, w)$: $\mathrm{B} \subseteq \mathrm{A}$;

overlap—$\forall \mathrm{A} \in \mathbf{r}_{\mathcal{C}}(P, w)$: $\mathrm{A} \cap \boldsymbol{P} \neq \emptyset$ implies $\exists Q \in \mathcal{P}$: $\boldsymbol{Q}_w \subsetneq \boldsymbol{P}$, $\mathrm{A} \cap \boldsymbol{Q} = \emptyset$, and $\mathrm{A} \in \mathbf{r}_{\mathcal{C}}(Q, w)$.

Proof. The proposition holds as a corollary of Theorem 2 below, as explained in §3.5. □

Although **ec-fallibilism** only says that we are fallibilists for at least one proposition, an **r** function can satisfy the Five Postulates while being fallibilistic for (infinitely) many propositions, as shown by Theorem 2 below. (Theorem 2 also shows that for **enough**, we could require $\mathrm{A} \subseteq \mathrm{W}_w - \boldsymbol{P}$.)

It is important to understand why the multipath picture avoids an analogue of Proposition 1. Recall that the proof forced us to conclude in (11) that knowing the contingent proposition $P \vee Q$ does not require eliminating any scenarios; for if it did, then by contrast/enough it would require eliminating $(\neg P \wedge \neg Q)$-scenarios; but that would contradict TF-cover, because knowing P did not require eliminating any $\neg Q$-scenarios, by the very choice of Q as a proposition true in all of the relevant $\neg P$-scenarios. Fortunately, the multipath picture does not lead to this contradiction. By **enough**, *one* path to knowing $P \vee Q$ is by eliminating all of the scenarios in some set of $(\neg P \wedge \neg Q)$-scenarios, which is nonempty by **noVK**. But in line with **TF-cover**, *another* path to knowing $P \vee Q$ is via knowing P, which may involve eliminating only $(\neg P \wedge Q)$-scenarios (note that here we use our rejection of both the single alternative set and contrast assumptions). All of these paths require eliminating scenarios, so we respect **noVK**. We have no problem of vacuous knowledge.

Contrast this account with those of Nozick (1981) and Lewis (1996). Let P be a mundane contingent proposition about the external world and S your favorite skeptical hypothesis incompatible with P. Recall that Nozick's tracking theory has the following problematic consequences: according to the theory, the logically weaker $P \vee \neg S$ may be much harder to know than the logically

[48] And it is *sufficient* together with certain assumptions on u, such as that of footnote 43.

stronger P; and the logically weaker $\neg S$ may be much harder to know than the logically stronger $P \wedge \neg S$. The reason is that on Nozick's theory, knowing P does not require eliminating skeptical $(\neg P \wedge S)$-scenarios, but knowing the *weaker* $P \vee \neg S$ *does* (where "elimination" for Nozick is understood as in §2.3); and knowing $P \wedge \neg S$ does not require eliminating skeptical S-scenarios, but knowing the weaker $\neg S$ does. This leads to the kind of extreme epistemic closure failures that illustrate the *problem of containment* from §2.4. As Vogel (2007, 76) explains:

> It seems hard to deny that one's epistemic position with respect to a logically weaker proposition (X or Y) is at least as good as one's epistemic position with respect to a logically stronger proposition X.... The tracking condition T improperly inverts that relation by making the conditions for knowing (X or Y) more stringent than the conditions for knowing X.... [S]atisfying T with respect to (X or Y) can require that one is right over a greater region of logical space than is required to satisfy T with respect to X. Therefore, one's epistemic position with respect to (X or Y) may be inferior to one's epistemic position with respect to X.

While Nozick thereby makes knowing something like $P \vee \neg S$ too hard, Lewis (1996) makes it too easy. On Lewis's theory, there will be many contexts in which an agent can know the *contingent* $P \vee \neg S$ without any requirement of eliminating scenarios, simply because it is true throughout the fixed set of relevant possibilities (recall §2.4). Nozick and Lewis are pushed to these extreme positions by their assumption that for each proposition Q, there is only a single alternative set for Q, containing only contrasting $\neg Q$-scenarios. By making such a single alternative set for $P \vee \neg S$ *nonempty*, Nozick avoids the problem of vacuous knowledge but saddles us with the problem of containment, whereas by making such a single alternative set for $P \vee \neg S$ *empty*, Lewis avoids the problem of containment but saddles us with the problem of vacuous knowledge.

We need not accept the Nozick–Lewis dilemma. In the multipath picture presented above, none of the alternative sets for the contingent $P \vee \neg S$ are empty, so there is no vacuous knowledge, and one of the alternative sets for $P \vee \neg S$ is from the path to knowing P via eliminating $(\neg P \wedge \neg S)$-scenarios, so there is no problem of containment arising from $P \vee \neg S$. Nor is there a problem of containment arising from $P \wedge \neg S$. Like Lewis's theory but unlike Nozick's, in the multipath picture presented above, an agent who knows $P \wedge \neg S$ has done enough empirical work to know $\neg S$.

By establishing the consistency of fallibilism, **noVK**, **TF-cover**, and the other principles, Proposition 2 shows that by adopting the multipath picture of knowledge, fallibilists can avoid the problems raised in §2.5, a significant positive result. Of course, fallibilists who adopt the multipath picture must address the question: where do the possibly multiple alternative sets for a proposition come from? It may seem that fallibilists working with the standard alternatives picture have an easier time saying where their single set of alternatives for a proposition comes from, e.g. by using a relevance or similarity ordering of scenarios to pick out the set of close(st) scenarios where the proposition is false. However, in §3.5 I show that the standard picture does not have an advantage in this respect.

3.5. *From Singlepath to Multipath*

The reason is that any standard alternatives function r determines a natural multipath alternatives function $\mathbf{r}^{\mathsf{r}}$; and if r satisfies a few conditions, which are satisfied by any r based on orderings of scenarios as in §2.3, then $\mathbf{r}^{\mathsf{r}}$ satisfies the Five Postulates of Proposition 2 and is fallibilistic in a way I will make precise. The alternative sets in $\mathbf{r}^{\mathsf{r}}_{C}(P, w)$ will depend on the *structure* of P. To keep things simple, I will first derive $\mathbf{r}^{\mathsf{r}}$ from r for propositions in an easy-to-handle *normal form*; then we can immediately derive $\mathbf{r}^{\mathsf{r}}$ for all propositions, using the fact that every proposition is TF-equivalent to one in normal form. To set this up, we need to review some basic logical concepts:

First, some notation and terminology. Assuming the structured propositions view of §1.1, let us write "p", "q", "r", etc. for TF-atomic propositions. A *TF-basic* proposition is a TF-atomic proposition or the negation thereof. Let **basic-singlepath** and **basic-contrast** be the conditions **singlepath** and **contrast** from §3.2 and §3.3 applied to TF-basic propositions only. A *clause* is a disjunction of TF-basic propositions: e.g. $(p \vee \neg q \vee r)$. I assume that permutation and repetition of disjuncts does not matter, so "$(p \vee \neg q \vee r)$" and "$(\neg q \vee p \vee p \vee r)$" represent the same clause. A clause is *nontrivial* if it does not contain both p and $\neg p$ for any p. If a clause C' can be obtained by adding zero or more disjuncts to C, then C' is a *superclause* of C, and C is a *subclause* of C': e.g. $(p \vee \neg q \vee r)$ is a superclause of $(p \vee \neg q)$ and a subclause of $(p \vee \neg q \vee \neg s \vee r)$. The set $sub(P)$ of TF-subpropositions of P is defined recursively: $sub(p) = \{p\}$ for p a TF-atomic proposition; $sub(\neg P) = sub(P) \cup \{\neg P\}$; $sub(P \# Q) = sub(P) \cup sub(Q) \cup \{P \# Q\}$ for any binary truth-functional connective $\#$, and so on for n-ary connectives. Finally, let $at(P)$ be the set of TF-atomic propositions in $sub(P)$.

Second, a fact: each proposition P (that is not a TF-tautology) is TF-equivalent to a proposition P' in *canonical conjunctive normal form* (CCNF), which is a conjunction of nontrivial clauses such that for each $q \in at(P')$, each clause in P' contains q or $\neg q$. Here is one way to calculate a CCNF of a proposition P. First, make a truth table for $at(P)$. Second, for each row of the truth table that makes the proposition *false*, write down a conjunction of TF-basic propositions describing that row; for example, the rows that make $p \wedge q$ false are described by: $(\neg p \wedge \neg q)$, $(\neg p \wedge q)$, and $(p \wedge \neg q)$. Third, write down a conjunction saying that we are not in any of those rows that make the proposition false: $\neg(\neg p \wedge \neg q) \wedge \neg(\neg p \wedge q) \wedge \neg(p \wedge \neg q)$. Finally, drive the negations inside: $(p \vee q) \wedge (p \vee \neg q) \wedge (\neg p \vee q)$. Thus, we obtain a CCNF equivalent of $p \wedge q$. What is important for our purposes is that each proposition P (that is not a TF-tautology) is TF-equivalent to a P' in CCNF *with* $at(P) = at(P')$ that is *unique up to reordering of the conjuncts and disjuncts* (see Theorem 1.29 of Cori and Lascar 2000). Since order will not matter, let us associate with each such P a unique $\mathrm{CCNF}(P)$ in CCNF. If P is a TF-tautology, let us stipulate that $\mathrm{CCNF}(P) = (p \vee \neg p)$ for some atomic p.

Third, a definition using the notions above: for P in CCNF (not a TF-tautology), define $\mathfrak{c}(P)$ to be the set of all subclauses C of conjuncts in P such that every nontrivial superclause C' of C with $at(C') = at(P)$ is a conjunct of P. This implies that every conjunct of P is in $\mathfrak{c}(P)$, but there may be other clauses in $\mathfrak{c}(P)$. For example, if P is $(p \vee q) \wedge (p \vee \neg q)$, then $\mathfrak{c}(P) = \{(p \vee q), (p \vee \neg q), p\}$; and if P is the conjunction of $(p \vee q \vee r)$, $(\neg p \vee q \vee r)$, $(p \vee \neg q \vee r)$,$(p \vee q \vee \neg r)$, and $(p \vee \neg q \vee \neg r)$, then $\mathfrak{c}(P)$ contains all of the conjuncts of P as well as $(p \vee q)$, $(p \vee r)$, $(q \vee r)$, and p. It turns out that $\mathfrak{c}(P)$ is the set of all nontrivial clauses C with $at(C) \subseteq at(P)$ that are TF-consequences of P.

Finally, some new notions related to fallibilism: a multipath function $\mathbf{r}$ is **fallibilistic** in $\mathcal{C}$ at w *with respect to* P iff there is some $\mathrm{A} \in \mathbf{r}_{\mathcal{C}}(P, w)$ with $\mathrm{A} \subsetneq \mathrm{W}_w - \boldsymbol{P}$; a standard alternatives function r is fallibilistic in $\mathcal{C}$ at w *with respect to* P iff $\mathsf{r}_{\mathcal{C}}(P, w) \subsetneq \mathrm{W}_w - \boldsymbol{P}$; and r is plurally fallibilistic in $\mathcal{C}$ at w with respect to a set $\{P_1, \dots, P_n\}$ of clauses iff there are subclauses $P_1', \dots, P_n'$ of $P_1, \dots, P_n$ such that the *union* of all $\mathsf{r}_{\mathcal{C}}(P_i', w)$ sets is a strict subset of $\mathrm{W}_w - (\boldsymbol{P_1} \wedge \dots \wedge \boldsymbol{P_n})$.

We are now ready to derive a multipath function $\mathbf{r}^{\mathsf{r}}$ from each standard alternatives function r. I will present the construction and the main result about the construction at the same time:

Theorem 2 (Multipath Theorem). Given a standard alternatives function r, define a multipath alternatives function $\mathbf{r}^{\mathsf{r}}$ as follows: for any clause C, define

$$\mathbf{r}^{\mathsf{r}}_{\mathcal{C}}(C, w) = \{\mathsf{r}_{\mathcal{C}}(C', w) \mid C' \text{ is a subclause of } C\}; \tag{14}$$

for any CCNF conjunction $C_1 \wedge \dots \wedge C_n$ of clauses with $\mathfrak{c}(C_1 \wedge \dots \wedge C_n) = \{D_1, \dots, D_m\}$, define

$$\mathbf{r}^{\mathsf{r}}_{\mathcal{C}}(C_1 \wedge \dots \wedge C_n, w) = \left\{ \mathrm{A} \subseteq \mathrm{W} \mid \exists \mathrm{A}_1 \in \mathbf{r}^{\mathsf{r}}_{\mathcal{C}}(D_1, w) \dots \exists \mathrm{A}_m \in \mathbf{r}^{\mathsf{r}}_{\mathcal{C}}(D_m, w) : \mathrm{A} = \bigcup_{1 \le i \le m} \mathrm{A}_i \right\}; \tag{15}$$

and if P is not in CCNF, define

$$\mathbf{r}^{\mathsf{r}}_{\mathcal{C}}(P, w) = \mathbf{r}^{\mathsf{r}}_{\mathcal{C}}(\mathrm{CCNF}(P), w).^{49} \tag{16}$$

Then $\mathbf{r}^{\mathsf{r}}$ satisfies **basic-singlepath** and **TF-cover**; if r satisfies r-RofA, then $\mathbf{r}^{\mathsf{r}}$ satisfies **r-RofA**; if r satisfies contrast, then $\mathbf{r}^{\mathsf{r}}$ satisfies **basic-contrast**, **enough**, and **overlap**; if r satisfies noVK, then $\mathbf{r}^{\mathsf{r}}$ satisfies **noVK**; for any clause P, if r is fallibilistic in $\mathcal{C}$ at w with respect to P, then $\mathbf{r}^{\mathsf{r}}$ is **fallibilistic** in $\mathcal{C}$ at w with respect to P; and for any P in CCNF, if r is plurally fallibilistic in $\mathcal{C}$ at w with respect to $\mathfrak{c}(P)$, then $\mathbf{r}^{\mathsf{r}}$ is **fallibilistic** in $\mathcal{C}$ at w with respect to P.

[49] Note that neither (14) nor (15) depend on the order of the disjunct or conjuncts, so the particular choice of $\mathrm{CCNF}(P)$ among equivalent but permuted CCNFs does not matter for $\mathbf{r}^{\mathsf{r}}_{\mathcal{C}}(P, w)$.

Proof. See Appendix B of Holliday 2014a. □

The idea behind the definition of $\mathbf{r}^{\mathsf{r}}_{c}$ is simple: (14) says that any path to knowing a subclause of a clause is a path to knowing the clause, a generalization of the idea that any path to knowing a disjunct of a disjunction is a path to knowing the disjunction; and (15) says that knowing a conjunction of clauses requires doing enough epistemic work to know each of the clauses that are TF-consequences of the conjunction. Note that for TF-basic propositions L, (14) implies $\mathbf{r}^{\mathsf{r}}_{c}(L,w) = \{\mathsf{r}_c(L,w)\}$, so the derived multipath function $\mathbf{r}^{\mathsf{r}}$ differs from the input function r only for complex propositions.

For complex propositions P, it is not guaranteed that $\mathsf{r}_c(P,w) \in \mathbf{r}^{\mathsf{r}}_{c}(P,w)$. To see this, one can check that for all $\mathrm{A} \in \mathbf{r}^{\mathsf{r}}_{c}(p \wedge q, w)$, $\mathsf{r}_c(p,w) \cup \mathsf{r}_c(q,w) \subseteq \mathrm{A}$, whereas there is no guarantee that $\mathsf{r}_c(p,w) \cup \mathsf{r}_c(q,w) \subseteq \mathsf{r}_c(p \wedge q, w)$, especially if r is based on orderings of scenarios. This is the source of the notorious problem for sensitivity theories that an agent may know that *the building is a barn and the building is red* ($p \wedge q$), despite not knowing that *the building is a barn* (p). By contrast, according to $\mathbf{r}^{\mathsf{r}}$, an agent knows a conjunction only if she has done enough to know each conjunct.

The definition of $\mathbf{r}^{\mathsf{r}}$ is best understood by example. Let us calculate the alternatives sets for $p \vee (q \wedge r)$. First, we calculate $\mathrm{CCNF}(p \vee (q \wedge r))$ as above. The rows of the truth table for p, q, and r in which $p \vee (q \wedge r)$ is false are described by $(\neg p \wedge \neg q \wedge r)$, $(\neg p \wedge q \wedge \neg r)$, and $(\neg p \wedge \neg q \wedge \neg r)$, so

$$\mathrm{CCNF}(p \vee (q \wedge r)) = (p \vee q \vee \neg r) \wedge (p \vee \neg q \vee r) \wedge (p \vee q \vee r).$$

One can then verify using the definition of $\mathfrak{c}$ that

$$\mathfrak{c}(\mathrm{CCNF}(p \vee (q \wedge r))) = \{(p \vee q \vee \neg r), (p \vee \neg q \vee r), (p \vee q \vee r), (p \vee q), (p \vee r)\}.$$

Call the members of this set D_1–D_5. By (15), every $\mathrm{A} \in \mathbf{r}^{\mathsf{r}}_{c}(p \vee (q \wedge r))$ is of the form $\mathrm{A}_1 \cup \mathrm{A}_2 \cup \mathrm{A}_3 \cup \mathrm{A}_4 \cup \mathrm{A}_5$, where by (14) each A_i is $\mathsf{r}_c(C_i, w)$ for some subclause C_i of D_i.[50] The important choices of the subclauses C_1–C_5 of D_1–D_5 are: each C_i is p; each C_i is q or r; each C_i is $(p \vee q)$ or $(p \vee r)$. These yield the following alternative sets in $\mathbf{r}^{\mathsf{r}}_{c}(p \vee (q \wedge r))$: $\mathsf{r}_c(p,w)$ for the path via knowing p; $\mathsf{r}_c(q,w) \cup \mathsf{r}_c(r,w)$ for the path via knowing $q \wedge r$; and $\mathsf{r}_c(p \vee q) \cup \mathsf{r}_c(p \vee r)$ for a path to knowing the disjunction without necessarily knowing either disjunct individually.

If r is based on orderings of scenarios, as in §2.3, then taking the $\mathsf{r}_c(p \vee q) \cup \mathsf{r}_c(p \vee r)$ path means eliminating the closest $(\neg p \wedge \neg q)$-scenarios and the closest $(\neg p \wedge \neg r)$-scenarios. By contrast, the singlepath picture with r says that there is only one path to knowing $p \vee (q \wedge r)$, by eliminating the closest $\neg(p \vee (q \wedge r))$-scenarios, i.e. the closest $(\neg p \wedge (\neg q \vee \neg r))$-scenarios. Note that each of these

[50] In this case, there are $7^3 \times 3^2$ ways of choosing C_1–C_5. But this does not mean that there are $7^3 \times 3^2$ distinct sets in $\mathbf{r}^{\mathsf{r}}_{c}(p \vee (q \wedge r))$, since many ways of choosing the C_i may result in the same union $\mathrm{A}_1 \cup \mathrm{A}_2 \cup \mathrm{A}_3 \cup \mathrm{A}_4 \cup \mathrm{A}_5$, and even distinct unions may be redundant because they contain others (see footnote 44).

scenarios is either a closest $(\neg p \wedge \neg q)$-scenario or a closest $(\neg p \wedge \neg r)$-scenario, so $\mathsf{r}_{\mathcal{C}}(p \vee (q \wedge r)) \subseteq \mathsf{r}_{\mathcal{C}}(p \vee q) \cup \mathsf{r}_{\mathcal{C}}(p \vee r)$.[51]

Let us now prove Proposition 2. Recall from Theorem 1 that if r is such that for all $\mathcal{C}$, P, and w, $\mathsf{r}_{\mathcal{C}}(P, w)$ is the set of closest $\neg P$-scenarios according to an ordering $\leqslant_w^{\mathcal{C}}$ as in §2.3, then r satisfies **r-RofA**, **contrast**, and **noVK**. Hence by Theorem 2, $\mathbf{r}^{\mathsf{r}}$ satisfies the Five Postulates of Proposition 2. Moreover, $\mathbf{r}^{\mathsf{r}}$ is highly fallibilistic if r is (not with respect to *every* proposition that r is, but with respect to those that meet the conditions in the theorem). To establish *expressible* contrast fallibilism, **ec-fallibilism**, let us make an extremely weak assumption about expressibility: for some TF-basic proposition L and proposition Q, Q entails $\neg L$, but not vice versa: $\boldsymbol{Q}_w \subsetneq \mathrm{W}_w - \boldsymbol{L}$. Then there exists an ordering $\leqslant_w^{\mathcal{C}}$ such that $Closest_{\leqslant_w^{\mathcal{C}}}(\neg L) = \boldsymbol{Q}_w \subsetneq \mathrm{W}_w - \boldsymbol{L}$; so by the fact that $Closest_{\leqslant_w^{\mathcal{C}}}(\neg L) = \mathsf{r}_{\mathcal{C}}(L, w) \in \mathbf{r}^{\mathsf{r}}_{\mathcal{C}}(L, w)$, Q is a witness for the fact that $\mathbf{r}^{\mathsf{r}}$ satisfies **ec-fallibilism**.

Although I have focused on the idea of deriving $\mathbf{r}^{\mathsf{r}}$ from a function r based on the familiar qualitative orderings of scenarios, it is not necessary that the input function r be based on such orderings. If $w \notin P$, then we can assume $\mathsf{r}_{\mathcal{C}}(P, w) = \{\{w\}\}$, so **r-RofA** is satisfied. If $w \in P$, then perhaps $\mathsf{r}_{\mathcal{C}}(P, w)$ is some function of the probability, or cost-weighted probability, or other value of each $\neg P$-scenario, so that the $\neg P$-scenarios with relatively substantial probability, or cost-weighted probability, or whatever, relative to other $\neg P$-scenarios, are in $\mathsf{r}_{\mathcal{C}}(P, w)$, where what "relatively substantial" means may depend on $\mathcal{C}$, P, or w.[52] These options would also satisfy **contrast** and **noVK**,[53] so the resulting

[51] The reason is that each of the closest $\neg(C_1 \wedge \cdots \wedge C_n)$-scenarios is a closest $\neg C_i$-scenario for some i:

$$Closest_{\leqslant_w^{\mathcal{C}}}(\neg(C_1 \wedge \cdots \wedge C_m)) \subseteq \bigcup_{1 \leq i \leq m} Closest_{\leqslant_w^{\mathcal{C}}}(\neg C_i),$$

so

$$\mathsf{r}_{\mathcal{C}}(C_1 \wedge \cdots \wedge C_m, w) \subseteq \bigcup_{1 \leq i \leq m} \mathsf{r}_{\mathcal{C}}(C_i, w),$$

and it is easy to check from (15) and (14) that that there is some $\mathrm{A} \in \mathbf{r}^{\mathsf{r}}_{\mathcal{C}}(C_1 \wedge \cdots \wedge C_m)$ such that

$$\bigcup_{1 \leq i \leq m} \mathsf{r}_{\mathcal{C}}(C_i, w) \subseteq \mathrm{A}.$$

[52] It would be more natural to think in terms of the probability of something more coarse-grained than scenarios, such as the *Alternatives* mentioned in §2.1 and footnote 10, but I skip over these details here. I also skip over the kind of probability in question, whether probability on the agent's evidence—in which case the suggestion in the text might blur the roles of the r and u functions (recall footnotes 33 and 14)—or a kind of objective probability or, in the spirit of contextualism, probability for the attributors. It is noteworthy here that Vogel (1999, 163) argues that probability cannot provide a sufficient condition for relevance. Roughly, the argument runs as follows: if a *proposition* S is an irrelevant "alternative" to P, but a *proposition* Q is probable enough to be a relevant "alternative" to P, then of course $Q \vee S$ is also probable enough to be a relevant "alternative" to P; but then since on Vogel's view, ruling out $Q \vee S$ requires ruling out S, it follows that S is a relevant "alternative" that must be ruled out for knowledge of P after all. Contradiction. Of course, this argument assumes the propositional view of alternatives that I rejected in §2.1 because it violates the disjointness condition on alternatives in a context.

[53] They would satisfy **noVK** because there are always some $\neg P$-scenarios with maximal probability, cost-weighted probability, or whatever, *relative to* other $\neg P$-scenarios.

$\mathbf{r}^{\mathsf{r}}$ would have the properties given by Theorem 2. I will not go into the details here. My main point in this section is that the multipath picture is not at a disadvantage relative to the singlepath picture with respect to having more alternative sets for which to account.

In my view, the construction of $\mathbf{r}^{\mathsf{r}}$ from r above provides a kind of "lower bound" on what a multipath function should look like if derived from a singlepath function r: if $\mathrm{A} \in \mathbf{r}^{\mathsf{r}}_{\mathcal{C}}(P,w)$, then A should be an alternative set for P in w relative to $\mathcal{C}$ according to any reasonable multipath function derived from r. In §4, I will consider whether a reasonable multipath function should provide even more alternative sets—even more paths—for knowing some propositions.

4. MORE PATHS?

In §3, we saw what might be called the "conservative" version of the multipath picture. On the conservative version, the source of additional paths to knowledge of a proposition is the structure of the proposition itself; this is why the single alternative set and contrast assumptions are rejected for *complex* propositions. Let us now consider the questions: Are there additional paths to knowledge of a proposition that do not come from the structure of the proposition? Should the single alternative set and contrast assumptions be rejected in general, not just for complex propositions?

4.1. *Inductive Closure*

Recall that my motivating examples for the multipath picture in §3.2 involved cases where some of the multiple paths to knowing a complex proposition—such as a disjunctive or existential proposition—went via knowing logically *stronger* propositions—a disjunct or an instance. Might there be multiple paths to knowing a proposition via knowing logically *weaker* propositions? Anyone who thinks that *inductive* knowledge is possible is committed to an affirmative answer. Although so far I have concentrated on closure principles where the relation **R** (recall §2.1) is a deductive relation, one can also consider closure with respect to inductive relations, asking whether an agent who knows the empirical premises of a "good" inductive argument has thereby done enough empirically to know the conclusion. Let us see how the multipath picture of knowledge bears on this issue.

To use a standard (oversimplified) example of enumerative induction, let $\mathbf{E} = \{e_1, \ldots, e_n\}$ be the set of the first n emeralds, by distance, from some location; for any $e \in \mathbf{E}$, let G_e be the proposition that *e is green*; and let G be $\bigwedge_{e \in \mathbf{E}} G_e$, a conjunctive version of *all emeralds in* **E** *are green*. According to some fallibilists, for large n one can come to know G by observing just the emeralds in some strict subset $\mathrm{E}_1 \subsetneq \mathbf{E}$. Since the proposition $\bigwedge_{e \in \mathrm{E}_1} G_e$ is logically weaker than G, this answers the second part of the question above; but what about the multiplicity of paths? Presumably believers in inductive knowledge do not think that G can only be known by observing the emeralds in just *one* set $\mathrm{E}_1 \subsetneq \mathbf{E}$; instead, there should be many sets $\mathrm{E}_i \subsetneq \mathbf{E}$ (with $\mathrm{E}_i \not\subseteq \mathrm{E}_j$ for $i \neq j$) such that if the agent observes all of the emeralds in one of them, she has done enough to know G. Hence for each such E_i, there will be an alternative set $\mathrm{A}_i \in \mathbf{r}_c(G, w)$ such that for every $e \in \mathrm{E}_i$ there is some $\mathrm{B}_e \in \mathbf{r}_c(G_e, w)$ with $\mathrm{B}_e \subseteq \mathrm{A}_i$. Assuming $\mathrm{A}_i \not\subseteq \mathrm{A}_j$ for $i \neq j$, this gives us the multiple alternative sets, answering the first part of the question above. Indeed, this provides another reason to accept the multipath picture for fallibilists who wish to make room for the possibility of knowledge by enumerative induction.

Note that on the assumption of closure under single-premise TF-consequence, someone who comes to know G by observing the emeralds in some E_i should also be able to know that the so-far-unobserved emerald $b \in \mathbf{E}$ in my back pocket is green by observing those *other* emeralds in E_i ($b \notin \mathrm{E}_i$). If this is correct, then it seems there may be multiple paths to knowing even the TF-atomic proposition G_b: by eliminating the $\neg G_b$-scenarios in $\mathsf{r}_c(G_b, w)$ or by eliminating the scenarios in some $\mathrm{A}_i \in \mathbf{r}_c(G, w)$. These paths will be genuinely distinct if $\mathsf{r}_c(G_b, w) \not\subseteq \mathrm{A}_i \not\subseteq \mathsf{r}_c(G_b, w)$ (see footnote 44). If so, then **basic-singlepath** cannot hold for **r**. Moreover, if A_i contains some G_b-scenarios, e.g. $(G_b \wedge \neg G_e)$-scenarios for some $e \in \mathrm{E}_i$, then **basic-contrast** cannot hold for **r** either. Thus, fallibilists who wish to maintain the possibility of inductive knowledge and single-premise logical closure may be led to reject **basic-singlepath** and **basic-contrast**. One may then wonder whether such fallibilists can extend a singlepath function r to a multipath function **r** as in §3.5. To do so, they must modify (14) in the construction for Theorem 2 in order to allow for some extra inductive paths to knowing some TF-basic propositions.[54] In the current example, to say just how many or which emeralds must be observed in the various paths to knowing G_b inductively, to determine the extra alternative sets in $\mathbf{r}_c(G_b, w)$, is a topic for a theory of inductive knowledge. In general, presumably only propositions involving certain kinds of objects and ("projectible") properties admit such extra inductive paths, which is again a topic for a theory of inductive knowledge to explain.[55]

[54] Note that if for each $e \in \mathbf{E}$, the inductive path to knowing G_e by observing the emeralds in the set E_i is included in $\mathbf{r}'_c(G_e, w)$ by a modified version of (14), then the inductive path to knowing the conjunction G by observing the emeralds in the set E_i will be included in $\mathbf{r}'_c(G, w)$ by (15).

[55] It is important that Vogel's (1999, §4) arguments, according to which a certain version of the standard alternatives picture cannot handle inductive knowledge, do not apply to the multipath picture. In short, Vogel attacks a view according to which knowing a proposition like G involves eliminating a single set of $\neg G$-worlds that resemble the actual world. On such a view, it seems difficult to explain why after observing a few emeralds, one has not eliminated the right $\neg G$-worlds resembling the actual world, but after observing more emeralds, one has. The solution in the multipath picture is to reject the view that the only path to knowing a proposition like G involves eliminating "close" $\neg G$-scenarios; instead, an agent can take one of the A_i paths described above, which involves coming to know G_e for each $e \in \mathrm{E}_i$, where this may involve eliminating close $\neg G_e$-scenarios. One might object that this response just assumes inductive knowledge is possible, rather than deriving its possibility from first principles. But I do not see this as an objection. It is just an observation that the multipath picture by itself is not a theory of inductive knowledge.

4.2. Metaphysical and Multi-Premise Closure

Theorem 2 shows how closure under single-premise TF-consequence fits with the conservative view that additional paths to knowledge of a proposition come from the structure of the proposition itself. However, in order to guarantee more controversial closure principles, consistently with the Five Postulates of §3.4, one must go beyond the conservative view. This is easiest to see in the case of closure under single-premise metaphysical entailment, which requires the following assumption:

M-cover—if $\boldsymbol{P}_w \subseteq \boldsymbol{Q}$, then $\forall \mathrm{A} \in \mathbf{r}_c(P,w)\ \exists \mathrm{B} \in \mathbf{r}_c(Q,w)$: $\mathrm{B} \subseteq \mathrm{A}$,

which is the multipath generalization of

M-cover—if $\boldsymbol{P}_w \subseteq \boldsymbol{Q}$, then $\mathsf{r}_c(Q,w) \subseteq \mathsf{r}_c(P,w)$.

M-cover says that if P entails Q as a matter of (deep) metaphysical necessity, then any path to knowing P covers a path to knowing Q, regardless of the structures of P and Q or what kinds of objects and properties they involve. Since the construction of $\mathbf{r}^{\mathsf{r}}$ in Theorem 2 only looks at the structure of P for extra paths to knowing P other than $\mathsf{r}_c(P,w)$, it does not guarantee that **M-cover** will hold for $\mathbf{r}^{\mathsf{r}}$ if M-cover does not hold for r. Moreover, by the impossibility result in Proposition 1, M-cover cannot hold for r together with the other conditions in the theorem, since M-cover implies TF-cover. In order to guarantee **M-cover**, along with the Five Postulates in §3.4, one must modify $\mathbf{r}^{\mathsf{r}}$ to allow extra paths to knowing P, not given by the structure of P or by $\mathsf{r}_c(P,w)$.

Before discussing modifications, let us consider the desirability of the **M-cover** assumption. Dretske (1970; 2005) famously argues that it can take more epistemic work to know a Q metaphysically entailed by P than to know P itself, when Q has a "heavyweight" status compared to the "lightweight" status of P.[56] One of the Dretskean concerns is that **M-cover**/M-cover will lead to radical skepticism about knowledge. Let us try to derive this result in the standard alternatives picture and the multipath picture. As discussed in §2.1, for many empirical propositions P,

$$\mathsf{u}_c(P,w) \cap (\mathrm{W}_w - \boldsymbol{P}) \neq \emptyset, \tag{17}$$

so there are some uneliminated $\neg P$-scenarios. Hence it is reasonable to assume that there is some proposition S (think of a "skeptical counter-hypothesis") such that

$$\emptyset \neq \boldsymbol{S} \subseteq \mathsf{u}_c(P,w) \cap (\mathrm{W}_w - \boldsymbol{P}). \tag{18}$$

If for every set of scenarios there is a proposition true in exactly that set of scenarios, then (18) is immediate from (17). Now let us suppose that for at

[56] In one of Dretske's (2005) examples, P is the proposition that *there are cookies in the jar*, and Q is the proposition that *idealism is false*. As Dretske quips, "Looking in the cookie jar may be a way of finding out whether there are any cookies there, but it isn't—no more than kicking rocks—a way of refuting Bishop Berkeley" (15).

least one of the propositions S as in (18), $\neg S$ is what could be called a *semi-contrast proposition*, in the sense that knowing $\neg S$ requires eliminating at least one S-scenario:[57]

$$\mathsf{r}_{\mathcal{C}}(\neg S, w) \cap \boldsymbol{S} \neq \emptyset; \tag{19}$$

$$\forall \mathrm{B} \in \mathbf{r}_{\mathcal{C}}(\neg S, w) : \mathrm{B} \cap \boldsymbol{S} \neq \emptyset. \tag{20}$$

It follows from $\boldsymbol{S} \subseteq \mathrm{W}_w - \boldsymbol{P}$ in (18) that $\boldsymbol{P}_w \subseteq \neg \boldsymbol{S}$, so M-cover/**M-cover** implies

$$\mathsf{r}_{\mathcal{C}}(\neg S, w) \subseteq \mathsf{r}_{\mathcal{C}}(P, w); \tag{21}$$

$$\forall \mathrm{A} \in \mathbf{r}_{\mathcal{C}}(P, w) \exists \mathrm{B} \in \mathbf{r}_{\mathcal{C}}(\neg S, w) : \mathrm{B} \subseteq \mathrm{A}. \tag{22}$$

Together (21) and (19) imply $\mathsf{r}_{\mathcal{C}}(P, w) \cap \boldsymbol{S} \neq \emptyset$, which with (18) implies $\mathsf{r}_{\mathcal{C}}(P, w) \cap \mathsf{u}_{\mathcal{C}}(P, w) \neq \emptyset$. Thus, by (Knows), the agent in w *does not know* P relative to $\mathcal{C}$. Similarly, (22) and (20) imply that for every $\mathrm{A} \in \mathsf{r}_{\mathcal{C}}(P, w)$, $\mathrm{A} \cap \boldsymbol{S} \neq \emptyset$, which with (18) implies $\mathrm{A} \cap \mathsf{u}_{\mathcal{C}}(P, w) \neq \emptyset$. Since this holds for *every* $\mathrm{A} \in \mathbf{r}_{\mathcal{C}}(P, w)$, by **(Knows)** the agent in w *does not know* P relative to $\mathcal{C}$. Since P, w, and $\mathcal{C}$ were arbitrary, we seem to be left with radical skepticism about empirical knowledge.

Essentially the same argument for skepticism can be given using other closure principles. I will demonstrate this in the multipath picture, leaving the singlepath case as an exercise for the reader. First, consider closure under *metaphysical equivalence* and the principle $\mathrm{K}(P \wedge Q) \Rightarrow (\mathrm{K}P \,\&\, \mathrm{K}Q)$:[58]

> **M-equiv**—if $\boldsymbol{P}_w = \boldsymbol{Q}_w$, then $\forall \mathrm{A} \in \mathbf{r}_{\mathcal{C}}(P, w)\ \exists \mathrm{B} \in \mathbf{r}_{\mathcal{C}}(Q, w)$: $\mathrm{B} \subseteq \mathrm{A}$;
> **concover**—$\forall \mathrm{A} \in \mathbf{r}_{\mathcal{C}}(P \wedge Q)\ \exists \mathrm{B} \in \mathbf{r}_{\mathcal{C}}(P, w)\ \exists \mathrm{B}' \in \mathbf{r}_{\mathcal{C}}(Q, w)$: $\mathrm{B} \cup \mathrm{B}' \subseteq \mathrm{A}$.

[57] It follows from (18) that $\neg S$ is not deeply necessary, $\neg \boldsymbol{S}_w \neq \mathrm{W}_w$, which with **noVK** implies $\emptyset \notin \mathbf{r}_{\mathcal{C}}(\neg S, w)$, which with **semi-contrast** implies (20).

[58] A similar argument is given by Hawthorne (2004, 41), employing a principle similar to **M-equiv**, namely closure under a priori equivalence, though Hawthorne uses the argument for different dialectical purposes. See Sherman and Harman 2011 for an argument against Hawthorne's equivalence principle, which also applies to **M-equiv**. Another reason to worry about the equivalence principle and **M-equiv** is that these principles seem to commit the mistake of what Perry (1989) calls "losing track of subject matter": losing track of what propositions are *about*, considering only the possibilities in which they are true. Barwise and Perry (1983) have argued that losing track of subject matter leads to serious problems in semantics, and the same may be true in epistemology. If the range of alternatives that one must eliminate in order to know a proposition may depend not only on the structure of the proposition, as I have argued, but also on what the proposition is *about*, then it is not clear why knowing $P \wedge \neg S$, from the example below in the text, should not require eliminating more alternatives than knowing P—even if (it is a priori that) P metaphysically entails $P \wedge \neg S$. Given a typical skeptical hypothesis S, an ordinary proposition P does not, in the terminology of Barwise (1981, 395), *strongly imply* $P \wedge \neg S$, i.e. it is not the case that every *situation* that supports P supports $P \wedge \neg S$, since supporting $P \wedge \neg S$ requires supporting $\neg S$, which brings in extra subject matter (nor, of course, is $P \wedge \neg S$ a truth-functional consequence of P). (By contrast, $P \wedge Q$ strongly implies P, and P strongly implies $P \vee Q$.) The move to a framework that includes partial situations is fully compatible with the multipath picture of knowledge, though I cannot go into details here (see Holliday 2014b). The general idea that the range of alternatives that one must eliminate in order to know some Q depends on what Q is *about*, contrary to Hawthorne's equivalence principle, is due to Yablo (2011; 2012; 2014). However, his specific view of the connection between closure and subject matter disagrees with some of the views about closure in this paper (recall footnote 40).

M-equiv says that if P and Q are equivalent as a matter of (deep) metaphysical necessity, then any path to knowing P covers a path to knowing Q; and **concover**, which follows from **TF-cover**, says that any path to knowing a conjunction covers paths to knowing each conjunct. It follows from $\boldsymbol{S} \subseteq \mathrm{W}_w - \boldsymbol{P}$ in (18) that $\boldsymbol{P}_w = (\boldsymbol{P} \wedge \neg \boldsymbol{S})_w$, which with **M-equiv** and **concover** implies (22).[59] The rest of the skeptical argument goes exactly as before.

Finally, the argument can be given with closure under multi-premise TF-consequence:

> **Multi**—if Q is a TF-consequence of $\{P_1, \ldots, P_n\}$, then
> $\forall \mathrm{A}_1 \in \mathbf{r}_c(P_1, w) \ldots \forall \mathrm{A}_n \in \mathbf{r}_c(P_n, w)\ \exists \mathrm{B} \in \mathbf{r}_c(Q, w)$: $\mathrm{B} \subseteq \bigcup_{1 \leq i \leq n} \mathrm{A}_i$,

so any paths to knowing $P_1, \ldots, P_n$ together cover a path to knowing Q. It follows from $\boldsymbol{S} \subseteq \mathrm{W}_w - \boldsymbol{P}$ in (18) that $(\neg \boldsymbol{P} \vee \neg \boldsymbol{S})_w = \mathrm{W}_w$, so the disjunction $\neg P \vee \neg S$ is a (deeply) necessary truth. Some would conclude that knowing $(\neg P \vee \neg S)$ does not require empirically eliminating scenarios, i.e. $\emptyset \in \mathbf{r}_c(\neg P \vee \neg S)$, but let us only make the weaker assumption that there is a path to knowing $(\neg P \vee \neg S)$ that does not require eliminating alternatives for $\neg S$,[60] i.e. some $\mathrm{A}_1 \in \mathbf{r}_c(\neg P \vee \neg S)$ that does not overlap with any $\mathrm{B} \in \mathbf{r}_c(\neg S, w)$. Since $\neg S$ is a TF-consequence of $(\neg P \vee \neg S)$ together with P, **Multi** with $P_1 = (\neg P \vee \neg S)$ and $P_2 = P$ implies that for every $\mathrm{A}_2 \in \mathbf{r}_c(P, w)$, there is some $\mathrm{B} \in \mathbf{r}_c(\neg S, w)$ such that $\mathrm{B} \subseteq \mathrm{A}_1 \cup \mathrm{A}_2$ and hence $\mathrm{B} \subseteq \mathrm{A}_2$ by the choice of A_1; and this implies (22), which leads to skepticism as before.

What are our options for avoiding this kind of argument for radical skepticism?

I have already mentioned the Dretskean option of denying closure under single-premise metaphysical entailment. The same considerations about lightweight propositions entailing heavyweight propositions suggest that Dretske would reject closure under metaphysical *equivalence* as well; for if $\neg S$ is a heavyweight proposition compared to the lightweight P, then surely $P \wedge \neg S$ is heavyweight as well. The construction in Theorem 2 is compatible with this view: without further assumptions about r or about how $\mathbf{r}^{\mathsf{r}}$ arises from r, there may be an alternative set in $\mathbf{r}^{\mathsf{r}}_c(P, w)$ that does not cover any in $\mathbf{r}^{\mathsf{r}}_c(P \wedge \neg S, w)$,[61] even if P and $P \wedge \neg S$ are metaphysically equivalent. If we had assumed that propositions are sets of metaphysically possible scenarios or worlds, then **M-equiv** would basically be unavoidable, but for generality I have not assumed such a view (recall §1.1).

[59] This argument (like that of Hawthorne 2004, 41) reflects the fact, which should be obvious to students of modal logic, that together the principles $(\mathrm{K}P \,\&\, \Box(P \leftrightarrow Q)) \Rightarrow \mathrm{K}Q$ and $\mathrm{K}(P \wedge Q) \Rightarrow (\mathrm{K}P \,\&\, \mathrm{K}Q)$ (or $\mathrm{K}P \Rightarrow \mathrm{K}(P \vee Q)$) suffice to derive $(\mathrm{K}P \,\&\, \Box(P \rightarrow Q)) \Rightarrow \mathrm{K}Q$, where $\Box$ is a normal modal operator.

[60] Epistemologists typically assume that knowing a conditional $P \rightarrow \neg S$, where P is an ordinary proposition and S is a metaphysically incompatible skeptical hypothesis, does not require eliminating skeptical S-scenarios.

[61] One can easily verify this by comparing the the CCNFs of p and $p \wedge \neg s$ for TF-atomic p and s.

As for multi-premise closure, without further assumptions about r or about how $\mathbf{r}^{\mathrm{r}}$ arises from r, Theorem 2 does not guarantee that $\mathbf{r}^{\mathrm{r}}$ satisfies **Multi** or, as a special case, $(\mathrm{K}P \,\&\, \mathrm{K}P') \Rightarrow \mathrm{K}(P \wedge P')$. The reason is that someone who knows P according to $\mathbf{r}^{\mathrm{r}}$, so has done enough work to know every $C \in \mathfrak{c}(P)$,[62] and knows P' according to $\mathbf{r}^{\mathrm{r}}$, so has done enough work to know every $C' \in \mathfrak{c}(P')$, has not necessarily done enough work to know $P \wedge P'$ according to $\mathbf{r}^{\mathrm{r}}$, because there may be some $D \in \mathfrak{c}(P \wedge P')$ that is not a superclause of anything in $\mathfrak{c}(P)$ or $\mathfrak{c}(P')$; in Dretskean terms, D may be a new "heavyweight" TF-consequence of $P \wedge P'$, which neither P nor P' had individually. Theorem 2 does guarantee that if an agent knows two TF-atomic (or TF-basic) propositions p and p', then she has done enough empirically to know $p \wedge p'$; every $D \in \mathfrak{c}(p \wedge p')$ is a superclause of something in $\mathfrak{c}(p)$ or $\mathfrak{c}(p')$, so the problem of new heavyweight consequences does not arise.[63] But if P and P' are complex, then the aggregation principle is not guaranteed without further assumptions, given the possibility of new heavyweight consequences coming from the combination of P and P'. On this view, it is not necessarily harmless to combine P and P' with $\wedge$; the impression that nothing more is required to know $P \wedge P'$ ("just put the $\wedge$ in between!") may be an illusion induced by too much focus on syntax. The same points apply to closure under known implication:[64] p and $p \to q$ together have what may be a heavyweight TF-consequence, q, that neither has individually.[65]

Much more could be said about views that limit the scope of closure. But let us change gears: is it possible to reject the skeptical argument while defending the strong closure principles? The only real way to do so is to maintain that for *every* S as in (18) for a known P, $\neg S$ can be known without a requirement of eliminating S-scenarios. In the standard alternatives picture, this would force defenders of strong closure to say that every such contingent $\neg S$ can be known without *any* requirement of eliminating scenarios, i.e. $\mathsf{r}_C(\neg S, w) = \emptyset$, which is the problem of vacuous knowledge from §2.4. For if any scenarios had to be eliminated, they would be S-scenarios according to the contrast/enough condition that I have argued is built in to the standard alternatives picture (§2.5).

62 Here I mean $\mathfrak{c}(\mathrm{CCNF}(P))$, but I will write "$\mathfrak{c}(P)$" for convenience.

63 This assumes that p and p' do not have further structure, ignored by the truth-functional analysis given here, such that $p \wedge p'$ has new heavyweight consequences.

64 Of course, there is a close connection between the multi-premise principles of closure under known implication, $(\mathrm{K}P \,\&\, \mathrm{K}(P \to Q)) \Rightarrow \mathrm{K}Q$, and $(\mathrm{K}P \,\&\, \mathrm{K}P') \Rightarrow \mathrm{K}(P \wedge P')$. First, the former essentially guarantees the latter: by closure under known implication, an agent who knows P and the tautology $P \to (P' \to (P \wedge P'))$ has done enough empirical work to know $P' \to (P \wedge P')$, so if the agent also knows P', then by closure under known implication again she has done enough empirical work to know $P \wedge P'$. Second, the latter guarantees the former assuming **TF-cover**: if $(\mathrm{K}P \,\&\, \mathrm{K}P') \Rightarrow \mathrm{K}(P \wedge P')$ holds, then an agent who knows P and knows $P \to Q$ must have done enough empirical work to know $P \wedge (P \to Q)$, which by **TF-cover** requires that she has done enough empirical work to know Q. Thus, assuming single-premise logical closure, the two multi-premise principles stand or fall together.

65 Can a *TF-consequence* Q of P have a "heavyweight" status compared to the "lightweight" status of P, requiring more epistemic work to know? I don't think so. See the Appendix for a related discussion.

However, in the multipath picture, defenders of strong closure can say that knowing $\neg S$ *does* require eliminating scenarios: a hard path to fulfilling this requirement is to eliminate some nonempty set of skeptical S-scenarios, in line with **enough** and **noVK** from §§3.3–3.4; but another path is to go via an ordinary proposition P that entails $\neg S$, eliminating all of the scenarios in some set $\mathrm{A} \in \mathbf{r}_{\mathcal{C}}(P, w)$ of $(\neg P \wedge \neg S)$-scenarios, rejecting **semi-contrast** for $\neg S$, but consistently with **overlap** from §3.3. That's why not just *anyone* gets to know the contingent $\neg S$, but someone who did the epistemic work to know an ordinary P that entails $\neg S$ can.[66] This is certainly an improvement over the vacuous knowledge story. What it shows, I think, is that the issue of how far closure holds ultimately comes down to the question of how far **contrast/semi-contrast** fails. In particular, since there is no guarantee that S will be complex, defenders of strong closure must reject **basic-contrast, basic-semi-contrast**, and **basic-singlepath**.

We have seen that defenders of strong closure benefit from the multipath picture. Can they also view a multipath function $\mathbf{r}^{\mathsf{r}}$ as arising from a singlepath function r? The simplest way to do so is to replace

$$\mathbf{r}^{\mathsf{r}}_{\mathcal{C}}(C, w) = \{\mathsf{r}_{\mathcal{C}}(C', w) \mid C' \text{ is a } \textit{subclause} \text{ of } C\} \quad (14)$$

from Theorem 2 with

$$\mathbf{r}^{\mathsf{r}}_{\mathcal{C}}(P, w) = \{\mathsf{r}_{\mathcal{C}}(P', w) \mid \boldsymbol{P}'_w \text{ is a } \textit{subset} \text{ of } \boldsymbol{P}\}.^{67} \quad (23)$$

Then clearly $\mathbf{r}^{\mathsf{r}}$ satisfies **M-cover**; if r satisfies r-RofA, then $\mathbf{r}^{\mathsf{r}}$ satisfies **r-RofA**; if r satisfies contrast, then $\mathbf{r}^{\mathsf{r}}$ satisfies **enough** and **overlap**; if r satisfies noVK, then $\mathbf{r}^{\mathsf{r}}$ satisfies **noVK**; and if r satisfies alpha (recall §2.3), then $\mathbf{r}^{\mathsf{r}}$ satisfies the analogous multipath principle,

$$\textbf{alpha}\text{—}\forall \mathrm{A}_1 \in \mathbf{r}_{\mathcal{C}}(P, w)\ \forall \mathrm{A}_2 \in \mathbf{r}_{\mathcal{C}}(Q, w)\ \exists \mathrm{B} \in \mathbf{r}_{\mathcal{C}}(P \wedge Q, w)\text{: } \mathrm{B} \subseteq \mathrm{A}_1 \cup \mathrm{A}_2$$

for $(\mathrm{K}P \,\&\, \mathrm{K}P') \Rightarrow \mathrm{K}(P \wedge P')$. It follows that $\mathbf{r}^{\mathsf{r}}$ guarantees full single- and multi-premise closure.

I will not try to decide here between the two positions on closure outlined above. In essence, defenders of strong closure think that knowledge is easier to come by than do defenders of limited closure. Does the former camp make knowledge too cheap? Without answering this question, we can say

[66] This is compatible with the super-shifty contextualist view (recall §2.4) that whenever we mention or think about S, we shift the context from $\mathcal{C}$ to a $\mathcal{C}'$ relative to which the agent does not count as knowing $\neg S$. The benefit to super-shifty contextualists of adopting the multipath picture is that they are no longer forced to say that relative to $\mathcal{C}$, the agent counted as knowing $\neg S$ no matter what epistemic work she had done; instead, the reason she could count as knowing $\neg S$ relative to $\mathcal{C}$ is that she did the epistemic work required to know the P that entails $\neg S$.

[67] This way of achieving strong closure bears some resemblance to the more sophisticated *recursive tracking theory* of Roush (2005; 2012). So does the recursive definition in Theorem 2, although in Theorem 2 the alternative sets for a proposition P are determined by the structure of the proposition itself and the alternative sets for its parts, rather than the alternative sets for all propositions that (are known to) entail P.

this much: at least in the multipath picture, no fallibilist need be committed to the cheapest knowledge of all—the vacuous knowledge of the standard alternatives picture in §2.4.

5. CONCLUSION

There are multiple paths to knowing some propositions about the world. This sounds like a truism, but it has yet to be fully appreciated in the theory of knowledge. According to the *standard alternatives picture* assumed in so much fallibilist epistemology, knowing a proposition involves eliminating a single set of alternatives. Proposition 1 in §2.5 suggests that this picture is fundamentally flawed. In its place, I proposed a *multipath picture* of knowledge for fallibilists, according to which knowing a proposition involves eliminating all of the alternatives in one of the proposition's alternative sets, of which there may be many. Proposition 2 in §3.4 showed that this picture solves the problems raised by Proposition 1 for the standard alternatives picture. Moreover, the Multipath Theorem in §3.5 showed how the multiple alternative sets for a proposition may emerge out of the standard alternatives picture in a way that depends on the structure of the proposition. Unlike the standard alternatives picture, the multipath picture allows fallibilists to maintain uncontroversial (single-premise, logical) epistemic closure principles without having to make extreme assumptions about the ability of humans to know empirical truths without empirical investigation. It also offers benefits to those who endorse more controversial (multi-premise and metaphysical) closure principles, thereby taking a more liberal attitude about paths to knowledge. Hard questions remain about how far fallibilists should claim that closure goes. But nobody ever said being a fallibilist was easy.

APPENDIX: NEGATION, CONTRAST, AND CLOSURE

The rejection of the single alternative set assumption in §3.1 and the contrast assumption in §3.3 can help us make sense of an otherwise puzzling feature of Nozick and Dretske's views on closure, concerning the following closure principles:

$$\mathrm{K}P \Rightarrow \mathrm{K}\neg(\neg P \wedge S); \tag{24}$$

$$\mathrm{K}\neg P \Rightarrow \mathrm{K}\neg(P \wedge S). \tag{25}$$

Beginning with Nozick (1981, 228f.), he explicitly rejects (24): "it is possible for me to know *p* yet not know the denial of a conjunction, one of whose conjuncts is not-*p*. I can know *p* yet not know ... not-(not-p & SK) .... However, we have seen no reason to think knowledge does not extend across known logical equivalence." Only a page later Nozick (1981, 230) writes: "It seems that a person can track 'Pa' without tracking 'there is an x such that Px'. But this apparent nonclosure result surely carries things too far. As would the apparent result of nonclosure under the propositional calculus rule of inferring 'p or q' from 'p'..."[68] Let us write the latter principle as $\mathrm{K}P \Rightarrow \mathrm{K}(P \vee Q)$. What is interesting is that Nozick's views in these passages are inconsistent.[69] Surely Nozick knows that $P \vee \neg S$ is logically equivalent to $\neg(\neg P \wedge S)$, so given his endorsement of closure under known logical equivalence, if he knew $P \vee \neg S$ then he would know $\neg(\neg P \wedge S)$. But he says he does not know $\neg(\neg P \wedge S)$, so he must not know $P \vee \neg S$. But he also says he knows P and accepts $\mathrm{K}P \Rightarrow \mathrm{K}(P \vee Q)$, so he should know $P \vee \neg S$. (We can assume Nozick makes the relevant inferences.)

I do not think this inconsistency was simply a mistake. Instead, I suspect that it reflects an intuition that Nozick shares with others, including Dretske. While Nozick explicitly endorses $\mathrm{K}P \Rightarrow \mathrm{K}(P \vee Q)$ and explicitly rejects (24), Dretske explicitly endorses $\mathrm{K}P \Rightarrow \mathrm{K}(P \vee Q)$ and is committed to rejecting (25). First, Dretske (1970) says that "it seems to me fairly obvious that if someone... knows that P is the case, he knows that P or Q is the case" (1009). Second, Dretske (1970, 1015–16) claims in his famous zebra that the zoo visitor does not know that the animal in the zebra cage is not a mule (M) disguised to look like a zebra (D): $\neg\mathrm{K}\neg(M \wedge D)$. But I assume that as a strong fallibilist, Dretske will allow that in ordinary cases of observing zebras at the zoo, people who know the difference between zebras and mules know that the zebras are not mules: $\mathrm{K}\neg M$. But together these commitments force Dretske to deny (25). Then since Dretske endorses $\mathrm{K}P \Rightarrow \mathrm{K}(P \vee Q)$, an instance of which is $\mathrm{K}\neg M \Rightarrow \mathrm{K}(\neg M \vee \neg D)$, Dretske must deny $\mathrm{K}(\neg M \vee \neg D) \Rightarrow \mathrm{K}\neg(M \wedge D)$.

Thus, Dretske must deny the "De Morgan" closure principle $\mathrm{K}(\pm P \vee \pm Q) \Rightarrow \mathrm{K}\neg(\mp P \wedge \mp Q)$,[70] and there is pressure for Nozick to do the same to resolve the inconsistency in his views.

In my view, (24) and (25) seem problematic because their consequents claim knowledge that something is *not* the case, and this *negation* brings with it the idea of **contrast** that I have argued fallibilists should not accept in general. In particular, I argued that **contrast** can fail for disjunctions like $P \vee \neg S$; for I agree with Dretske, Nozick, and Kripke that one path to knowing $P \vee \neg S$ is via knowing P, and I agree with fallibilists in general that coming to know P may not require eliminating $(\neg P \wedge S)$-scenarios. But can one come to know $\neg(\neg P \wedge S)$ without eliminating $(\neg P \wedge S)$-scenarios?

With the negated conjunction, I expect some people's intuitions to shift in favor of **contrast**, perhaps because the processing of negations in non-epistemic contexts in natural language involves the construction of contrast classes (see Oaksford and Stenning 1992). In Dretske's example, when considering a disjunction like $\neg M \vee \neg D$, one may recognize that knowing $\neg M$ provides a path to knowing the disjunction; but when considering the equivalent $\neg(M \wedge D)$, one might have a competing intuition in favor of **contrast** and the thought

[68] The second quoted sentence is from the footnote to the first sentence.

[69] Kripke (2011, 199) also discusses the inconsistency, pointed out to him by Assaf Sharon and Levi Spectre.

[70] Notation: $\pm P$ is either P or $\neg P$; if $\pm P$ is P, then $\mp P$ is $\neg P$; if $\pm P$ is $\neg P$, then $\mp P$ is P.

that $(M \wedge D)$-scenarios must be eliminated, which fallibilists would not insist on for knowing $\neg M$. (One might also have the mistaken intuition that $\neg D$ follows from $\neg(M \wedge D)$, so D-scenarios must be eliminated.)[71] There are three ways the explanation might go from here, depending on the kind of significance assigned to these intuitions:

1 *Epistemic*: $\mathrm{K}(\pm P \vee \pm Q) \Rightarrow \mathrm{K}\neg(\mp P \wedge \mp Q)$ is not a valid principle even for a fixed context, because **contrast** may apply to $\mathbf{r}_c(\neg(\mp P \wedge \mp Q), w)$ without applying to $\mathbf{r}_c(\pm P \vee \pm Q, w)$.
2 *Pragmatic*: $\mathrm{K}(\pm P \vee \pm Q) \Rightarrow \mathrm{K}\neg(\mp P \wedge \mp Q)$ is valid, but when an attributor claims that an agent knows a negated proposition N, this has a tendency to pragmatically trigger the (mistaken) intuition that **contrast** must hold for $\mathbf{r}_c(N, w)$.
3 *Contextual*: $\mathrm{K}(\pm P \vee \pm Q) \Rightarrow \mathrm{K}\neg(\mp P \wedge \mp Q)$ is valid for a fixed context, but when an attributor claims that an agent knows a negated proposition N, this has a tendency to change the context to one in which **contrast** holds for $\mathbf{r}_{c'}(N, w)$ (cf. DeRose 1995).

In my view, the pragmatic or contextual explanations are more plausible than the epistemic, although I will not argue for this here.[72] The point I wish to make is that the multipath picture has the potential to explain divergent intuitions concerning knowledge of disjunctions and knowledge of equivalent negated conjunctions in terms of the keys ideas of §3.1 and §3.3.

REFERENCES

Alspector-Kelly, Marc. 2011. Why Safety Doesn't Save Closure. *Synthese*, 183(2): 127–42.

Arrow, Kenneth J. 1959. Rational Choice Functions and Orderings. *Economica*, 26 (102):121–7.

Barwise, Jon. 1981. Scenes and Other Situations. *Journal of Philosophy*, 78(7):369–97.

Barwise, Jon and John Perry. 1983. *Situations and Attitudes*. MIT Press.

Blome-Tillmann, Michael. 2009. Knowledge and Presuppositions. *Mind*, 118(470): 241–94.

Bordes, Georges. 1976. Consistency, Rationality and Collective Choice. *The Review of Economic Studies*, 43(3):451–7.

[71] Wright (2014) also warns his reader not to confuse the likes of $\neg D$ and $\neg(M \wedge D)$: "we don't have a visual warrant for thinking that *those animals have not been cleverly disguised in a visually undetectable way*, but we do, in the relevant circumstance, have a visual warrant for thinking that *those animals are not* ***mules*** *that have been so disguised*. Maybe we are confused by the operation of some kind of implicature here: maybe saying, or thinking, 'It is not the case that those animals are cleverly disguised mules' somehow implicates, in any context of a certain (normal) kind, that 'Those animals have not been cleverly disguised'. But anyway, it doesn't entail it: not-(P&Q), dear reader, does not entail not-Q!" (234–5).

[72] See Roush 2010 for arguments to the effect that if an agent knows $\pm P$, then she can know $\neg(\mp P \wedge S)$ for any S you like, and that intuitions to the contrary can be explained away.

Cohen, Stewart. 1988. How to Be a Fallibilist. *Philosophical Perspectives*, 2:91–123.

Cohen, Stewart. 1998. Contextualist Solutions to Epistemological Problems: Scepticism, Gettier, and the Lotttery. *Australasian Journal of Philosophy*, 76(2):289–306.

Cohen, Stewart. 2000. Contextualism and Skepticism. *Philosophical Issues*, 10:94–107.

Comesaña, Juan. 2007. Knowledge and Subjunctive Conditionals. *Philosophy Compass*, 2(6):781–91.

Cori, René and Daniel Lascar. 2000. *Mathematical Logic: A Course with Exercises Part I*. Oxford University Press.

Davies, Martin and Lloyd Humberstone. 1980. Two Notions of Necessity. *Philosophical Studies*, 38(1):1–30.

DeRose, Keith. 1995. Solving the Skeptical Problem. *The Philosophical Review*, 104 (1):1–52.

DeRose, Keith. 2000. How Can We Know that We're Not Brains in Vats? *The Southern Journal of Philosophy*, 38(S1):121–48.

DeRose, Keith. 2009. *The Case for Contextualism*. Oxford University Press.

Dretske, Fred. 1970. Epistemic Operators. *The Journal of Philosophy*, 67(24):1007–23.

Dretske, Fred. 1971. Reasons, Knowledge, and Probability. *Philosophy of Science*, 38 (2):216–20.

Dretske, Fred. 1981. The Pragmatic Dimension of Knowledge. *Philosophical Studies*, 40(3):363–78.

Dretske, Fred. 2005. The Case against Closure, in Matthias Steup and Ernest Sosa, editors, *Contemporary Debates in Epistemology*, pages 13–25. Wiley-Blackwell.

Dretske, Fred. 2006. Information and Closure. *Erkenntnis*, 64:409–13.

Evans, Gareth. 1979. Reference and Contingency. *The Monist*, 62(2):161–89.

Goldman, Alvin I. 1976. Discrimination and Perceptual Knowledge. *The Journal of Philosophy*, 73(20):771–91.

Hawthorne, John. 2004. *Knowledge and Lotteries*. Oxford University Press.

Heller, Mark. 1989. Relevant Alternatives. *Philosophical Studies*, 55(1):23–40.

Heller, Mark. 1999. Relevant Alternatives and Closure. *Australasian Journal of Philosophy*, 77(2):196–208.

Holliday, Wesley H. 2012. *Knowing What Follows: Epistemic Closure and Epistemic Logic*. PhD thesis, Stanford University. Revised Version, ILLC Dissertation DS-2012-09.

Holliday, Wesley H. 2014a. Fallibilism and Multiple Paths to Knowledge (Extended Version). URL <http://philosophy.berkeley.edu/holliday>.

Holliday, Wesley H. 2014b. Epistemic Closure and Epistemic Logic II: A New Framework for Fallibilism. Manuscript.

Holliday, Wesley H. Forthcoming. Epistemic Closure and Epistemic Logic I: Relevant Alternatives and Subjunctivism. *Journal of Philosophical Logic*. Published online <http://link.springer.com/article/10.1007/s10992-013-9306-2?sa_campaign=email/event/articleAuthor/onlineFirst>, DOI 10.1007/s10992-013-9306-2.

Ichikawa, Jonathan. 2011. Quantifiers and Epistemic Contextualism. *Philosophical Studies*, 155(3):383–98.

King, Jeffrey C. 2007. What in the World Are the Ways Things Might Have Been? *Philosophical Studies*, 133(3):443–53.

Kripke, Saul. 1980. *Naming and Necessity*. Harvard University Press.

Kripke, Saul. 2011. Nozick on Knowledge. In Saul Kripke, editor, *Collected Papers*, volume 1, pages 162–224. Oxford University Press.

Lawlor, Krista. 2013. *Assurance: An Austinian View of Knowledge and Knowledge Claims*. Oxford University Press.

Lewis, David. 1973. *Counterfactuals*. Basil Blackwell.

Lewis, David. 1979. Attitudes *De Dicto* and *De Se*. *The Philosophical Review*, 88(4): 513–43.

Lewis, David. 1996. Elusive Knowledge. *Australasian Journal of Philosophy*, 74(4): 549–67.

Luper-Foy, Steven. 1984. The Epistemic Predicament: Knowledge, Nozickian Tracking, and Scepticism. *Australasian Journal of Philosophy*, 62(1):26–49.

MacFarlane, John. 2005. The Assessment-Sensitivity of Knowledge Attributions. *Oxford Studies in Epistemology*, 1:197–233.

MacFarlane, John. 2014. *Assessment Sensitivity: Relative Truth and its Applications*. Oxford University Press.

Murphy, Peter. 2005. Closure Failures for Safety. *Philosophia*, 33:331–4.

Murphy, Peter. 2006. A Strategy for Assessing Closure. *Erkenntnis*, 65(3):365–83.

Nozick, Robert. 1981. *Philosophical Explanations*. Harvard University Press.

Oaksford, Mike and Keith Stenning. 1992. Reasoning with Conditionals Containing Negated Constituents. *Journal of Experimental Psychology*, 18(4):835–54.

Perry, John. 1986. From Worlds to Situations. *Journal of Philosophical Logic*, 15(1): 83–107.

Perry, John. 1989. Possible Worlds and Subject Matter. In Sture Allén, editor, *Possible Worlds in Humanities, Arts and Sciences: Proceedings of Nobel Symposium 65*, pages 124–37. Walter de Gruyter.

Roush, Sherrilyn. 2005. *Tracking Truth: Knowledge, Evidence, and Science*. Oxford University Press.

Roush, Sherrilyn. 2010. Closure on Skepticism. *The Journal of Philosophy*, 107(5): 243–56.

Roush, Sherrilyn. 2012. Sensitivity and Closure. In Kelly Becker and Tim Black, editors, *The Sensitivity Principle in Epistemology*, pages 242–68. Cambridge University Press.

Rysiew, Patrick. 2001. The Context-Sensitivity of Knowledge Attributions. *Noûs*, 35(4):477–514.

Schiffer, Stephen. 2004. Skepticism and the Vagaries of Justified Belief. *Philosophical Studies*, 119(1/2):161–84.

Sen, Amartya K. 1971. Choice Functions and Revealed Preference. *The Review of Economic Studies*, 38(3): 307–17.

Sherman, Brett and Gilbert Harman. 2011. Knowledge and Assumptions. *Philosophical Studies*, 156(1):131–40.

Sosa, Ernest. 1999. How to Defeat Opposition to Moore. *Noûs*, 33(13):141–53.

Stalnaker, Robert. 1986. Possible Worlds and Situations. *Journal of Philosophical Logic*, 15(1):109–23.

Stanley, Jason. 2005. Fallibilism and Concessive Knowledge Attributions. *Analysis*, 65(2):126–31.

Stine, G. C. 1976. Skepticism, Relevant Alternatives, and Deductive Closure. *Philosophical Studies*, 29(4):249–61.

Vogel, Jonathan. 1987. Tracking, Closure, and Inductive Knowledge. In S. Luper-Foy, editor, *The Possibility of Knowledge: Nozick and His Critics*, pages 197–215. Rowman and Littlefield.

Vogel, Jonathan. 1999. The New Relevant Alternatives Theory. *Noûs*, 33(s13): 155-80.

Vogel, Jonathan. 2007. Subjunctivitis. *Philosophical Studies*, 134(1):73–88.

White, Roger. 2006. Problems for Dogmatism. *Philosophical Studies*, 131(3):525–57.

Williamson, Timothy. 2000. *Knowledge and its Limits*. Oxford University Press.

Wright, Crispin. 2004. Warrant for Nothing (and Foundations for Free)? *Aristotelian Society Supplementary Volume*, 78(1):167–212.

Wright, Crispin. 2014. On Epistemic Entitlement (II): Welfare State Epistemology. In Dylan Dodd and Elia Zardini, editors, *Scepticism and Perceptual Justification*, pages 213–47. Oxford University Press.

Yablo, Stephen. 2011. Achieving Closure. Kant Lecture delivered at Stanford University on May 20.

Yablo, Stephen. 2012. Aboutness Theory. Online paper accessed August 26, 2013. URL <http://www.mit.edu/~yablo/home/Papers.html>.

Yablo, Stephen. 2014. *Aboutness*. Princeton University Press.

Yalcin, Seth. 2011. Nonfactualism about Epistemic Modality. In Andy Egan and Brian Weatherson, editors, *Epistemic Modality*, pages 295–332. Oxford University Press.

5. New Rational Reflection and Internalism about Rationality

Maria Lasonen-Aarnio

1. DOING WHAT ONE TAKES TO BE RATIONAL

Does rationality require that there be a match between one's opinions about what it would be rational to do in one's situation, and what one does in that situation? For instance, could it be rational to simultaneously judge that rationality requires one to __ (*to believe a proposition p; to have a certain credence distribution P; to desire, intend, or be motivated to ϕ; to ϕ*, etc.), while failing to__ ? Or, could it be rational to judge that rationality forbids one to __ , while nevertheless going on to __ ? Numerous views, appearing in different subfields of philosophy, affirm the existence of interesting connections between opinions, or at least rational opinions, about what is rational and what is in fact rational.

For instance, many ethicists have argued that someone who believes that a rational agent in her situation would be motivated to ϕ, while failing to be motivated to ϕ, fails to be rational.[1] To some it has seemed almost uncontroversial that something similar holds for belief: believing that her fully rational self would believe *p*—or simply, that it is rational to believe *p*—while failing to believe *p*, is a paradigmatic case of irrationality.[2] Similar-sounding principles have been defended for other doxastic states. Consider, for instance, the principle that if a possible rational agent is certain exactly what degrees of belief she ought, rationally, to have, then she has those degrees of belief.[3] And cases where one judges that it would be *irrational* to __ , while nevertheless going on to __ , might seem like even more egregious violations of rationality. Many have argued that it is irrational to believe (or have high confidence in)

[1] For instance, an assumption along these lines plays a pivotal role in Smith's defence of internalism in ethics, and, relatedly, his proposed solution of the so-called moral problem (see, for instance, Smith (1994: 65)). Arpaly (2000) notes that it is 'almost a universal assumption in contemporary philosophy' that it is never rational to act against one's best judgement about what it would be best for one to do in a given situation. If best judgements are judgements about what it would be rational to do, this assumption also manifests the kind of trend under issue.

[2] See, for instance, Smith (1994: 178). Titelbaum (Forthcoming) argues that a subject who makes *any* mistakes about the requirements of rationality—whether about how they apply to one's own situation or to someone else's situation—is irrational.

[3] Elga (2013) finds this principle, which he calls *Certain*, almost self-evident, and I suspect that a lot of epistemologists would agree.

p, while believing (or having high confidence) that it is not rational to believe *p*, or that one's evidence does not support *p*.[4]

The aforementioned views, or variants thereof, are often assumed as uncontroversial premises for further arguments. Defences often appeal to the intuition that there is something incoherent about a subject who fails to match her own states with those she takes to be rational. Such an agent fails to be *rational by her own lights*, and failing to be rational by one's own lights, it is urged, is a paradigm failure of rationality.[5] Some even argue that such incoherence involves blatantly inconsistent states.[6] Examples abound of subjects who fail to follow their own advice about what is rational, and who seem clearly irrational: subjects who believe that the evidence does not support astrology while nevertheless believing in astrology;[7] subjects who stick to their evaluations of a body of evidence even after coming to believe that those evaluations are flawed due to fatigue or some other cognitive impediment;[8] subjects who believe that flying is safe while believing that their plane will crash,[9] etc.

In effect, many of the views mentioned above are committed to something stronger than merely denying the rationality of a mismatch between the states one takes to be rational and the states one is in. For they are committed to the view that certain beliefs—or at least certain rationally permissible or rationally required beliefs—about what states it is (or is not) rational for one to be in are factive. Here is an instance of the kind of view I have in mind: if I rationally believe that it is (or is *not*) rational for me to believe *p*, then it is (or is *not*) rational for me to believe *p*. It is impossible to hold false but rational beliefs about what it is rational for one to believe. But let me first take a step back. Consider the following wide-scope schema:

> Rationality requires that, if I believe that rationality requires that I __ , then I __ .

Assume that I believe that rationality requires that I __ . The wide-scope schema requires that I avoid the mismatch, but the mismatch can be avoided by not believing that rationality requires that I __ . Rationality does not require that I __ . But even if belief about what rationality requires doesn't come out as factive, it is difficult to prevent a wide-scope schema from generating narrow-scope ones. As an example, consider a straightforward argument from the above schema to the factivity of *rationally required* belief about what is rational.

Assume that rationality requires that I believe that rationality requires that I __ . The above wide-scope schema, together with the K axiom, entails that

[4] See, for instance, Chisholm (1989: 6), Scanlon (1998: 25), Elga (2005), Hazlett (2012). See Horowitz (Forthcoming) and Greco (2014) for discussions of *epistemic akrasia*.

[5] See, for instance, Smith (1994: 65; 178) and Elga (2005).

[6] See Greco (2014).

[7] Elga (2005).

[8] Horowitz (Forthcoming).

[9] Greco (2014).

rationality requires that I __ .[10] The following reasoning strikes me as very convincing: According to the wide-scope requirement, rationality requires of me that *either* I __ , or that I fail to believe that rationality requires that I __ . But if rationality requires that I believe that rationality requires that I __ , then the only way in which I can respect the wide-scope requirement is if I __ . At least rationally required beliefs about what rationality requires come out as factive. Now, Broome (2013) rejects the K axiom, and he might even reject its present application on the following grounds. Assume that I fail to believe what rationality requires, failing to believe that rationality requires that I __ . Then, rationality does not require that I __ .[11] I disagree, but the issue is neither here nor there, for the question was whether rationally required *belief* about what is rational is factive. If it is rationally required that I believe that rationality requires that I __ , and I *do* in fact believe this, then the above kinds of counterexamples are irrelevant. It is difficult to see how a proponent of the wide-scope schema could block the conclusion that at least a certain kind of belief about what is rational is factive.[12]

It is not surprising, then, that numerous authors explicitly express their views as narrow-scope requirements of rationality, or as the view that certain beliefs about what it is rational for one to believe are factive.[13] Moreover, these authors don't take their views to follow from the general factivity of rationally permissible or rationally required belief or certainty. And therein lies the main challenge: if even rationally required belief, opinion, or certainty is not in general factive, why is it in the special cases under issue? The challenge can be sharpened by considering the fact that beliefs about whether it is rational to believe a proposition *p*, on the one hand, and beliefs about whether *p*, on the other, seem to be about distinct subject matters. The first is about an epistemic subject matter—namely, the rationality of certain beliefs. The second is about the matter of whether *p*—perhaps, for instance, the matter of whether it will rain today. In so far as what opinions it is rational to hold is a matter of

[10] Letting 'Op' stand for 'it ought to be the case that *p*', the K axiom is the following distribution axiom: $O(p \rightarrow q) \rightarrow (Op \rightarrow Oq)$. In the present context we can read 'Op' as 'rationality requires that *p*'.

[11] Broome (2013: 120) gives the following counterexample to the K axiom. Suppose that you enter a marathon. As a result, prudence requires you to exercise hard every day. Prudence also requires that, if you exercise hard every day, you eat heartily. But assume that you are lazy, failing to exercise hard, even though you could. Broome thinks that in this case it does not follow that prudence requires that you eat heartily.

[12] Now, someone might object that *all* requirements of rationality are wide-scope, and that rationality could never require such a thing. However, such a view is in tension with the idea that at least sometimes one is rationally required to hold beliefs that reflect one's evidence by, for instance, believing a proposition that is overwhelmingly likely on the evidence.

[13] For instance, Hazlett's (2012) *Undercutting principle* is a narrow-scope requirement. Greco (2014) takes opponents of epistemic akrasia to be committed to the claim that justified false beliefs about what one ought to believe are impossible . Smith (1994: 148) defends the following narrow-scope principle, 'If an agent believes that she has a normative reason to ϕ, then she rationally should desire to ϕ,' where, on Smith's view, believing that one has a normative reason to ϕ is just believing that one would desire to ϕ if one were fully rational.

what one's evidence supports, denying the rationality of opinions that are mismatched in the way under issue is to pose constraints on the kind of evidence that it is possible for rational subjects to have.[14]

I will refer to the loose trend sketched as *internalism about rationality*. *Internalism*, because all of the views mentioned affirm some connection between one's internal perspective concerning what is rational—that is, one's (rational) opinions about what is rational—and what is in fact rational. This is not to say that what is rational is fixed by one's internal perspective: even if, for instance, rationally believing that believing *p* is rational (irrational) entails that believing *p* is rational (irrational), it doesn't follow that whenever rationally (irrationally) believing *p*, one must hold some opinion about whether it is rational to believe *p*. The entailment need not go both ways. The idea is just that when one *does* hold (rational) opinions about what is rational, these opinions constrain what is in fact rational. Though the view bears similarities to various internalisms, the usage is by no means standard. For instance, within ethics, 'internalism' is often used to refer to a view on which there is a connection between one's judgements about what it is right to do (or what there is a normative reason to do) and what one is *motivated* to do. And in epistemology 'internalism' often refers to a view on which epistemic rationality (or justification, or ...) supervenes on one's internal states.

My main focus in this paper will not be on the kinds of internalist views about rationality just mentioned but rather, on how to extend the gist of these views to contexts involving uncertainty about what is rational. Consider, in particular, situations in which it is uncertain what credences or degrees of belief it is rational for one to have. I will assume that such uncertainty can be perfectly rational. Consider a rational subject, Sophie, who has a credence distribution over what credence distributions it would be rational for her to have, or which credence distributions best reflect her evidence. (Unless otherwise indicated, when speaking about which credence distribution would be rational, I will mean which distribution is *rationally required*.)[15] Perhaps, for instance, she is 1/3 confident in each of the following hypotheses about the rational credence that it will rain today: the rational credence is 0.3; the rational credence is 0.5; the rational credence is 0.7. What credence in rain would reflect or match Sophie's opinions about the rational credence in a way that respects the spirit of internalism about rationality?

What should Sophie's credence in rain be, *conditional* on 0.3 being the rational credence that it will rain today? Well, it seems that it should be 0.3!

[14] Greco (2014) spells out this objection forcefully. For a lengthier discussion of such a puzzle, see Lasonen-Aarnio (unpublished).

[15] For simplicity I will focus on cases in which a subject has a credence distribution over a partition of hypotheses about the rationally required credence distribution and, hence, cases in which the subject is certain that there is a uniquely rational distribution. Considering such cases will, I am hoping, teach us lessons that can be extended to cases in which a subject is uncertain whether such uniqueness obtains, and where at least some of the hypotheses she considers state that there is such-and-such a range of rationally permissible credence distributions.

And what should her credence be, conditional on 0.5 being the rational credence? 0.5. And so on. In this sense, Sophie should treat the rational credence function as an *expert* function. Then, her credence in rain should be the weighted average of the candidate rational credences, weighted by her confidence in their rationality. Her credence in rain ought to be $1/3 \times 0.3 + 1/3 \times 0.5 + 1/3 \times 0.7 = 0.5$. I will call the resulting principle, following David Christensen, *Rational Reflection*:

Rational Reflection
$P_t(h|p_{\text{RATIONAL},t}(h) = r) = r$[16]

Here P_t is the credence function of a rational subject. *Rational Reflection* entails that a rational subject's *expectations* about what the rational credences are match her credences in a very straightforward manner: her credence in a proposition p equals her expectation of the rational credence in p. A narrow-scope requirement could be derived from *Rational Reflection* in the way sketched above. Assume that Sophie has a *rationally required* credence distribution over hypotheses about the rational credence in p. Because this distribution is rationally required, giving it up is not an option. And because *Rational Reflection* itself states a rational requirement, it follows that Sophie is rationally required to assign to p the credence that equals her expectation of the rational credence in p. This cannot be taken to follow from rational expectations in general being factive, for they are not. Again, we have a special connection between (rational) opinions about what is rational and what is in fact rational. And again, a challenge arises: why, in the special case at hand, do we have an exception to the non-factivity of rational expectations?

Rational Reflection has recently come under a line of attack.[17] However, many still feel that there is something right about it. Rational opinions about what opinions are rational have got to constrain what is rational in some way that the principle tries, imperfectly, to capture. I will argue against this: *Rational Reflection* fails, and we shouldn't expect there to be any fix. In section 2 I briefly look at the problem with *Rational Reflection*, and a natural-seeming revision that has been proposed by Adam Elga, resulting in *New Rational Reflection* (NRR). In section 3 I discuss a line of argument for NRR employing as a premise a principle stating that if a rational agent is certain what credences it is rational for her to have, then she has those credences (*Certain*). I also introduce some formal apparatus to facilitate seeing what is problematic about the argument. I then argue against NRR more directly. In section 4 I introduce a simple class of formal frames in which the principle fails, as well as describing candidate counterexamples. In section 5 I say why the principle seems to compromise intuitive motivations prompting the search for rational

[16] In denoting probability functions, I will adopt the convention of using lower-case letters for non-rigid designators. So, '$p_{\text{RATIONAL},t}(h) = r$' should be read as 'the ideal, rational credence in h (at t) is r'.

[17] Williamson (2011) spells out a counterexample to the principle. See also Christensen (2010) and Elga (2013).

reflection principles. In section 6 I briefly discuss a puzzling consequence of reflection principles, namely, that if a subject who is uncertain what credences it is rational for her to have is to satisfy such principles, then it looks like she will sometimes be required to hold a credence while being certain that it is irrational. Section 7 draws together some conclusions.

2. THE PROBLEM WITH RATIONAL REFLECTION: KNOWING MORE THAN THE EXPERT

As initially plausible as it may seem, *Rational Reflection* faces serious problems. These have been forcefully spelled out by others, so I won't spend too much time on them.[18] What I say below will constitute an additional line of argument against the principle, for I sketch counterexamples with a different structure from those that have been recently discussed. But before seeing what is wrong with the principle, it may be useful to look at a slightly different formulation of the basic idea behind *Rational Reflection*.

Consider a subject who has a credence distribution over a partition of hypotheses about which (entire) credence function is rational for her. As above, I am using lower-case letters to stand in for non-rigid designators, and upper-case letters for rigid designators. So, for instance, 'P_t' and 'P^*' rigidly name credence functions. Claims like '$P^* = p_{\text{RATIONAL},t}$' should be read as '$P^*$ is the rational, ideal credence function at t.'[19] Then, let *Old Rational Reflection* be the following principle:

Old Rational Reflection (ORR)
$P_t\left(h \mid P^* = p_{\text{RATIONAL},t}\right) = P^*(h)$[20]

If a subject is rational, then her conditional credence in any proposition h, conditional on some function P^* being rational, ought to be whatever P^* assigns to h.

For a simple way to see what is wrong with ORR, consider the following entailment of the principle:

$$P_t\left(P^* = p_{\text{RATIONAL},t} \mid P^* = p_{\text{RATIONAL},t}\right) = P^*\left(P^* = p_{\text{RATIONAL},t}\right)$$

By the probability calculus, $P_t\left(P^* = p_{\text{RATIONAL},t} \mid P^* = p_{\text{RATIONAL},t}\right) = 1$. The upshot is that any probability function P^* that is, from the perspective of P_t, possibly the ideally rational function, must be certain that it is the rational function.[21] Hence, ORR rules out its ever being rational to think that

[18] See Williamson (2011), Christensen (2010), Elga (2013).

[19] Also, I will assume that a subject who assigns a credence to the propositions expressed is capable of thinking about the relevant credence function P^* in such a way that it is transparent to her exactly what credences P^* assigns to various propositions. So, for instance, she is not merely capable of thinking about the function through descriptions like 'the credence function my older sister has at the moment'.

[20] This is how Elga (2013) formulates *Rational Reflection*.

[21] Elga (2013) makes this observation.

it even *might* be rational to be uncertain of one's own rationality. Note also that as long as P_t assigns some non-zero credence to P_t itself being the rational credence function, it must be certain that it is the rational credence function. In other words, ORR requires any credence function that is not certain of its own irrationality to be certain of its own rationality! Such entailments are especially bad given that our search for rational reflection principles was propelled by the observation that it is not always rational to be certain about what the rational thing to do in one's situation is.

I think these observations provide ample reasons to reject ORR, and there is no need to reiterate the counterexamples that have been recently discussed. Those unconvinced by the observations made still need to explain away the seeming counterexamples.[22] But maybe there is a core motivation for rational reflection principles that can be retained, and that *Old Rational Reflection* simply fails to capture. Here is an idea. Assume that you learn that Sophie is an expert with a perfectly rational credence function P*, but you also know that you have information that Sophie lacks (Sophie's total evidence is a subset of your total evidence). Then, it seems clear that you ought not to simply adopt Sophie's credence function. Instead, perhaps you ought to adopt the function that results from conditionalizing Sophie's function on all the extra evidence you have. But note that merely learning that Sophie is an expert might give you evidence that Sophie lacks, for she might not know that she is an expert. Elga (2013) argues that this motivates the following thought: your credences (at *t*), conditional on Sophie's credence function being the expert function (at *t*), should be Sophie's credences, conditional on her credence function being the expert function (at *t*). Here, then, is what might seem like a natural way of revising ORR:

New Rational Reflection (NRR)
$$P_t\left(h|P^* = P_{\text{RATIONAL},t}\right) = P^*\left(h|P^* = P_{\text{RATIONAL},t}\right)$$

NRR looks promising: it doesn't require thinking that all credence functions that might be rational are certain of their rationality, and it doesn't fall prey to the recently discussed counterexamples to ORR. Note that NRR is analogous to the fix to another expert principle, Lewis's *Principal Principle*, suggested by Ned Hall and Michael Thau.[23] Moreover, the motivations given by Elga for *New Rational Reflection* are strikingly similar to those given by Hall for the *New Principal Principle*. (Ultimately, I suspect that the *New Principal Principle* is

[22] Williamson (2011), Christensen (2010), and Elga (2013) discuss a particular kind of counterexample to ORR involving 'clock beliefs'. What is essential to such counterexamples is that which credence it is rational to assign to propositions about the position of the minute hand of a clock seems to depend on what the minute hand in fact points to. Then, information about what credences it is rational for one to have gives one information about the position of the minute hand of the clock.

[23] Hall (1994), Thau (1994).

susceptible to problems that are structurally similar to the problems encountered by NRR, to be spelled out below, but this is not the place to pursue these worries.)

I will first discuss an argument from a principle mentioned above, *Certain*, to NRR. I then argue against NRR more directly, spelling out counterexamples. (Note that those who want to skip the next section can easily do so.)

3. AN ARGUMENT FOR NEW RATIONAL REFLECTION

The following principle might seem almost self-evidently correct:

> *Certain*
> Whenever a possible rational agent is certain exactly what degrees of belief she ought rationally to have, she has those degrees of belief.[24]

Now, one might think, as Elga does, that not only does *Certain* motivate rational reflection principles in the sense that there had better be a way of generalizing the principle to situations involving uncertainty about what is rational, but that at least in certain special cases *Certain* in fact entails instances of NRR. Assume that as a result of acquiring new evidence, I become certain what I ought to believe right now. If I ought to end up respecting *Certain* as a result of conditionalizing on this evidence, then a constraint is imposed on my initial state of mind. If the constraint is that my initial credences respect NRR, then NRR can be derived as a special case. This would seem to lend at least some support to NRR. An argument along these lines is not Elga's main way of motivating NRR. But it will be instructive to see why, even in these special cases, one cannot derive instances of NRR. For at first sight, the argument has a fair bit of intuitive appeal.

Here is the argument.[25] Assume that at a time t_0 Chandra is certain that the rational updating procedure is conditionalization, and that he is about to conditionalize on information about what credence function is rational for him at t_0. Chandra is in fact rational, and has probability function P_0 at t_0. These conditions are to specify the special case in which *Certain* yields instances of NRR. At a slightly later time t_1 he learns that credence function P^* was rational for him at t_0. Then, at t_1, he has probability function $P_1 = P_0(- \mid P^* = p_{RATIONAL,t0})$, which is rational for him.[26] But Elga also thinks that it follows from the conditions stated that Chandra is certain, at t_1, that the function $P^*(- \mid P^* = p_{RATIONAL,t0})$ is rational for him right now, at t_1. If this is right, then by *Certain*, this function *is* rational for him. And then,

$$P_0(- \mid P^* = p_{RATIONAL,t0}) = P^*(- \mid P^* = p_{RATIONAL,t0}).$$

This is an instance of NRR.

[24] Elga (2013).

[25] See Elga (2013), who notes that the argument adapts an idea from Ross (2006: 277–99).

[26] As I am using notation here, $P_0(- \mid P^* = p_{RATIONAL,t0})$ is P_0 conditionalized on $P^* = p_{RATIONAL,t0}$.

The argument proceeds in two steps: first, $P_0(- \mid P^* = p_{RATIONAL,t0})$ is rational for Chandra at t_1, and second, at t_1 Chandra is certain that $P^*(- \mid P^* = p_{RATIONAL,t0})$ is rational for him at t_1. But for Chandra to have such certainty, it looks like he would have to be certain, at time t_1, of both (a) and (b):

(a) $P^* = p_{RATIONAL,t0}$
(b) All the evidence I have acquired since t_0 is that $P^* = p_{RATIONAL,t0}$

After all, if Chandra was merely certain of (a), but not of (b), how could he be rationally certain that now, at t_1, the rational probability function for him is P^* conditionalized on the information that $P^* = p_{RATIONAL,t0}$? Let us assume for now that (a) and (b) are distinct pieces of information: in conditionalizing on the former Chandra isn't automatically conditionalizing on the latter (just what it would be for (a) and (b) not to be distinct pieces of information is discussed below). Indeed, in general, learning a proposition p doesn't entail certainty that one learnt exactly p.[27]

Perhaps the problem could be fixed by conceding that Chandra needs to learn *both* (a) and (b). But that won't help. For even if we have now saved the second step of the argument (that is, Chandra is certain that $P^*(- \mid P^* = p_{RATIONAL,t0})$ is rational for him at t_1), the first step faces a problem (that is, that $P_0(- \mid P^* = p_{RATIONAL,t0})$ is rational for Chandra at t_1). If Chandra has learnt both (a) and (b), and these are distinct pieces of information for him, then it is *false* that all the evidence he has acquired since t_0 is that $P^* = p_{RATIONAL,t0}$. Hence, (b) is a proposition that is falsified as soon as Chandra learns it. If Chandra has learnt (b) by t_1, then it is not the case that the rational credence function for him at t_1 is $P_0(- \mid P^* = p_{RATIONAL,t0})$. After all, $P^* = p_{RATIONAL,t0}$ is only *part* of the information he has acquired since t_0, but the above argument really did rely on the assumption that *all* Chandra has learnt by t_1 is $P^* = p_{RATIONAL,t0}$. An instance of NRR cannot be generated.

The problem does not essentially rest on the fact that learning (b) is self-falsifying. I do not want to deny that it can ever be possible to know that one has learnt exactly such-and-such up to the present moment. Perhaps, for instance, it is possible to learn a self-referential proposition along the following lines:

(b*) All the evidence I have acquired since t_0 is that $P^* = p_{RATIONAL,t0}$ and this very proposition.[28]

But replacing (b) with (b*) still does not yield an instance of NRR. In so far as (a) and (b*) are genuinely distinct pieces of information for Chandra, in

[27] This isn't entailed even by the assumption that certainty is luminous, iterating in a trivial manner: If $P(p) = 1$, then $P(P(p) = 1) = 1$. One might have such iterated certainty in a proposition p, while being uncertain that p is exactly what one learnt at a given time.

[28] Thanks to a referee for prompting me to discuss this suggestion.

conditionalizing on (b*) he isn't just conditionalizing on the information that $P^* = p_{RATIONAL,t0}$. But then, the rational credence function for Chandra at t_1 is not $P_0(- \mid P^* = p_{RATIONAL,t0})$. Let p' be the conjunction of (a) and (b*). Even if we still get the result that $P_0(- \mid p') = P^*(- \mid p')$, this is not an instance of NRR, nor, as far as I can see, does it entail one.

Perhaps the argument still works on the assumption that (a) and (b) are not genuinely distinct pieces of information for Chandra. And perhaps there is a way of understanding the assumption that Chandra is certain, at t_0, that he is about to conditionalize on information about what credence function is rational for him at t_0, on which this is the case. In particular, assume that though it may be possible for rational agents to conditionalize on (i.e. learn) false propositions, Chandra is certain that he is not going to conditionalize on a falsehood about which credence function is rational for him at t_0. For any of the probability functions P that Chandra considers as possibly rational at t_0, he is certain (at t_0) that P is rational at t_0 if and only if at t_1 he learns exactly the proposition that P is rational at t_0. Then, in conditionalizing on proposition (a) above, isn't he automatically also conditionalizing on proposition (b)—and doesn't he then become certain that $P^*(- \mid P^* = p_{RATIONAL,t0})$ is rational for him at t_1? But even given these assumptions, there is a glitch in the argument. To spell out the problem, it will help to introduce some tools from epistemic logic that I will employ more below, and that will help make explicit the assumption that (a) and (b) are not genuinely distinct pieces of information. My approach will closely parallel that of Williamson (2011).

A probabilistic frame $<W, R, P_{PRIOR}>$ will consist, first, of a set W of situations or cases. These need not be thought of as entire worlds. It might be more useful to think of them as something like centred worlds or cases. R is a relation between members of W which we can informally think of as that of epistemic accessibility: wRw' just in case the subject in w cannot epistemically rule out being in w'. We might express this by saying that for all she knows, she is in w'. But more generally, we can simply maintain that w' is epistemically accessible from w just in case being in w' is compatible with the subject's evidence at w. Note that it is *not* being assumed that epistemic accessibility must be reflexive. The formal framework can accommodate a view on which false propositions can constitute evidence. At least in the finite case, a rational subject assigns some non-zero credence to a case just in case it is epistemically accessible for her. P_{PRIOR} is a prior probability distribution over subsets of W. I won't assume the distribution to be uniform, assigning to each member of W the same weight. Now, we will also need to assign probabilities to propositions at members of W. To get the probability function P_w at a world w, we conditionalize P_{PRIOR} on the set of worlds accessible from w. Informally, we can think of this set as one's total evidence at w. Note that given this framework, two worlds have the same probability function just in case they can access exactly the same worlds.

In effect, what this apparatus enables us to model is a view that assumes updating to happen by Bayesian conditionalization on evidence thought of as sets of worlds or cases.[29] Given the apparatus introduced, here is what the assumption that (a) and (b) are not distinct pieces of information for Chandra amounts to: the set of accessible worlds (worlds accessible for Chandra at t_0) at which P* is rational at t_0 is exactly the set of accessible worlds at which Chandra learns just this proposition at t_1. But care must be taken in spelling out just what the relevant proposition is. Since we are thinking about propositions as sets of worlds, to avoid confusion it is best to just explicitly talk about such sets. Let the set of accessible worlds (accessible for Chandra at t_0) at which function P* is rational at t_0 be S. Then, the assumption is that S is identical to the set of accessible worlds at which Chandra conditionalizes exactly on set S at t_1. As long as this is the case, the kind of problem raised above dissipates: in conditionalizing on S, Chandra is simultaneously conditionalizing (a) and (b) (or (a) and (b*) above), for from Chandra's perspective, both propositions are identical to set S. But there is still a problem.

At t_1 Chandra learns that P* is rational at t_0. That is, he conditionalizes on set S. By the assumptions made, it is true at every world in S that P* is rational at t_0, and it is also true at every world in S that since t_0 Chandra has conditionalized exactly on set S. Then, let us grant that Chandra is certain that *right now* (at t_1) the following function is rational for him:

$$P^*(-|S).$$

But since S just *is* the proposition that P* is rational at t_0, doesn't this entail that Chandra is certain that *right now* (at t_1) the following function is rational for him?

$$P^*(-|P^* = P_{RATIONAL,t0})$$

It would be too quick to draw this conclusion, for just which set the proposition that P* is rational at t_0 is identical to depends on one's perspective: from the perspective of P_0 the proposition is set S, but it does not follow that from the perspective of P*, the proposition is also set S. This follows only if we are operating with frames that satisfy the following condition: the set of worlds at which P* is rational at t_0 and that are accessible for Chandra (at t_0) is *identical to* the set of worlds at which P* is rational at t_0 and that are accessible for worlds at which P* is rational at t_0.[30] This condition guarantees the desired instance of NRR, without any reliance on *Certain*.[31] The condition would follow from the assumption that Chandra can only become certain of and conditionalize on

[29] Since the above argument from *Certain* to instances of NRR assumes updating to happen by conditionalization, employing the framework should not bring in any obviously illicit assumptions.

[30] To see how the argument fails if this assumption is not made, consider a frame in which there is a world w^* at which P* is rational at t_0, but that Chandra cannot access at t_0. Nevertheless, assume that (all) worlds at which P* is rational at t_0 can access w^*. Then, from the perspective of worlds at which P* is rational at t_0, the set of worlds identical to the proposition that $P^* = P_{RATIONAL,t0}$ will contain w^*. And then, $P^*(-|P^* = P_{RATIONAL,t0}) \neq P^*(-|S)$.

[31] Assume that the set of worlds at which P* is rational at t_0 and that are accessible for Chandra at t_0 (that is, accessible from worlds at which P_0 is rational at t_0) is *identical* to the set of worlds at which P* is rational at t_0 and that are accessible from worlds at which P* is rational

true propositions (and hence, it must be the case that $P_0 = P^*$), but this wasn't supposed to be assumed by the above argument.[32]

Hence, even assuming that (a) and (b) are not distinct pieces of information for Chandra, the argument from *Certain* to special cases of NRR does not go through. Of course, one could still argue for NRR using the kind of condition on frames just sketched as a starting point. But that would be an altogether different argument. Let me now turn to frames in which NRR fails, and to direct counterexamples to NRR.

4. A SIMPLE FRAME AND COUNTEREXAMPLES TO NRR

Let a proposition that *unifies* two probability functions P and P* be a proposition *p* such that the function that results from conditionalizing P on *p* just *is* the function that results from conditionalizing P* on *p*. In other words, conditionalizing the two probability functions on *p* produces the same function. In effect, what *New Rational Reflection* says is that for any probability functions P and P*, the proposition that P* is rational unifies P and P* so long as both P and P* assign some non-zero probability to P* being rational (and note that this is required for the relevant conditional probabilities to be defined in the first place).

Using the tools from epistemic logic introduced above, I will first describe a simple model (or, more precisely, frame) in which NRR fails. I will then argue that there are cases with the structure of the frame described, cases that constitute counterexamples to NRR.

A Simple Frame

Again, the frame $\langle W, R, P_{PRIOR} \rangle$ will consist of a set W of situations or cases, of a relation R between members of W which we can informally think of as that of epistemic accessibility, and of a prior probability distribution P_{PRIOR} over subsets of W. To get the probability function P_w at a world *w*, we conditionalize P_{PRIOR} on the set of worlds accessible from *w*. Informally, this set can be thought of as one's total evidence at *w*. Given this framework, two worlds have the same probability function just in case they can access exactly

at t_0. (Note that P* may or may not be identical to P_0.) Let this set be S. Now consider $P_0(- \mid S)$ and $P^*(- \mid S)$. Let one's total evidence at worlds in which P_0 is rational at t_0 be E_0, and let one's total evidence at worlds in which P* is rational be E*. Then, $P_0(- \mid S) = P_{PRIOR}(- \mid E_0 \cap S)$, and $P^*(- \mid S) = P_{PRIOR}(- \mid E^* \cap S)$. Given the assumption that each world in S is accessible for both P_0 and P*, S is a subset of both E_0 and E*. But then, $P_{PRIOR}(- \mid E_0 \cap S) = P_{PRIOR}(- \mid S)$, and $P_{PRIOR}(- \mid E^* \cap S) = P_{PRIOR}(- \mid S)$ and hence, $P_0(- \mid S) = P^*(- \mid S)$. Because, from the perspective of both P_0 and P*, S just *is* the proposition that P* is rational at t_0, we get the following instance of NRR: $P_0(- \mid P^* = p_{RATIONAL,t0}) = P^*(- \mid P^* = p_{RATIONAL,t0})$.

32 If Chandra only conditionalizes on truths, we can generate an instance of NRR by a much simpler argument. Assume that P_0 is rational for Chandra at t_0, and at t_1 he learns that P* is rational at t_0. Then, $P_0(- \mid P^* = p_{RATIONAL,t0})$ is rational for Chandra at t_1. Because Chandra only conditionalizes on truths, $P^* = P_0$. The resulting instance of NRR is trivial, and lends no support to NRR in its full generality.

the same worlds. Here, then, is a simple, three-world frame in which NRR fails:[33]

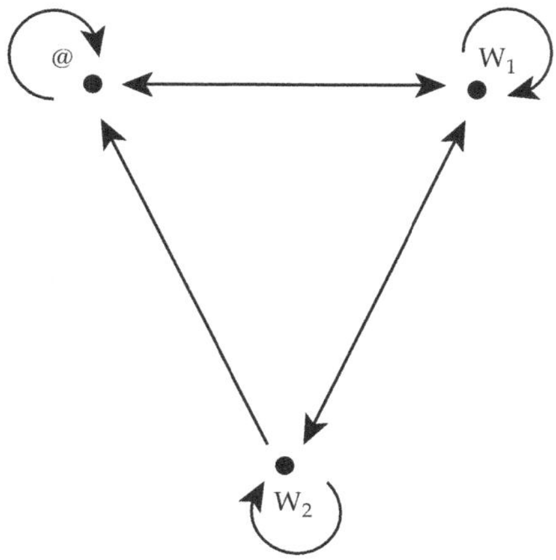

$W = \{w_1, w_2, @\}$

$R = \{<w_1, w_1>, <w_2, w_2>, <@, @>, <w_1, w_2>, <w_1, @>, <w_2, w_1>,$
$\times <w_2, @>, <@, w_1>\}$

W consists of three worlds or cases, @ being the actual one. w_1 and w_2 can access all cases, @ can access itself and w_1. Because w_1 and w_2 can access exactly the same worlds, they have the same probability function. Let P be the probability function (that is rational) at @, and P* the probability function (that is rational) at w_1 and w_2. Then, the proposition that P is rational is just {@}, and the proposition that P* is rational is $\{w_1, w_2\}$. To see that NRR fails for frames for which the above holds, consider the following:

$$P(\{w_1\} | P^* \text{ is rational}) = 1$$
$$P^*(\{w_1\} | P^* \text{ is rational}) < 1$$

This will hold as long as P_{PRIOR} assigns to each world a non-zero weight and, hence, as long as each world assigns a non-zero weight to each world that it can access.

In a nutshell, the reason why NRR fails is that the set of worlds in which P* is rational that are accessible from @ is different from the set of worlds in which P* is rational that are accessible from the worlds in which P* is in fact rational. (Note that it was precisely this feature that also created trouble for the argument, discussed above, from *Certain* to instances of NRR.) As a result, the proposition that P* is rational is not a unifier of the two probability

[33] There is another kind of three-world frame as well, but I focus on this one, as it gives the structure of the kinds of examples described below.

functions. Note that *Old Rational Reflection* (ORR) also fails in frames of the sort described, for:

$$P(\{w_1\} \,|\, P^* \text{ is rational}) = 1$$
$$P^*(\{w_1\}) < 1.$$

Those who are already convinced that epistemic accessibility is neither symmetric nor transitive should be surprised if we *couldn't* find examples with the structure of the simple frame described above. This—more so than the examples I describe below—is what ultimately convinces me that NRR has got to fail. By contrast, those already convinced that epistemic accessibility is transitive and symmetric will deny that the above frame is ever realized. But note that transitivity fails in the kinds of counterexamples to *Old Rational Reflection* recently discussed.[34] Still, a proponent of *New Rational Reflection* could argue, at the very outset, that at least symmetry holds. Now, I think that there are good, independent reasons to reject such symmetry: for instance, we can ordinarily rule out the possibility that we are brains in vats deceived about the world around us, but were we such brains in vats, we would not be able to rule out the possibility that we are in a good, ordinary case. But in so far as brains in vats and ordinary subjects can be in the very same internal states, those with internalist commitments about evidence will disagree. Below I discuss candidate examples of symmetry failure that cannot be dismissed just by appeal to internalism about evidence, for they don't involve pairs of cases in which one is in the very same internal states. I will also discuss a framework on which epistemic accessibility is symmetric for the reason that acquiring new evidence doesn't involve ruling out possibilities or worlds in the first place, but merely redistributing one's credences. As will become clear, insisting that updating happens by something like Jeffrey conditionalization will not, on its own, suffice to block counterexamples to NRR.

To start out with, assume the kind of framework discussed on which updating happens by Bayesian conditionalization on evidence thought of as sets of worlds or cases.

The Ages of Agnes

We often estimate peoples' ages based on how they look. On several occasions I have been way off in my estimates. In particular, on several occasions I have presumed that a friend or acquaintance is in their thirties, later finding out, to my utter amazement, that they are in fact nearly fifty. And on at least one of these occasions, I have at some point, after finding out the real age of my friend, seen a picture of them at thirty, thinking, 'Wow, she looks *much* younger in that picture—I can now see how her face is that of a fifty-year old.' By contrast, had I met only the younger, thirty-year-old version of my friend,

[34] Here I have in mind the 'clock belief' types of cases described by Williamson (2011), Christensen (2010), and Elga (2013).

I would never have thought that she might be fifty. Hence, at least in some cases, and for some faces, the following asymmetry obtains: the younger face cannot easily be mistaken for a significantly older one, but the older face can be mistaken for a younger one. (I am not denying that this could sometimes go the other way.) Consider, then, the following case.[35]

You are at the birthday party of your good friend Agnes, a sprightly centenarian. In order to mark the occasion, Agnes has hung up pictures of herself taken exactly ten years apart, starting from birth, in each room of her house. You know that the floor you are in has three rooms, with pictures of Agnes as a thirty-, forty-, and fifty-year-old. You are about to walk into a room, but don't know which. We can assume that you have not seen any pictures of Agnes at these ages (and you have only known Agnes since she was, say, ninety), though you do know that Agnes has always been youthful, and that at fifty she was frequently mistaken for a thirty-year-old. Let @ be the case in which you step into a room and see a picture of Agnes at thirty, w_1 the case involving a picture of Agnes at forty, and w_2 a case involving a picture of Agnes at fifty.

You, in fact, walk into the room with a picture of Agnes at thirty. Now, the following strikes me as perfectly plausible. You look at the picture, seeing the thick, black hair, porcelain-like skin, and hardly a wrinkle on her face: you know that this is not Agnes at fifty. However, you cannot rule out her being forty—after all, you know she never looked her age. At w_1 you see a forty-year-old Agnes. Here you could go either way: you can't rule out her being thirty, but neither can you rule out her being fifty. At w_2 you see a picture of a fifty-year old Agnes. You see what looks like pretty thick black hair, a tanned face, and mountains in the background. There are some wrinkles around the eyes and mouth, but what you are looking at could be the face of a thirty-year-old who has spent lots of time outdoors, her face being exposed to the elements. This could be a picture of Agnes at thirty, forty, or fifty. You just cannot tell.

Instead of describing what you know in each of the three cases, I could simply have talked about which possibilities the evidence you gain by looking at the photographs allows you to rule out: when looking at the picture of a thirty-year-old Agnes, your evidence rules out the possibility that you are looking at a fifty-year-old Agnes, but not vice versa. All that really matters is that the view of evidence one is working with allows for the kind of asymmetry (and intransitivity) that the example relied on. Note that the kind of asymmetry described is compatible with internalist views of evidence, for one is not in the same internal states in any of the three cases described. And the precise numbers, of course, don't matter. What matters is just that the following holds: when looking at a picture of Agnes at a younger age of n years, you can (just) rule out her being some age $n + i$, whereas if you were instead looking at a picture of Agnes at $n + i$ years, you couldn't rule out her being n years.

[35] I am indebted to Ville Aarnio.

Let P be the probability function at @. Since w_1 and w_2 can access exactly the same worlds (namely, all three), they have the same probability function. Let this function be P*. Now consider:

$$P(\{w_1\} \mid P^* \text{ is rational}) = 1 \neq P^*(\{w_1\} \mid P^* \text{ is rational}) < 1.$$

This is a failure of NRR. Before moving on to discuss an alternative framework, let me sketch a couple of recipes for generating more counterexamples to NRR.

Reals, Fakes, and Diminishing Discriminative Abilities

I take the following to be a pretty common phenomenon: coming across a fake, one can mistake it for the real thing, but when one sees (feels, hears, tastes, smells) the real thing, one can tell that it is not a fake. This is one reason why verbal directions are bad. It is easy to get lost by taking the little, barely visible path, where the guidebook tells one to take the *obvious* path to the left, and not to get distracted by tiny, barely visible paths. When seeing one of the small paths, it is far from obvious that it is not obvious—after all, it is a path. Upon finally finding the right path, one immediately recognizes that this is what the guidebook was talking about. This phenomenon can be exploited to construct more counterexamples to NRR. Assume, for instance, that the guidebook says that there is a really obvious path that will lead to the lake, a slightly smaller but still clear path that will lead to the mountain, and, finally, a small, barely visible path that will lead nowhere. Looking at the obvious path, one knows that it is not the barely visible path. But one cannot rule out its being merely the slightly smaller path. Looking at the slightly smaller path, perhaps one cannot rule out its being either. And looking at the smallest path, one cannot rule out its being one of the larger paths—for it is clearly a path, and someone *could* refer to a path like that as really obvious.

Cases with the following feature provide resources for constructing yet further counterexamples to NRR: as we move along some parameter, our discriminative abilities get worse, and our margins of error grow. For instance, the more wine I drink, the worse I am telling how many glasses I have had.[36] The further I move from an object, the worse I am at telling how distant it is from me. The more works by Alexandre Dumas I read, the worse I am at telling how many I have already read. And as the minutes go by while I wait for my doctor's appointment, the worse I get at telling just how long I have waited. Consider the following example.

You are participating in an experiment. You know that you will be sat in a room for a period of three minutes (@), five minutes (w_1), or seven minutes (w_2), after which a bell will ring, and the experiment will be over. During your time in the room, you will be asked to focus on a menial task preventing you from counting seconds in your head. The precise numbers don't matter, but it

[36] Thanks to Brian Weatherson for this example.

seems plausible that we could choose numbers in such a way as to generate a case with the sort of structure described above—and all we need, of course, is that there is some possible rational being for whom we could choose such numbers. The idea is that if you are in the case in which the bell sounds at three minutes (@), you know, upon hearing the bell, that seven minutes have not passed. But you cannot rule out the possibility that five minutes have passed. If you are in a case in which five minutes have passed (w_1), you can rule out neither the possibility that only three have passed, nor the possibility that seven have passed. And if you are in a case in which seven minutes have passed, you can rule out neither five nor just three minutes having passed, for your margin for error expands with time.

While I am convinced by the above examples, some of the assumptions made could be resisted. Consider again *The Ages of Agnes*. You undergo different experiences at w_1 and w_2, since you are looking at different pictures of Agnes, taken ten years apart. But then, wouldn't it be rational to have different credence functions at these two cases? Part of the objection may be resistance to the idea that you can only access three worlds. If more worlds are at play—as in more complex and perhaps more realistic examples—pairs of cases involving different experiences will typically be able to access different worlds. It is a challenge to make sure that we really have an example involving only three cases. But I suspect that the real source of the objection is resistance to the idea that updating on new evidence should always take the form of straight conditionalization: even if your evidence at w_1 or w_2 doesn't rule out any of the three cases, one might insist that some information is surely gained, information that should impact your probability functions at these worlds in different ways.[37] For instance, at w_1 you should be more confident that you are at w_1 than that you are at w_2, and vice versa. A related objection is that your evidence at @ doesn't allow completely ruling out w_2, even if you *know* that you are not at w_2. In particular, combining an internalist view of evidence with the idea that we nevertheless know a lot of the things we think we know leads to the thought that a subject can know a proposition p, even if not-p is not ruled out by her evidence. If this is right then, again, we cannot think of updating as happening by straight conditionalization on an evidence-proposition.

Let me now discuss the possibility of constructing counterexamples to NRR within a framework that assumes updating to happen by a more general form of conditionalization. As will become clear, such counterexamples can be resisted. But resisting them requires defending assumptions that are far from being self evident.

Jeffrey Conditionalization

As a starting point, assume that facts about a subject's evidence (and about which credence function is rational for her) supervene on her internal states, so long as it is not always perfectly luminous or transparent to a subject just

37 Thanks to David Christensen for pressing this objection.

what internal state she is in. So, for instance, even if a subject is in a different internal state in some case w' from the state she is actually in, she may not be able to rationally rule out being in w'. My diagnosis of *The Ages of Agnes* was at least compatible with such an internalist view of evidence.[38] Second, if evidence at least sometimes allows the ruling out of possibilities, it is not clear why such ruling out couldn't be asymmetric in a way that allows for the above kinds of counterexamples to arise. So let me discuss a view on which updating doesn't happen by ruling out worlds in the first place. We can no longer think of the probability function at a world as resulting from conditionalizing a prior probability function on the set of accessible worlds, thought of as one's total evidence at that world. This means giving up the kind of formal framework sketched above.

Instead of Bayesian conditionalization, think of updating as happening by Jeffrey conditionalization. When one is at a world w, rather than there being an evidence proposition that is true in all the worlds accessible from w, there is a probability distribution over the worlds accessible from w, and we can think of the distribution as indicating what one's evidence is. Hence, being at a world is like undergoing a Jeffrey experience. Because straight conditionalization is a special case of Jeffrey conditionalization, the latter does, of course, allow the ruling out of worlds. But what I am envisaging is a view on which undergoing new experiences never makes it epistemically appropriate to completely rule out a possibility.

The counterexamples described above (cases such as *The Ages of Agnes*) relied on different worlds accessing different sets of worlds. But counter-examples to NRR can also be constructed that rely not on accessing different worlds, but on weighting the accessible worlds differently. Let S* be a set of worlds in which one is in some internal state E*, and let the probability function that is rational in all of these worlds be P*. If there is a probability function P (rational in some different set of worlds S in which, instead of E*, one is in some internal state E) that assigns different relative weights to the worlds in S* than P* itself does, NRR is in trouble. Assume for now that if one undergoes different experiences at two worlds w and w', it is not possible for the same probability function to be rational at these worlds. Then, counter-examples to NRR could be blocked by insisting that the partitions with respect to which Jeffrey updates happen never cut across experiential boundaries (in a way I discuss below). I will first discuss this constraint on partitions; I then say why the assumption that different experiences always mean different probability functions is implausible.

Consider a very simple example. As part of an experiment, you are sitting in a pitch-dark room. When a bell chimes, there is a 2/3 chance that a sprig of lavender will placed right before you, and a 1/3 chance that there will be a sprig of mint. You also know that half of the sprigs of lavender have been

[38] I leave it open what counts as an internal state. Perhaps, for instance, all non-factive mental states are internal in this sense.

genetically modified to smell of mint. Hence, just before the bell chimes, you consider the following three possibilities as equally likely: there will be a sprig of lavender that smells of lavender (@), a sprig of lavender that smells of mint (w_1), and a sprig of mint that smells of mint (w_2). The bell chimes, and you catch the distinct smell of lavender—hence, you are at @. How should you update your credences?

@	w_1	w_2
Smell lavender	*Smell mint*	*Smell mint*
Lavender	*Lavender*	*Mint*

Your experience doesn't allow you to completely rule out any of the three cases: though you are in perceptual contact with the sprig of lavender before you, you cannot be certain that it is lavender (rather than mint), or even that you are experiencing the olfactory sensation you seem to be experiencing (perhaps you are imagining things). There are four different partitions with respect to which you might perform a Jeffrey update: {{@}, {w_1}, {w_2}} is the maximally fine-grained partition; {{@, w_1}, {w_2}}, which divides the worlds into those at which there is lavender before you and those at which there is mint; {{@}, {w_1, w_2}}, which divides the worlds according to whether or not you experience the olfactory sensation as of lavender; and, finally, there is {{@, w_2}, {w_1}}, the seemingly least natural partition, which divides according to whether or not the sprig before you is genetically modified, or whether or not your experience is misleading.

If, upon smelling the lavender at @, you are permitted to deploy the partition {{@, w_1}, {w_2}}, which divides worlds into those at which there is lavender and those at which there is mint, NRR is in trouble. Let the probability function that is rational at @ be P, and let the probability function that is rational at w_1 and w_2 be P*. First, how should P* weight w_1 and w_2? If the bell chimes and there is a smell of mint, you may not be able to rule out the possibility that you are catching a whiff of lavender after all, but you shouldn't think that it is likelier that there is lavender disguised as mint than that there is mint. That is, $P^*(w_1) \leq P^*(w_2)$. By contrast, consider your situation in @. You have a distinct olfactory experience as of lavender. You should raise your credence that there is a sprig of lavender before you. If you do a Jeffrey update by deploying the partition {{@, w_1}, {w_2}}, then because the probability of {@, w_1} goes up from 2/3, that of {w_2} goes down from 1/3, and @ and w_1 are equally probable (as they previously were), $P(w_1) > P(w_2)$. Then,

$$P(\{w_1\}|P^* \text{ is rational}) \neq P^*(\{w_1\}|P^* \text{ is rational}).$$

Again, NRR fails. The point is not restricted to the extremely simple model described. As long as one can at least sometimes deploy partitions that divide up worlds in accordance with how things stand in the external world, it looks like opportunities for failures of NRR arise.[39]

If acceptable partitions always divide worlds according to the experiences one undergoes at those worlds, then the above reasoning is blocked. $\{\{@\},\{w_1, w_2\}\}$ is such a partition. If w_1 and w_2 are lumped together, then a Jeffrey update cannot change the ratios of the probabilities of these worlds. More generally, one might defend a principle according to which the ratios of probabilities of worlds in which one undergoes the same experiences, or in which one is in the same total internal states, are always preserved.[40] In defence of $\{\{@\},\{w_1, w_2\}\}$—and against $\{\{@, w_1\}, \{w_2\}\}$—one might complain that there is something very awkward about raising one's confidence in w_1 as a result of smelling lavender, even though one does not undergo any such experience w_1 (recall that in w_1 one has an olfactory experience as of mint, not lavender). But we shouldn't assume, without further argument, that smelling lavender couldn't directly give one information about whether there is a sprig of lavender before one, without doing so via information about the experience one is undergoing.[41]

The question of what partitions ought to be deployed for the purposes of Jeffrey updates goes beyond the scope of this paper. But I very much doubt that the question can be resolved in a motivated way just by comparing intuitions. For instance, insisting on experience-partitions while denying scepticism seems to have consequences that are far from intuitive.[42] Assume that you open your eyes, and upon seeing your hands, have a perceptual experience as of hands. You are in normal conditions and there are no defeaters. The result we want is that it is rational for you to end up reasonably confident that you have hands. But assume that you perform a Jeffrey update by partitioning worlds according to which experience you are having. Given that you are undergoing a paradigm perceptual experience as of hands, it would seem rational for you to increase your confidence that you undergo such an experience (at least if, prior to opening your eyes, you were not already confident just what experience you would have). But if you increase your confidence that you are undergoing an experience as of hands, you will have to increase your confidence that you are a handless brain in a vat undergoing an experience as of hands. Your new confidence that you have hands

[39] One objection to the kind of three-world model described is that if acquiring new evidence never justifies ruling out worlds, then one could never be in a model with only three worlds. But the point of using a three-world model was just to point to the kind of structural feature that allows for failures of NRR—namely, there being a set of worlds S* such that the probability function that is rational in those worlds weights the worlds (or subsets of worlds in S*) differently from some other probability function, rational in a different set of worlds.

[40] It is worth remarking again that such a principle would not suffice if the following was possible: the same credence function is rational at two worlds w and w', even though one undergoes *different* experiences at the two worlds.

[41] It is worth noting that Jeffrey himself didn't think that the relevant partitions should concern one's experiences.

[42] The reasoning here relies on White's (2006) discussion of dogmatism. See Pryor (2013) for further discussion, including ways in which the reasoning could be resisted.

must be lower than your old confidence that you are not such a handless brain in a vat.[43] To avoid scepticism, it looks as though the proponent of experience-partitions will have to concede that it is rational for you to be very confident *a priori* that you are not a handless brain in a vat undergoing an experience as of hands. But this is far from intuitive.

It is worth reiterating the commitments of a view that can hope to block counterexamples to NRR. Updating on new evidence takes the form of Jeffrey conditionalization, and one is never justified in completely ruling out a possibility. What one's evidence is at a world—and hence, which credence function is rational at that world—is fixed by one's internal state in that world. If one undergoes different experiences at two worlds, then different credence functions are rational at these worlds. Finally, the partitions one deploys in Jeffrey updates divide up worlds in accordance with the internal states one is in, or the experiences one undergoes, at those worlds.

Even if one granted that Jeffrey conditionalization should always happen with respect to experience-partitions, the assumption that different experiences always make for different probability functions is far from uncontroversial. It is very plausible that what credences it is rational for a subject to have based on an experience she undergoes often depends on her discriminative abilities. But then, it should be possible for the same credences to be rational for two subjects, even if they undergo slightly different experiences. That overall experiential states fix rational credences does not entail that rational credences fix experiential states. Let w_1 and w_2 be two worlds in which one undergoes different experiences, but that have the same probability function. Then, even if Jeffrey conditionalization always happens with respect to experience-partitions, this would not be enough to guarantee that counterexamples for NRR never arise, since the probability function that is rational at some third world w^* might weight w_1 and w_2 differently from the probability function that is rational at these worlds.

At first sight it didn't seem like the appeal of rational reflection principles rested on a specific way of thinking about evidence and updating. Just as the kind of synchronic constraint on rational credence functions posed by the *Principal Principle* seems to flow from the nature of chance, the kind of synchronic constraint on rational credence functions posed by rational reflection principles seemed, at first sight, to flow from the nature of what it is for

[43] Here is a sketch of the reasoning yielding this conclusion. Upon having an experience as of hands, you do a Jeffrey update by deploying a partition that concerns your experiences. Let S be the set of worlds at which you undergo an experience as of hands. Let h be the proposition that you have hands, and let b be the proposition that you are a handless brain in a vat undergoing an experience as of hands. h entails not-b. As a result of your experience, you increase your confidence in S, decreasing your confidence in at least some other members of the relevant partition. Each world at which b is true is a member of S—hence, b is a subset of S. But then, by increasing your confidence in S, you increase your confidence in b, and decrease your confidence in not-b. Because h entails not-b, the probability of h is at most the probability of not-b. Putting everything together, your new confidence in h is lower than your old confidence in not-b.

something to be a rational credence function. But it turns out that *New Rational Reflection* rests on a controversial view about evidence and updating.

Let me now discuss a further set of problems for NRR, which is that it fails to respect internalism about rationality. The problems raised cast doubt on the whole project of trying to capture such internalism within a context of uncertainty about what is rational by means of some rational reflection principle.

5. RATIONAL REFLECTION AND INTERNALISM ABOUT RATIONALITY

As we saw, NRR can be motivated by the rough idea that if I am rationally certain that Sophie is an expert (i.e. an ideally rational agent), but I have information that Sophie lacks, then I ought to defer to what Sophie thinks, *conditional on all the extra evidence I have.* I now want to discuss a further problem with NRR, which is that it doesn't respect the kind of internalism about rationality that was supposed to motivate rational reflection principles in the first place. Ironically, when I am uncertain just what my extra evidence is, the idea that I ought to adopt Sophie's credence function, conditionalized on all the extra evidence I have, goes against the rough idea that I ought to have the attitudes that I take a rational agent in my situation to have.

Assume that though I have extra evidence E (evidence that Sophie does not have), I have excellent reasons to believe that I have extra evidence E^*. Moreover, I know that while telling Sophie E would make her very confident in a proposition h, telling her E^* would make her very confident in not-h. Then, I have excellent reason to think that in my situation the perfectly rational agent Sophie would be confident in not-h. But if I ought to follow the rule of adopting the credence function that results from conditionalizing Sophie's credence function on the extra evidence I in fact have, then I ought to be confident that h is true. Hence, it is not at all clear whether NRR respects internalism about rationality within a context of uncertainty about what is rational in one's situation. Both *Old* and *New Rational Reflection* adhere to the internal perspective of a subject when it comes to the question of which credence function is rational, but not when it comes to the question of what the subject's extra evidence is.

This might prompt a search for a new rational reflection principle. Here is a thought. Assume that it is time t, and shortly thereafter at t' I learn (become certain) that P* is rational for me at t. What should my credence function at t' be? The problem was that $P^*(- \mid P^* = p_{RATIONAL,t})$ isn't necessarily the function I *take* to be rational for me, as I may not be certain (at t') that what I have learnt between t and t' is that $P^* = p_{RATIONAL,t}$. Assume that at t' I am thus uncertain about my evidence, assigning a probability of 0.5 to having learnt E since t, and a probability of 0.5 to having learnt E'. Now, one might think that the following *would* respect internalism about rationality:

$$P_{t'}(-) = P^*(- \mid E) \times 0.5 \ + \ P^*(- \mid E') \times 0.5.$$

This suggests a more general thought, which can be heuristically expressed as follows. What should my credence in a proposition h be (at a time t), conditional on some probability function P* being rational for me (at t)? Well, what would I take myself to have just learnt were I to learn that P* is rational at t? My credence should just be P* conditionalized on that evidence. And what if, upon learning that P* is rational at t, I wouldn't be sure exactly what I just learnt? Well, then I should adopt the weighted sum of P*-conditional credences on the various pieces of evidence I think I might just have acquired, weighted by my credences that those are my new pieces of evidence. If $E_1,\ldots,E_n$ are my candidates for the pieces of evidence acquired, and $r_1,\ldots,r_n$ are my credences, for each piece of evidence, that I just acquired it, then we can formulate the new principle as follows:

Internalist Rational Reflection (IRR)

$$P_t(-\mid P^* = P_{RATIONAL,t}) = P^*(-\mid E_1)\times r_1 + \cdots + P^*(-\mid E_n)\times r_n.$$

But it is far from clear whether this rough idea is workable.

First, just what are $E_1,\ldots,E_n$ and $r_1,\ldots,r_n$? I made use of counterfactual talk to express the gist of the new rational reflection principle IRR: 'What would I take myself to have just learned, were I to learn that P* is rational at t?' It would be wise to clean up such talk, but there is no obvious way to do so. To get a glimpse of the problem of determining the values of $r_1,\ldots,r_n$, assume that there is some motivated way of deciding what $E_1,\ldots,E_n$ are. Consider the following candidates for determining the weights $r_1,\ldots,r_n$:

(i) $r_i = P_t$ (I learn exactly E_i between t and t')
(ii) $r_i = P^*$ (I learn exactly E_i between t and t')
(iii) $r_i = P^*$ (I learn exactly E_i between t and $t' \mid P^* = P_{RATIONAL,t}$)
(iv) $r_i = P_t$ (I learn exactly E_i between t and $t' \mid P^* = P_{RATIONAL,t}$)

(i) won't work, for the issue isn't what my opinions at t are concerning how likely it is that I will acquire various pieces of evidence at t'. What we are trying to get at (again, using the heuristic way of talking introduced above) is the opinions I would have were I to learn that P* is rational at t. (ii) is problematic for similar reasons. And conditionalizing probability function P* on the information that that P* is rational at t wouldn't seem to help, for we are interested in *my* perspective on what my evidence is, upon learning that P* is rational at t, not the perspective of P*. (iv) looks much better: instead of talking about the opinions I would have were I to learn that P* is rational at t, why not just talk of my present opinions, conditional on the proposition that P* is rational at t? But within the present context, the problem with (iv) should be obvious: plugging it into IRR creates a regress, for precisely what we are trying to do is to come up with a rational reflection principle that tells us just which function $P_t(-\mid P^* = P_{RATIONAL,t})$ is.

Second, I wonder whether even a principle along the lines of IRR is internalist enough. The above assumed that I am at least certain that conditionalization is the rational updating procedure: the strategy was to arrive at the probability function $P_t(- \mid P^* = p_{RATIONAL,t})$ not by conditionalizing P* on what I would in fact learn were I to learn that P* is rational at *t*, but on what I would *take* myself to have learnt. But I may doubt not only whether I just learnt that P* is rational at *t*, but also whether I should take into account evidence by conditionalizing on—and if a lot of recent literature on higher-order defeat is on the right track, it may be rational for me to do so. Then, even if the sort of idea discussed above could be made to work, the resulting internalism would still seem half-baked.

One reply to such a worry is that in raising it I am bending the rules of the game, for it was simply being assumed that rational agents are certain of the rationality of conditionalization. But recall that the need for rational reflection principles arose within a context of uncertainty about just what is rational, and one way such uncertainty can arise is if one is uncertain about just what the right way of updating on one's evidence is. I see no motivated way of drawing a line between acceptable and unacceptable kinds of uncertainty.

Before concluding, let me briefly discuss a further issue, which is that the kinds of internalist principles and requirements discussed at the very beginning of this paper seem to conflict with rational reflection principles.

6. A POTENTIAL CONFLICT

In the very beginning I mentioned several views committed to the idea that certain kinds of mismatch between the states one takes to be rational and the states one is in are themselves irrational. It was noted that numerous epistemologists think that it is always irrational to be in a state of epistemic akrasia: one believes (or has high confidence in) a proposition *p*, while believing (or having high confidence) that it is irrational to believe *p*.[44] Consider now the state a subject is in if she assigns to *p* a credence of *r*, while being certain that it is irrational for her to assign to *p* a credence of *r*. Such a state is at least a close cousin of epistemic akrasia, and it might seem irrational for similar kinds of reasons:

> *Weak self-confidence*
> If a possible rational agent assigns credence *r* to *p*, then she is not certain that it is irrational to assign credence *r* to *p*.

This sounds like a pretty modest requirement—it doesn't even require rational agents to be moderately confident in the rationality of the opinions they hold. The requirement could be defended by a strategy mentioned at the very beginning. Consider a subject who assigns to the proposition that it

[44] While many mean by 'epistemic akrasia' the state a subject is in when she believes a proposition *p*, while believing that it is irrational to believe *p*, there is no completely standard usage. For instance, Horowitz (Forthcoming) means having high confidence in *p*, while having high confidence that one's evidence does not support *p*.

will rain today a credence of 0.9, despite being certain that a credence of 0.9 is irrational. Such a subject holds a credence that is, by her own lights, certain to be irrational. Doesn't she exhibit a kind of incoherence that is rationally forbidden?[45] However, *Weak self-confidence* seems to be in tension with the very idea of a rational reflection principle. Assume that Sophie is 0.5 confident that right now the rational credence to assign to the proposition that it will rain today is 0.3, and 0.5 confident that the rational credence is 0.7 (perhaps an epistemology oracle just told her so). The only way in which Sophie can respect *Old Rational Reflection* is by being 0.5 confident that it will rain. But because Sophie is certain that 0.5 is an irrational credence, by having that credence she would violate *Weak self-confidence*. And however ORR is revised, I see no reason why the new reflection principle would urge Sophie to assign to p either a credence of 0.3 or 0.5—in fact, such a recommendation would go against the spirit of rational reflection principles.

Further, it is at least unclear whether rational reflection principles are compatible with the view that it can never be rational to believe p, while believing that it is irrational for one to believe p—that is, the view that epistemic akrasia is always irrational. Assume a non-maximal threshold view of belief: to believe a proposition is to assign some credence above a value r to that proposition, where r is less than 1. Assume that the threshold for belief is 0.9, and that you know this. Assume that you have the following rational credences: your credence that the rational credence in p is 0.89 is 0.9, and your credence that the rational credence in p is 0.99 is 0.1. Then, your expectation of the rational credence is 0.9. If you respect *Old Rational Reflection*, your credence in p is 0.9. Given the 0.9 threshold for belief, you believe p. But you also believe that it is not rational to believe p. Hence, you are in a state of epistemic akrasia. Now, it is true that the reasoning relied on ORR. But there doesn't seem to be anything about NRR that would block the problem from arising—indeed, in many cases the two principles pose exactly the same constraint on one's credences.

Rational reflection principles sometimes recommend being in akrasia-like states. It is not clear whether this is a problem for rational reflection or the anti-akrasia principles discussed. But in any case, it does seem to put some pressure on the idea that rational reflections exemplify internalism about rationality within a finer-grained context of uncertainty about what is rational. Let me now draw together some conclusions.

7. CONCLUSIONS: GIVING UP ON RATIONAL REFLECTION

My starting point was the rough idea that rationality requires that there be a match between one's opinions about what states it is rational to be in, on the

[45] I intend to set aside a subject who is certain that a credence of 0.9 in rain is irrational while having that credence, but who is confident that her credence in rain is, instead, some value other than 0.9. The issue could be avoided by adding to the antecedent of *Weak Self-Confidence* a clause specifying that the subject has sufficient epistemic access to her own credences.

one hand, and the states one is in, on the other. I dubbed this idea *internalism about rationality*. I took the search for rational reflection principles to be spurred by an attempt to implement such internalism within a context of uncertainty about what is rational. Whatever one's favoured rational reflection principle is, any such principle seems to entail the following: if I have a rational credence distribution over hypotheses about which of $P_1, \ldots, P_n$ is the rational credence function for me, then that credence distribution fixes just which credence function is in fact rational for me. There is a very tight connection between one's opinions about what is rational and what is in fact rational.

While *Old Rational Reflection* faces serious problems, many still think that there is something right about rational reflection. Elga's *New Rational Reflection* (NRR) seems like a well-motivated fix that elegantly avoids the main problems and counterexamples for the old principle. The fix parallels a popular fix to another expert principle, Lewis' *Principal Principle*. I have argued against *New Rational Reflection*. First, one might hope to motivate the principle by showing that in certain special cases instances of it have to hold if one is to comply with a highly plausible principle, *Certain*. *Certain* denies that it could be rational to be certain that a credence function P is rational, while failing to have P. But, I argued, the argument from *Certain* to instances of NRR fails. I also argued that NRR faces counterexamples. Such counterexamples can be avoided, but this requires adopting views about evidence and updating that are far from self-evident. Finally, I discussed ways in which NRR fails to implement internalism about rationality—for while the principle defers to a subject's perspective when it comes to the question of what probability functions are rational for her, it does not defer to her perspective when it comes to the question of what her evidence is.

I think the above discussion warrants drawing at least some general lessons. First, NRR looks like a promising and well-motivated revision, but I hope to have shown that its viability rests on buying into somewhat controversial assumptions about evidence and updating. Such assumptions could be defended, but so could alternative ones. There is reason to think that the same will be true of any proposed reflection principle. However, the initial appeal of reflection principles seemed to rest on the nature of being a rational credence function. Just as it seemed like one ought to treat the chance function as an expert, it seemed that one ought to treat the rational credence function (or an agent with such a credence function) as an expert. What I am suggesting is that whether or not the rational credence function should be thought of as an expert function depends on very substantial epistemological issues.

Second, if the search for rational reflection principles is an attempt to implement internalism about rationality, there is reason to be sceptical about whether any such attempt *could* succeed. Any such principle will have to assume that certain facts about rationality cannot be rationally doubted: as we saw, even *Internalist Rational Reflection* cannot respect the internal perspective of a subject who doubts whether conditionalization is the rational updating procedure. Given that the need for rational reflection principles arose within a context of uncertainty about what is rational, this raises the challenge of saying why we should only take certain kinds of uncertainty about what is rational seriously, and not others.[46]

REFERENCES

Arpaly, Nomy (2000) 'On Acting Rationally against One's Best Judgment', *Ethics*, 110.3: 488–513.

Broome, John (2013) *Rationality through Reasoning*, Oxford: Wiley Blackwell.

Chisholm, Roderick (1989) *Theory of Knowledge*, 3rd edition, Englewood Cliffs, NJ: Prentice-Hall.

Christensen, David (2010) 'Rational Reflection', *Philosophical Perspectives*, 24.1: 121–40.

Elga, Adam (2005) 'On Overrating Oneself ... and Knowing It', *Philosophical Studies*, 123: 115–24.

Elga, Adam (2013) 'The Puzzle of the Unmarked Clock and the New Rational Reflection Principle', *Philosophical Studies*, 164.1: 127–39.

Greco, Daniel (2014) 'A Puzzle about Epistemic Akrasia', *Philosophical Studies*, 167.2: 201–19.

Hall, Ned (1994) 'Correcting the Guide to Objective Chance', *Mind*, New Series, 103.412: 505–17.

Hazlett, Allan (2012) 'Higher-order Epistemic Attitudes and Intellectual Humility', *Episteme*, 9.3: 205–23.

Horowitz, Sophie (Forthcoming) 'Epistemic Akrasia', *Noûs*, published online May 2013.

Lasonen-Aarnio, Maria (ms). Enkrasia or Evidentalism? Unpublished manuscript.

Pryor, Jim (2013) 'Problems for Credulism', in Chris Tucker (ed.), *Seemings and Justification: New Essays on Dogmatism and Phenomenal Conservatism*, Oxford: Oxford University Press, pp. 89–132.

Ross, Jacob (2006) *Acceptance and Practical Reason*, PhD thesis, Rutgers University, October.

Scanlon, Thomas (1998) *What We Owe to Each Other*, Cambridge: Belknap Press.

Smith, Michael (1994) *The Moral Problem*, Oxford: Blackwell.

Thau, Michael (1994) 'Undermining and Admissibility', *Mind*, 103: 491–504.

Titelbaum, Michael (Forthcoming) 'Rationality's Fixed Point (or: In Defense of Right Reason)', *Oxford Studies in Epistemology*, 5, Oxford: Oxford University Press.

White, Roger (2006) 'Problems for Dogmatism', *Philosophical Studies*, 131: 525–57.

Williamson, Timothy (2011) 'Improbable Knowing', in T. Dougherty (ed.), *Evidentialism and its Discontents*, Oxford: Oxford University Press, pp. 147–64.

46 I am extremely indebted to conversations with Ville Aarnio, Brian Hedden, David Christensen, Cian Dorr, Dmitri Gallow, Jim Joyce, Bernhard Salow, Josh Schechter, Brian Weatherson, and Tim Williamson, as well as two *superb* referees for *Oxford Studies in Epistemology*. I also want to especially thank Alex Worsnip for exceptionally detailed and helpful comments on the paper.

6. Time-Slice Epistemology and Action under Indeterminacy

Sarah Moss

There is a question that unifies several recent debates in epistemology, namely whether there are any essentially diachronic norms of rationality, or whether all fundamental norms of rationality are temporally local.[1] Let us say that fans of temporally local norms advocate *time-slice epistemology*, where at a first pass, we define this theory as the combination of two claims. The first claim: what is rationally permissible or obligatory for you at some time is entirely determined by what mental states you are in at that time. This supervenience claim governs facts about the rationality of your actions, as well as the rationality of your full beliefs and your degreed belief states. The second claim: the fundamental facts about rationality are exhausted by these temporally local facts. There may be some fact about whether you are a rational person, for instance. But this fact is a derivative fact, one that just depends on whether your actions and opinions at various times are rational for you at those times.

Suppose that perdurantism is correct, i.e. that objects have temporal parts located at different times, just as they have spatial parts located in different places. Then we can restate time-slice epistemology as a theory that concerns time slices, or instantaneous temporal parts of objects. The first claim: what is rationally permissible or obligatory for a time slice is entirely determined by the mental states of that time slice. The second claim: the fundamental facts about rationality are exhausted by facts about the rationality of time slices. There may also be derivative facts about whether temporally extended agents are rational. For instance, we could say that you are rational just in case you are composed only of rational time slices, or just in case most of your time slices are rational. The point is just that rationality is not fundamentally predicated of you, but of your time slices. In a nutshell: time slices are the fundamental subjects of epistemic evaluation.[2]

Time-slice epistemology may initially seem to have plenty of counterintuitive consequences. Say you see your friend Alice eat four scoops of ice

[1] Thank you to audiences at the University of Michigan and the 2013 Columbia-NYU Graduate Conference in Philosophy for helpful discussion of this paper. I am also especially grateful to Tom Dougherty, Brian Hedden, Susanna Rinard, Scott Sturgeon, Brian Weatherson, Robbie Williams, and an anonymous referee for extensive comments on earlier drafts.

[2] The initial statement of time-slice epistemology should make it clear that the theory is not committed to perdurantism. In the context of this paper, claims about time slices are merely convenient shorthand for claims about agents that instantiate certain properties at times, such as being in particular mental states or being permitted to do particular actions.

cream for lunch, and after lunch you form the belief that she has not eaten anything all day. This seems like a perfectly good example of an irrational belief. Say you yourself eat seven scoops of ice cream for lunch, even though you are going to regret your binge as soon as it is over. This seems like a perfectly good example of an irrational action. In both cases, it is tempting to say that you are irrational precisely because there is no connection between your past mental states and what you currently believe, or between your future mental states and what you are currently doing. Hence these cases may seem like counterexamples to time-slice epistemology.

But the cases are not counterexamples. The time-slice epistemologist agrees that you are irrational in these cases, just not that you are irrational in virtue of ignoring what you used to believe or what you will later desire. The idea behind the theory is that you are irrational in virtue of ignoring your *current* beliefs and desires. For instance, you currently remember seeing Alice eat ice cream, and that is why it is irrational for you to believe that she has not eaten anything. You currently care about whether you are happy later, which is why it is irrational for you to do something that you believe will make you unhappy later. In more generality, the idea behind time-slice epistemology is that the current normative import of your past and future mental states is entirely mediated by your current mental states.

This paper is programmatic in nature, with two very general goals. First, I want to tie together several epistemological theories by identifying them as theories that advance time-slice epistemology. This goal is addressed in the first section of the paper, where I define and motivate time-slice epistemology.

Second, I want to suggest that analogies with ethical claims can help us defend certain time-slice theories, namely time-slice theories of action under indeterminacy. In §2, I discuss several theories about how you should act when the outcome of your decision depends on some indeterminate claim. I start with Caprice, a theory of action under indeterminacy defended in Williams 2014. Caprice says how agents should act in isolated, one-off decision situations. Caprice follows from a more complete theory, Liberal, that also says how agents should act when they face multiple decision situations over time. Liberal is a time-slice theory. In §3, I defend it against objections. In particular, I defend Liberal against general objections to time-slice theories by comparing it with compelling ethical claims.

Although Caprice and Liberal are compelling, they are not perfect. In §4, I raise some objections to these theories. In light of these objections, I develop alternative theories of action under indeterminacy in §5. Here again, I rely on useful analogies with ethical claims as I develop more robust principles in support of time-slice epistemology.

1. DEFINING AND MOTIVATING TIME-SLICE EPISTEMOLOGY

Time-slice epistemology replaces diachronic norms of rationality with synchronic norms, norms that say what is rational for individual time slices when they have particular mental states. The most obvious target of the time-slice epistemologist is the classic updating norm introduced by Bayes 1763 and widely adopted by Bayesians, namely the claim that agents should update their credences according to Conditionalization. Conditionalization demands that your later credence in a proposition should match your earlier conditional credence in that proposition, conditional on any information you have since learned. The norm is at odds with time-slice epistemology, as it says your current credences are rationally constrained by your past credences, on which your current mental states need not supervene.

As time-slice epistemologists, our case against Conditionalization begins with the observation that, as Williamson 2000 put it, "forgetting is not irrational; it is just unfortunate" (219). There may be meaningful epistemic norms that require your memory to be perfect. But these are not norms of rationality in the ordinary sense. If you forgot what you ate for dinner last night, we might criticize you—but not by saying, "how *irrational* of you!" In the context of an argument about whether Alice is a rational person, it is not obviously relevant to mention that she is forgetful. Intuitively, rational requirements on evidence retention are more similar to requirements on evidence gathering. Being *negligent* about what you learn or remember may signal or constitute irrationality. This is especially the case for strategic negligence, e.g. if you selectively forget or fail to gather evidence that disconfirms your favorite theory. But just as you are not irrational merely for having imperfect powers of evidence gathering, you are not irrational for having imperfect powers of evidence retention.[3]

In more generality, the problem is that on our traditional understanding of Conditionalization, the norm requires that evidence is cumulative for rational agents, whereas intuitively, rationality does not impose such strict demands. In fact, sometimes it imposes contrary demands. In the Shangri La case in Arntzenius 2003, an agent is rationally required to violate Conditionalization, rather than remaining certain of a proposition for which she lacks sufficient evidence.

There are two very different strategies for responding to these challenges for Conditionalization. The first strategy is to simply modify the diachronic norm with some restrictions. Titelbaum 2013 suggests this approach when he says that "the domain of applicability of the Conditionalization-based framework is limited to the sorts of stories that originally motivated it: stories in which all the doxastic events are pure learning events" (124). In fact, some theorists claim that this restriction is already implicitly understood in traditional discussions of the updating norm. For instance, Schervish et al. 2004 respond to the examples in Arntzenius 2003 by complaining that certain "restrictions or limitations" are "already assumed as familiar" when Conditionalization is applied, and that these restrictions include the constraint that agents not lose information over time (316).

[3] I am grateful to an anonymous referee for suggesting this argument. See Talbott 1991 for a classic discussion of Conditionalization and memory loss.

The second strategy for responding to challenges for Conditionalization is to trade in this diachronic norm for a synchronic norm that will yield its intuitive verdicts, but also yield the right verdicts about cases of memory loss. For instance, Williamson 2000 argues that "a theory of evidential probability can keep separate track of evidence and still preserve much of the Bayesian framework" (220). According to Williamson, your current credences are not constrained by your past credences, but by your current evidence. At any given time, your current credence in a proposition should match the prior conditional probability of that proposition, conditional on your current evidence. The prior probability distribution is a distinguished measure of "something like the intrinsic plausibility of hypotheses prior to investigation" (211), and your current evidence is just your current knowledge (185). Since knowledge is not necessarily cumulative for rational agents, this proposal answers challenges involving rational memory loss. Since your current mental states include your current knowledge, this proposal advances time-slice epistemology.

The same strategies can be used to respond to other challenges to Conditionalization. For instance, it is a familiar observation that without forgetting anything, rational agents can start out certain that some *de se* proposition is false, and then later have some credence in that same proposition. For example: you may rationally be sure that it is not yet after midnight, and then later have some credence that it is after midnight. This sort of rational credal change is incompatible with Conditionalization as it is traditionally stated. Again, there are two strategies for responding to the challenge. Some theorists simply restrict the diachronic norm. For instance, Titelbaum 2013 endorses "Limited Conditionalization," which "looks exactly like Conditionalization, except that it applies only when [an agent] retains all certainties at the later time that she had at the earlier time" (124). By contrast, time-slice epistemologists will again trade in Conditionalization for a synchronic norm, one that yields the right verdicts about cases of *de se* updating. For instance, the theory of updating defended in Moss 2012 constrains your current credences in *de se* propositions using only facts about your current mental states, namely your current memories and your current opinions about the passage of time. To sum up so far: in response to counterexamples to diachronic norms, we can simply restrict those norms so that they do not apply in cases where they would yield bad results. Or we can come up with alternative synchronic norms that yield intuitive verdicts in the challenging cases.

All else being equal, the second strategy wins. From the point of view of theory building, the repeated restriction of diachronic norms is unsatisfying. The simplest diachronic norm would say that your present opinions should match your past opinions. But this norm fails when you have more information than your past self, or less information than your past self, or different *de se* information than your past self. The problematic cases for diachronic norms are exactly those cases where your past opinions do not have their usual effects on your current mental states. Usually you remember and trust what your past self believed. But when this connection fails, diachronic norms yield counterintuitive consequences. Time-slice epistemology is a natural response to this pattern of observations. Instead of restricting diachronic norms to cases where your past credences have their usual effects on your current mental states, we should admit that your current mental states are what determine whether your current credences are rational.

The same goes for the normative import of your future opinions and desires. The simplest future-oriented diachronic norm for belief would say that your present opinions should match your future opinions. This norm is false when you are not sure what you will later believe, when you are wrong about what you will believe, when you think you might get misleading evidence, or when you think you might later be irrational. The simplest diachronic norm for desire would say that your present desires should match your future desires. This norm fails when you are not sure what you will later desire, when you are wrong about what you will desire, or when you think you might come to have some despicable desires, either as a result of some rational change like joining another political party or some irrational change like being influenced by drugs. Usually you anticipate and trust what your future self believes, and you want what your future self desires. But when these connections fail, it is no longer intuitive to suppose that your future states should constrain your current beliefs and desires.

The foregoing discussion reveals two motivations for preferring time-slice theories over restricted diachronic norms. First, time-slice theories are generally stronger, yielding verdicts in cases where restricted norms are silent. Second, time-slice theories are generally simpler. They do not include ad hoc maneuvers around potential counterexamples. Instead they describe the indirect ways in which your past and future mental states do affect what you should currently believe and desire. To appreciate the contrast, consider analogous debates about interpersonal norms of rationality. The simplest interpersonal norm for belief would say that your opinions should match the opinions of your neighbor. This norm fails when you are not sure what your neighbor believes, when you are wrong about what they believe, when you think that your neighbor has misleading evidence, or when you think that your neighbor is irrational. In response, we could restrict interpersonal norms so that they do not apply in these cases. But it is far more natural to say that how your neighbor affects what you should believe depends on how much you trust that she has reasonable beliefs. The normative import of her beliefs is mediated by your opinion about her. The time-slice epistemologist makes just the same move in the intrapersonal case.

This reflection on interpersonal norms offers more than just a helpful analogy. In fact, it suggests a second definition of time-slice epistemology. Time-slice epistemologists say that what is currently rational for you does not fundamentally depend on your past and future attitudes. This insight can be developed in multiple ways. First, what is rational for you might depend on your current mental states. As we have seen so far, we can define time-slice epistemology as the claim that the fundamental norms of rationality are *synchronic*, grounding the rationality of your current states and actions in facts about your current mental states. Second, what is rational for you might depend on very general relations that hold between your current self and your past and future selves. In particular, these relations might be general enough that they hold not just between distinct temporal parts of a single agent, but between distinct agents. Hence we could alternatively define time-slice epistemology as the claim that the fundamental norms of rationality are *impersonal*, so that your rationality is grounded in facts about normative relations that hold between persons as well as between temporal parts of persons. The handle 'time-slice epistemology' is better suited for the first theory, but not entirely unfit for the second. Compare: the United States Supreme Court's ruling in *Citizens United v. Federal Election Commission* confers meaningful legal status on corporations by conceiving of them as subjects of legal norms traditionally reserved for persons. The second notion of time-slice epistemology confers meaningful epistemic status on temporal parts of persons by conceiving of them as subjects of epistemic norms traditionally reserved for persons. By featuring in fundamental epistemic norms, time slices play a more significant role in our theorizing about rational belief and action.

Like our first notion, our second notion of time-slice epistemology has been defended in recent literature. For instance, Christensen 1991 observes that distinct agents are not rationally required to have beliefs that cohere with each other simply because they are guaranteed to lose money otherwise, and he concludes that the same goes for distinct time slices of individuals: "the guaranteed betting losses suffered by those who violate Conditionalization have no philosophical significance" (246). Hedden 2013a expands on this conclusion, arguing that we should treat distinct time slices like distinct agents in many cases where their collective action is against their collective interest. Hedden 2013b advocates "moving to an independently motivated picture of rationality which treats any in*tra*personal requirements of rationality as deriving from more general requirements that apply equally in the in*ter*personal case" (36). Hedden goes on to advocate claims that resemble both definitions of time-slice epistemology, often defending them simultaneously.

Both definitions of time-slice epistemology have evolved as responses to the aforementioned challenges for traditional norms. It is not a coincidence that the theories are responsive to the same challenges. The first and second definitions are naturally related, since synchronic and impersonal norms are intimately connected. If the rational import of your past and future attitudes is mediated by your current opinions about those attitudes, then your opinions about the attitudes of other agents will often have that same import. Conversely, if the rational import of other agents is mediated by your opinions about their attitudes, then your opinions about your past and future attitudes will often have that same import.

However, it is important to recognize that while the first and second definitions of time-slice epistemology are connected, they are indeed independent claims. For starters, some norms are essentially synchronic and personal. The standard principle of Reflection defended by van Fraassen 1984 is one example. Reflection demands that your current credence in a proposition match your expected future credence in that proposition. The norm is synchronic, since your expectations of credences are among your current mental states. But Reflection is personal, as it assigns special rational import to your expectations of your own future credences, as opposed to the future credences of other agents. The Qualified Reflection norm defended by Briggs 2009 can helpfully direct us to impersonal revisions of Reflection. Qualified Reflection says roughly that if it is given that you will later have some particular credence in a proposition as a result of rationally updating on veridical evidence, then you should already have that very credence in that proposition. This norm naturally follows from an impersonal norm. Suppose that you are certain that some agent with all your evidence has rationally updated on some additional veridical evidence. Then given that she has some particular credence in a proposition, you should have that very credence in that proposition. This norm applies whether the agent in question is someone else or some other time slice of yourself. In developing this impersonal replacement for Reflection, we are not advancing our first notion of time-slice epistemology, as each of the norms just considered is synchronic. But we are advancing the second notion of time-slice epistemology, and thereby addressing many of the same concerns that motivated our earlier rejection of diachronic norms.

In addition to synchronic personal norms, some theorists may accept impersonal norms that are essentially diachronic in the intrapersonal case. Burge 1993 is one example. Burge argues that getting information from your past self is like getting information from other agents, as both memory and testimony involve "purely preservative" processes that directly transfer justification from one self to another. Suppose you have a justified belief and that I get this belief from you by testimony. Burge argues that my justification for my belief may have nothing to do with facts about me, such as the fact that I just heard you express the belief or the fact that I believe that you are reliable. The warrant may instead be just the same warrant that you have for your belief. In the same way, my inherited justification for remembered beliefs may have nothing to do with facts about my current mental states. Hence Burge may accept the second notion of time-slice epistemology without endorsing the first. The same goes for Lackey 2008 when she compares memory and testimony, while arguing that "it is not enough for testimonial justification or warrant that a hearer have even epistemically excellent positive reasons for accepting a speaker's testimony—the speaker must also do her part in the testimonial exchange by offering testimony that is reliable or otherwise truth-conducive" (155).

Both definitions of time-slice epistemology are natural and important. Neither has a stronger claim to fame. By contrast, it is worth considering a third definition that may initially seem attractive, namely the thesis that what is rational for your current time slice supervenes on its intrinsic properties. This definition is admirably simple, but ultimately less natural and compelling than those we have considered so far. For instance, the definition entails that time-slice epistemology is simply off-limits for many externalists, including anyone who accepts both that what you should believe depends on what knowledge you have and also that intrinsic duplicates can differ with respect to what knowledge they have. This is an unwelcome result, as the spirit of time-slice epistemology is intuitively independent of debates over epistemic externalism. In addition, it is not clear that the intrinsic properties of time slices even include the sort of mental properties on which normative facts are meant to supervene. It may well be that, strictly speaking, instantaneous temporal parts of objects have mental states only in virtue of having certain extrinsic properties. By contrast, the synchronic and impersonal definitions of time-slice epistemology merely require that there are facts about what mental states you are in at a particular time. These definitions allow that you may have a certain desire at a particular time partly in virtue of properties that you satisfy at other times, just as you may be painting a house at one instant in time partly in virtue of properties that you satisfy at nearby times. The same goes for your beliefs, memories, and other mental states.[4]

2. A CASE STUDY: NORMS FOR ACTION UNDER INDETERMINACY

Having addressed the first programmatic goal of this paper, I will now turn to assessing particular time-slice theories. Our central case study begins with a thought experiment from van Inwagen 1990:

> Suppose that a person, Alpha, enters a certain infernal philosophical engine called the Cabinet. Suppose that a person later emerges from the Cabinet and we immediately name him 'Omega'. Is Alpha Omega? ... Let us suppose the dials on the Cabinet have been set to provide its inmates with indeterminate adventures. (We need not agree on what would constitute an indeterminate adventure to suppose this. Let each philosopher fill in for himself the part of the story that tells how the dials are set.) Alpha has entered and Omega has left. It is, therefore, not definitely true or definitely false that Alpha is Omega. (243–4)

Following Williams 2014, we can use this story to raise questions about how agents ought to act in indeterminate decision situations.[5] Say that you are

[4] A referee worries that some norms govern "movements of mind," such as the norm requiring you to *change* your beliefs when they are inconsistent. But the point of time-slice epistemology is that we can derive such requirements from synchronic, impersonal norms. It is a fundamental fact that you are required to have consistent beliefs. This fact entails the less fundamental fact that you must reject some of your earlier beliefs when those beliefs were inconsistent, just as you must reject inconsistent beliefs held by others.

[5] For readers skeptical about whether there are any genuinely indeterminate decision situations, the arguments of this paper may be more felicitously applied to situations where agents have imprecise credences about facts relevant to their decisions (cf. §3).

Alpha. You are walking toward the Cabinet. It is indeterminate whether you will survive what is about to happen inside it. How should you make decisions about the future? Say that a broker offers you an investment: if you pay him ten dollars now, he will pay Omega twenty-five dollars when Omega comes out of the Cabinet.[6] Should you take the bet? Standard decision theory does not answer the question for you. Williams argues that a complete account of indeterminacy should tell you how to respond to the broker, i.e. which if any responses are permissible, and which if any are obligatory.

The decision theory defended in Williams 2014 yields a straightforward verdict about the broker case, namely that it is okay for you to take the investment, and okay for you to reject it. In fact, when it comes to isolated, one-off decision situations, the decision theory is simple. In any case where the supervaluationist says that some indeterminate claim has multiple sharpenings, it is permissible for you to act as if any of those sharpenings is certainly correct. In other words:

> CAPRICE:
> An isolated action is currently permissible for you just in case there is some sharpening such that the action has highest expected utility according to your current utility function and your current conditional credence function, conditional on that sharpening being correct.

In the Cabinet case, the indeterminate claim relevant for your decision is the claim that you are identical with Omega. The claim has two sharpenings: either you are Omega, or you are not. The first sharpening sanctions your taking the investment. The second sanctions your rejecting it. Hence we may conclude that either action is permissible for you.[7]

Caprice is restricted to isolated decisions, cases where an indeterminate claim is relevant for your decision but has never before been relevant for any others. Hence Caprice is not a complete decision theory, because it does not yield verdicts about diachronic decision cases. Say that after you accept or reject the investment, the broker offers you a loan. He will immediately pay you fifteen dollars, in exchange for charging Omega twenty-five dollars when Omega comes out of the Cabinet. The investment and the loan together constitute a great pair of bets. The payoffs are just the same as in the bets discussed in Elga 2010. If you accept both the investment and the loan, you will end up five dollars ahead: you immediately pay ten dollars and receive fifteen, while Omega pays and receives twenty-five dollars later. Caprice tells

[6] For simplicity, I assume throughout that agents are certain of the relevant details of their decision situations. In order to sidestep concerns about whether you could spend ten dollars before entering the Cabinet, we could imagine the broker adding or subtracting from your immediate felt pleasure rather than from your bank account.

[7] There is a further complication in the theory defended in Williams 2014, namely that strictly speaking, your action must be randomly chosen from among the actions sanctioned by sharpenings. This complication does not affect my arguments, and so I will set it aside for sake of simplicity.

you that you may accept or reject the investment. But it does not say what you should do about the loan.

There are multiple ways of extending Caprice into a more complete decision theory. For instance, our complete theory may say that your previous decision does not constrain your current rational actions. To state the norm precisely, let us say that an action is currently *sanctioned by a sharpening* for you just in case it maximizes utility according to your current utility function and your current conditional credence function, conditional on that sharpening. Then we may extend Caprice as follows:

> LIBERAL:
> An action is permissible just in case it is sanctioned by some sharpening.

This norm is implicitly indexed: an action is permissible for an agent at a time, and a sharpening sanctions an action for an agent at a time. According to Liberal, the condition mentioned in Caprice is necessary and sufficient for the permissibility of any action.

By contrast, we could instead expand Caprice by saying that if you rejected the investment, you cannot also reject the loan. This extension of Caprice is inspired by the following three claims in Williams 2014:

> [A]gents should strive to make their actions dynamically permissible. (16)

> We call an action *dynamically permissible* at time *t* just in case it maximizes utility on some sharpening live at the score at *t*. (16)

> When an action is carried out that is permissible on some but not all sharpenings, the score updates by eliminating those on which it is not permissible. (16)

Here is one interpretation of these passages: let us say that a sharpening is *live* for you just in case it sanctioned all of your past actions at the time at which you did them. Then we may extend Caprice as follows:

> RESTRICTIVE:
> An action is permissible just in case it is sanctioned by some live sharpening.

Like Liberal, Restrictive is implicitly indexed. In particular, a sharpening is live for an agent at a time. The idea is simple: if your previous actions were sanctioned by some sharpenings but not others, then an action is currently permissible for you just in case it is consistent with how you acted before. If you rejected the investment earlier, then also rejecting the loan is not sanctioned by a live sharpening, and so it is not dynamically permissible, and so you should not do it. According to Restrictive, what you did earlier constrains what actions are currently permissible for you.[8]

[8] If you have already rejected the investment, some theorists may say that rejecting the loan is permissible, while the joint act of rejecting the investment and the loan is not (cf. Caprice in

There are two natural ways of implementing Restrictive. The set of live sharpenings may be independent of your current mental states, in which case Restrictive will clearly conflict with our first notion of time-slice epistemology. By contrast, it may be that your current mental states are rationally constrained so that they determine which sharpenings are currently live for you. For instance, it may be that your opinions about indeterminate propositions are rationally constrained to evolve over time in a way that reflects how you have already acted on those propositions. In that case, Restrictive will be compatible with our first notion of time-slice epistemology, since which actions are permissible for you will supervene on facts about your current mental states. But that is only because we will have accepted another constraint incompatible with time-slice epistemology: which opinions are rationally permissible for you will not supervene on facts about your current mental states, but will be partly determined by independent facts about how you have acted before. Hence either way, fans of Restrictive will end up endorsing some norm at odds with our first notion of time-slice epistemology. The same goes for our second notion. Restrictive resembles Reflection. As you deliberate about what you should believe, Reflection demands that you assign a special normative status to certain future beliefs, simply because they are *your* future beliefs. In the same way, as you deliberate about how you should act, Restrictive assigns a special normative status to certain past actions, simply because they were *your* past actions.

To sum up the dialectic so far: both Liberal and Restrictive entail Caprice, while also saying something about how you may act in repeated decision situations. Liberal is a synchronic and impersonal norm. The set of sharpenings of some indeterminate claim need not have anything to do with your actions at other times, or indeed with you in particular. Hence according to Liberal, your rational options are independent of your earlier actions, just as they are independent of the actions of any other agent. In order to implement Restrictive, you must accept some norm at odds with time-slice epistemology. The choice between Liberal and Restrictive is an illuminating case study in the development of time-slice epistemology. I have already made some general remarks in favor of time-slice theories. In the next section, I will defend Liberal against specific objections often raised by advocates of norms like Restrictive.

3. ANSWERING ARGUMENTS AGAINST LIBERAL

Caprice, Liberal, and Restrictive are theories about how you should act when faced with indeterminacy. They have close cousins, namely theories about how you should act when your evidence is limited. For example, suppose that you are merely deeply ignorant of what happens inside the Cabinet.

Weatherson 2008 and Sequence in Elga 2010). If that is right, then intuitively agents should be interested in which actions are such that performing them will not make it the case that you have performed an impermissible sequence of actions, and readers may interpret 'permissible' in the text as denoting this property.

There is a determinate fact of the matter about whether you will survive, but your evidence does not support a particular precise credence about your survival. Perhaps many precise credence distributions are rational given your evidence, or perhaps your evidence licenses only imprecise credal states that contain many precise credence distributions as members. Either way, your evidence does not uniquely determine how you should act. The analog of Liberal for imprecise agents says that an action is permissible just in case it is sanctioned by some member of your imprecise credal state, while the analog of Restrictive says that an action is permissible just in case it is sanctioned by some member that also sanctions all your previous actions. These theories have been discussed in recent literature: Elga 2010 and White 2010 present challenges for both, while Weatherson 2008 explores a third option that combines attractive features of each. This paper focuses on Caprice and Liberal, but many of my arguments apply equally to decision theories for imprecise agents. The critical comments in this section are largely inspired by literature on the latter.[9]

There are a couple of arguments against Liberal that have nothing to do with betting. These arguments are scarce in print but common in conversation, and so they merit some discussion here. The first argument is that it could not be rational to first act according to one sharpening and next according to another, without some *reason* for changing how you act. If you first act as if you will survive the Cabinet, for instance, then you cannot start to act otherwise for no good reason. This argument may seem compelling, until we observe that similar reasoning yields conclusions that contradict standard principles of decision theory. It is widely accepted that if multiple actions each have maximal expected utility, then each of the actions is permissible, regardless of whether you have chosen between just these actions before. It is permissible to act one way and then another, with no reason for changing how you act. Since some groundless switching between alternatives is permissible, it cannot be that Liberal is incorrect merely in virtue of permitting some groundless switching.

In addition, it is not clear why the argument under consideration is any better than an analogous argument for the opposite conclusion, namely that you cannot rationally keep acting in the same way without some reason for *continuing* to act that way. The fact that you acted some way before has no intrinsic epistemic significance. Insofar as you must have reasons why you are acting according to some sharpening, the mere fact that you have acted on some sharpening before constitutes just as much reason as the mere fact that

[9] For all I have said, it may be that some cases of action under indeterminacy are best treated with theories that govern imprecise agents, while others deserve another treatment entirely. For instance, sometimes it may be *indeterminate* whether actions are permissible when those actions lead to indeterminate outcomes. Dougherty 2014 argues that cases of vagueness generate indeterminate ethical judgments. Rinard 2013 defends an alternative decision theory according to which it is often indeterminate whether actions are permissible for imprecise agents. I regret that I cannot explore these theories in more detail here.

you have never acted on some sharpening before. This point about epistemic reasons can be made more clear by comparing it with the same point about moral reasons. The fact that you have done some action before may be good evidence that the action is morally permissible, provided that you are a good person. But that fact itself does not usually have intrinsic moral significance in your deliberation about which actions are currently permissible for you.

The second common argument against Liberal is that the theory is impractical or even impossible to implement. The complaint is that Liberal requires that you reconsider every decision made in the face of indeterminacy, constantly reassessing actions from moment to moment. It is plausible that ordinary agents cannot constantly reassess their actions. If ought implies can, then we may conclude that Liberal is false. The same sort of objection applies to several other time-slice theories. It may not be possible for agents to reinvent their credences from moment to moment, for instance, as may seem required by the updating norms defended by Williamson 2000 and Moss 2012.

To respond: this objection misidentifies the subject matter of Liberal and other time-slice theories. The norms articulated in an epistemology classroom govern deliberating agents. Time slices that are not deliberating are simply not in the scope of Liberal. It may be true that people often chug along without deliberating, responding to any indeterminate claim as they did before, without reconsidering what sharpening they are acting on. It may even be true that people cannot survive without acting in this way. But this does not challenge norms that tell agents what they should do when they do deliberate. To compare: it may be true that people often fall asleep and hence fail to consider or assess any reasons at all, and it may even be true that people cannot survive without sleeping. But this fact about human nature does not challenge ordinary norms governing lucid agents.

The most compelling argument against Liberal is pragmatic. Recall that Liberal says that you can reject each of a pair of bets when accepting both would guarantee you sure money, and when you do not gain or lose any evidence about the bets as they are offered to you. In a recent paper about agents with imprecise credences, Elga 2010 claims that results like these are unacceptable. Elga explains: "rejecting both bets is worse for you, no matter what, than accepting both bets. And you can see that in advance. So no matter what you think about [whether you will survive the Cabinet], it doesn't make sense to reject both bets" (4).[10] This argument resembles a diachronic Dutch book argument. Liberal does not force agents to act on different sharpenings over time, and so agents following Liberal cannot be guaranteed to lose money. But it appears that rational agents should not only prefer keeping money over losing it, but also prefer gaining money over gaining nothing at all.

[10] Joyce 2010 develops a related argument against Liberal and endorses a consistency requirement that "ensures that at least one of Elga's bets will always be chosen by any agent using any reasonable decision rule" (316).

This pragmatic argument depends on the claim that it is impermissible for informed agents to forego sure money. But on reflection, this claim is not so clear. In fact, sometimes it seems that we ought to be forgiving of agents who forego sure gains. The most familiar examples of this phenomenon do not involve agents who are torn between beliefs, but agents who are torn between values.[11] To take an example from Sartre 1946: you must either join the Free French as a soldier in England, or stay home to care for your ailing mother. Suppose that after several days of agonizing reflection, you board a train for England. But on the train, you have a change of heart. The situation has not changed, i.e. you have just the same evidence and just the same values as you did the day before. But you regret joining the army, and you feel resolved to care for your mother. In this situation, it seems perfectly permissible for you to get off the train to England and head home. In the literature on moral dilemmas, several authors aim to predict this sort of result. Raz 1997 argues that when facing moral dilemmas, "we are within our rights to change our minds" (119). Broome 2000 argues that you may "make the best of a bad job" in such situations (34). It is true that if you return home, you could have done better overall. Instead of buying train tickets, you could have saved your money and definitely come out ahead. But that does not mean that you are inextricably bound to your decision from the moment you board any train.

The same goes for situations where you are deeply torn between options, not because you are torn between values, but because you are torn between beliefs. For example, suppose that your friend is vacationing out of the country, and that some horrible wildfires have started to destroy the city where he lives. There are several family photograph albums in his apartment and several valuable manuscripts in his university office. It is impossible for you to save both, and impossible for you to contact your friend to ask which he would prefer you to save. Suppose that after several minutes of agonizing reflection, you board a subway train for his office. But on the train, you change your mind. The situation has not changed, i.e. you have just the same evidence about your friend as you did before. But you regret heading for the manuscripts, and you have made up your mind that your friend would rather you save the photographs. In this situation, it seems perfectly permissible for you to head for his apartment. It is intuitively permissible even if it means that you end up eating the cost of your subway ticket. The ticket is a sunk cost, and you may rationally ignore it. If this intuitive judgment is right, it is not always impermissible for informed agents to forego sure money.[12]

[11] See Moss 2014 for a more detailed development of this response to the arguments in Elga 2010. For further connections between imprecise agents and agents with incommensurable values, see the discussion of insensitivity to evidential sweetening in Schoenfield 2012.

[12] To clarify the dialectic: some theorists accept that your evidence uniquely determines which precise credence function you should have. Fans of this uniqueness claim have a ready response to pragmatic arguments against Liberal, namely that rational agents are never in situations where Liberal recommends foregoing sure money. The point of the present discussion is that Liberal and similar time-slice theories can and should be accepted even by those who reject the uniqueness claim.

To sum up so far: there are many ways to be deeply torn between options. Agents may be torn between values or between beliefs, and they may be torn between beliefs because they have limited evidence or because they recognize that there is no fact of the matter about some question. It is not clear that agents in these situations are strictly forbidden from changing their minds. In fact, we are intuitively disposed to forgive some agents who forego sure money, even when their change of heart is not prompted by any change in their evidence. I do not mean to suggest that the moral dilemmas literature univocally supports our intuitions about the above cases. For instance, Chang 1997b argues that rational agents cannot have incommensurable values, precisely on the grounds that practical reason prohibits agents from being "merit pumps" (11). But it is telling that Chang focuses on an example in which an agent accepts unfortunate trades with no hesitation or reflection. In general, we are most inclined to reject apparent mind changing as irrational when it happens quickly, unreflectively, repeatedly, or for strategic reasons. These intuitions can be comfortably accommodated by a theory according to which changing your mind is not itself impermissible, namely because the salient features of these cases may provide evidence that they do not involve the same sort of genuine changes of mind exhibited by agents in the Sartre case and the wildfire case. By contrast, it is more difficult for blanket injunctions against mind changing to accommodate the intuition that changing your mind can sometimes be okay.[13]

From a third-person perspective, we sometimes forgive agents for changing their minds. But there is one special sense in which mind changing never seems permissible. From a situated first-person perspective, changing your mind may seem wrong for first-order reasons. For instance, it may seem that you should not change your opinion about a proposition because you would then have the wrong opinion about that proposition. The foregoing permissive theory of mind changing can accommodate this sort of judgment. If you are acting according to some sharpening, other actions will always seem wrong simply insofar as they are not sanctioned by that sharpening. If you are acting as if you will survive the Cabinet, then accepting the loan will seem wrong simply insofar as it does not have maximal expected utility for you conditional on your surviving. From a reflective third-person perspective, you may acknowledge that accepting the loan is sanctioned by some sharpening, which constitutes an important sense in which accepting the loan is just as permissible as rejecting it. But even as you reflect, you may act according to a sharpening that does not treat these actions as equally permissible. Here it is especially useful to think of different time slices as being like different

[13] The arguments here and in Moss 2014 are limited: rather than arguing that mind changing is always permissible, I defend mind changing against norms that say that it is never permissible. This defense is incompatible with blanket injunctions against mind changing, but compatible with more nuanced theories according to which rational agents can have mental states that preclude mind changing. For further discussion of such theories, see Hinchman 2003, Korsgaard 2008, Holton 2009, Bratman 2012, and Ferrero 2012.

agents. Say that you first reject the investment offered by the broker, and later reject the loan. Then your later slice will judge that your earlier slice acted incorrectly, and your earlier slice would have said the same about your later slice. But there is an important sense in which you may reflectively judge that this disagreement is faultless. This judgment is captured by our reflective endorsement of Liberal.

The notion of faultless disagreement is familiar from literature about agents with different tastes and values. For instance, we may simultaneously endorse something as beautiful or fun or valuable, while reflectively judging that other evaluations of it are not wrong. Here again, familiar literature about valuing can help us better understand what to say about acting under indeterminacy.

This connection is not an accident, but part of a larger pattern. It is reasonable to expect similarities between agents acting on incommensurable values and agents acting on indeterminate or imprecise credences. For one thing, some imprecise credences may themselves be the product of incommensurable values, namely incommensurable epistemic values. For example, it may be that different hypotheses about the Cabinet are supported by different prior probability distributions, where these priors encode incommensurable ways of balancing epistemic values such as strength, elegance, and simplicity. To make matters worse, it may ultimately prove difficult even to distinguish between the state of having incommensurable values and the state of believing that there is no precise fact of the matter about what is valuable. The mental states of believing and valuing may not function as independently as classical decision theory would have us believe, which could give us further reason to expect literature on incommensurable values to provide us with fruitful analogies for theories of action under indeterminacy.[14]

4. DEVELOPING ARGUMENTS AGAINST CAPRICE

Liberal is a promising theory of action under indeterminacy. But it is not perfect. Recall that Liberal entails Caprice, namely that an action is permissible in an isolated decision situation just in case it is sanctioned by some sharpening. Williams 2014 credits Elga with raising a problem for a close cousin of this latter principle.[15] Suppose that the broker offers you the same investment as before, only now he also offers you a third option. Instead of immediately accepting or rejecting the investment, you may choose to delay your decision. If you delay, the broker will offer you the investment again in five minutes, and pay you one dollar for waiting. Elga claims that in this situation, it is intuitively permissible for you to delay your decision.

[14] For further discussion, see Lewis 1988 and Price 1989.

[15] Elga actually raises a problem for the combination of a claim like Caprice and the claim that your action must be randomly chosen from among the actions sanctioned by sharpenings. Since I am interested in assessing Liberal, I will present a version of the problem that constitutes a challenge to Caprice itself.

Williams 2014 agrees that delaying is intuitively permissible: "After all, Alpha won't close off any of the rival options, and he'll gain a dollar whichever way he goes" (19). The problem is that Caprice entails that when you have some significant credence that you would act on a different sharpening later, it can be impermissible for you to delay your decision. In particular, any sharpening that you could act on will fail to sanction delaying whenever the small amount you would gain by waiting fails to outweigh the expected possible loss of your making the wrong decision about the investment later.

In response, Williams says that this problem for Caprice arises only if we neglect some standard assumptions about the rationality of agents. He says that when assessing the permissibility of actions, we standardly assume that agents are certain that they are rational and will remain rational. From this assumption, Williams concludes: "the credences induced by the sharpening [that entails that Alpha is Omega] will say that rationality requires investing; and hence (given that they assume that the agent will do what is rational) those credences will assign full probability to the agent investing" (25). In other words, the part of you that believes that you will survive the Cabinet also believes that you will accept the investment later if you delay, and the part of you that believes that you will not survive also believes that you will reject the investment if you delay. If we understand Caprice as saying that you may act according to any of these opinions, then it is permissible for you to delay your decision. Williams is cheered by this result, and ultimately this argument constitutes the response to Elga that he most prefers.

Unfortunately, this interpretation of standard rationality assumptions yields several unhappy consequences. For starters, the interpretation entails that you are rationally *compelled* to delay your decision, since delaying always has highest expected utility according to your conditional credence function, conditional on some sharpening together with the claim that you will later act according to that sharpening. But intuitively, it is sometimes permissible for you to just go ahead and accept the investment from the broker. Insofar as part of you can say, "I should accept this investment, namely because I will probably survive to collect on it," that part of you can also say, "I should accept this investment as soon as possible, namely because I will probably survive and there is a real chance that I will miss out on a great investment if I delay." This intuition becomes stronger as delaying is accompanied by smaller sure gains and larger possible losses.

The simplest response to this objection would be to weaken the assumption that you must act as if you are certain of some sharpening and also *certain* that you will always act according to that sharpening. Perhaps we should assume only that you must act as if you are certain of some sharpening and also have at least some threshold credence that you will later act according to it.[16] Here is a precise proposal that could replace Caprice: consider the constraint that it is both certain that some particular sharpening is correct and also fairly

[16] Robbie Williams suggested this response to me in personal communication.

likely that you will later act according to that sharpening. We could say that an action is permissible for you just in case it maximizes utility according to your utility function together with your credence function after it has been updated on some constraint of just this sort.

This tempered proposal would allow you to accept the investment from the broker. But the proposal shares some other unhappy consequences with the proposal that Williams defends. For example: intuitively, your decision about delaying could be informed by independent evidence about what would happen if you did delay. Let us suppose that if you were forced to accept or reject the investment, you would accept it. And suppose you know that when it comes to indeterminate questions about your survival, you tend to change your mind a lot. It is rare for you to form intentions and carry them out without vacillating. Then your self-aware opinions may recommend that you accept the investment immediately, while the artificial conditional credence functions relevant for the tempered proposal may recommend that you delay. In more generality: when faced with an indeterminate claim, you have an imaginary "mental committee" of sharp opinions about how you should act. The proposals considered so far mandate wishful thinking on the part of each of your mental committee members, e.g. each member is confident that you will do just the right thing if you delay your decision, accepting the investment if and only if you will survive the Cabinet. But that means your decisions may not be appropriately responsive to your evidence. Proposals that require you to act on sufficiently optimistic credence functions blunt the force of relevant information, such as facts about the likelihood and relative cost of your changing your mind.

The possibility of side bets raises a related problem. In the initial simple broker case, the proposal in Williams 2014 requires you to bet at any odds that you will either accept the investment and survive the Cabinet, or reject the investment and fail to survive. This requirement is counterintuitive; rationality should not mandate hubristic certainty about your decisions. It seems fine for you to hedge your bets, accepting significant side bets that pay off just in case you are making the wrong decision about investing. The permissibility of hedging is forcefully illustrated by other hypothetical decision situations. Suppose that instead of entering the Cabinet yourself, you are about to send your pet hamster Fluffy into the Cabinet. Fluffy is valuable to you, and ordinarily you would pay up to fifty dollars to ensure her survival. As Fluffy enters the Cabinet, you see that there is a lethal device attached to the exit door that will kill any creature that emerges. If it only costs fifty cents to disarm the device, it seems intuitively permissible for you to pay the fifty cents to save the creature that will emerge from the Cabinet. But if you are willing to pay the fifty cents, you are not intuitively also obligated to pay up to fifty dollars to disarm the device. As the price of disarming the device dramatically increases, you may eventually decide that it is not worth the money.

Another decision situation involving hedging highlights a final problem for Caprice. The situation comes from Weatherson 2008:

> An agent is told (reliably) that there are red and black marbles in a box in front of them, and a marble is to be drawn from the box. They are given the choice between three bets. α pays $1 if a red marble is drawn, nothing otherwise, β pays a certain 45 cents, and γ pays $1 if a black marble is drawn. (12)

Let us suppose that which color marble is drawn is fixed by some indeterminate claim, like the claim that you survive the Cabinet. Not everyone has clear intuitions about this decision situation. But some report that it does not seem irrational to prefer β over the other gambles. If that is correct, Caprice is incorrect. If you act on one sharpening, you act as if you are certain that a red marble was drawn, and so you prefer α over the other gambles. If you act on the other, you prefer γ. Hence accepting the certain forty-five cents is not sanctioned by any sharpening. If choosing β is rationally permissible, then we have yet another case where Caprice seems too uncompromising. Williams 2014 states that "action under indeterminacy does not tolerate compromise" (29). But our intuitions suggest that it should.

5. ALTERNATIVE THEORIES OF ACTION UNDER INDETERMINACY

In light of these problems for Caprice, it makes sense to look for alternative theories about making isolated decisions in the face of indeterminacy. Then we can extend these theories to get alternative time-slice theories of action under indeterminacy, improvements on the Liberal theory defended in §3 above.

Here is one alternative to Caprice: instead of identifying your mental committee members with particular sharpenings, we could identify them with nuanced opinions about sharpenings. This proposal would unify what we say about cases where your evidence is indeterminate and about other cases where your evidence fails to uniquely determine how you should act. The idea is that in the former cases, you should act as you do when you have imprecise credences about determinate propositions. This imprecise credal state is a set of credence distributions, each of which assigns some precise credence to each sharpening of the indeterminate proposition under discussion. The accompanying decision theory resembles Caprice, only with members of the imprecise credal state standing in for sharpenings. In more detail: an isolated action is currently permissible for you just in case it maximizes utility according to your current utility function and some member of the imprecise credal state. For example, in addition to acting as if you will certainly survive the Cabinet or acting as if you will certainly not survive, you may act as if you think those outcomes are equally likely, as long as the relevant imprecise credal state contains this third opinion. The corresponding analog of Liberal is straightforward: in any decision situation, an action is permissible just in case it maximizes utility according to your current utility function and some member of the imprecise credal state.

This alternative decision theory is neutral about the nature of the imprecise credal state that determines which actions are permissible for you. There may well be additional bridge principles that constrain this imprecise credal state in light of your opinions about the relevant indeterminate propositions. For example, it may well be that if you are certain that some proposition is indeterminate, then you should act as if your imprecise credal state contains every single precise credence distribution over sharpenings of that proposition. On the other hand, we may sometimes feel compelled to say things like "it is determinate that you might survive the Cabinet, and determinate that you might not survive it, even though it is not determinate whether you will survive," in which case we may feel justified in eliminating only very decisive precise opinions from the relevant imprecise credal state.[17]

The revised Liberal theory under discussion accommodates our intuitive verdicts about almost all of the §4 examples. For instance, some members of your imprecise credal state may be on the fence with respect to whether you should accept or reject the investment offered by the broker. Those members will prefer delaying the investment decision over making it, since you definitely gain something by waiting, and there is no cost associated with your making either particular decision later. Hence the revised Liberal theory allows that delaying your decision is sometimes permissible. In addition, the theory allows that simply accepting the investment is sometimes permissible, namely whenever your imprecise credal state contains members that are sufficiently confident that you will survive the Cabinet. Finally, it is permissible for you to merely pay fifty cents in the hamster case whenever your imprecise credal state contains members that have some credence that Fluffy will survive the Cabinet, but not enough credence to justify paying fifty dollars to save the creature that emerges. In fact, many members of your imprecise credal state will normally have this feature, which may be partly responsible for our intuition that being willing to pay fifty cents but not fifty dollars is an eminently reasonable disposition.

The revised Liberal theory calls for just one more point of clarification. Suppose that Weatherson 2008 is right that you may prefer getting a certain forty-five cents over getting one dollar just in case you survive the Cabinet, and over getting one dollar just in case you do not survive. This preference is not sanctioned by any precise credence about your survival, according to standard decision theory. The problem is two-fold: our model of your mental state is not fine-grained enough, and standard decision theory hastily condemns certain sorts of risk aversion as irrational. Instead of identifying members of your credal state with precise credences, we should identify them

[17] In light of such bridge principles, one might worry about whether we are failing to distinguish action under indeterminacy from action on insufficient evidence. Williams 2014 rejects some theories of the former on the grounds that they do not assign a distinctive cognitive role to uncertainty induced by indeterminacy. However, it is not clear that we should expect norms of rationality to distinguish action under indeterminacy from all other sorts of action. There are multiple reasons why your evidence could fail to determine the likelihood of some outcome. It may be that the likelihood relation is not well defined, or that the outcome itself is not well defined. The failure of your evidence may matter for the purposes of evaluating your action, while the source of that failure does not matter at all.

with pairs of precise credences and *subjective risk functions*, measurements of risk aversion defined as in Buchak 2013. Instead of sanctioning just those actions that maximize expected utility, members of your credal state should sanction actions that maximize *risk-weighted expected utility*.[18] For example, risk-averse agents may have exactly 0.5 credence that they will survive the Cabinet, and yet rationally prefer getting a certain forty-five cents over getting one dollar just in case they survive. Augmented in this way, the revised Liberal theory can accommodate our intuition that you may prefer the certain forty-five cents, namely since such hedging will be rationally permissible as long as your credal state contains some sufficiently risk-averse members with middling credences about your survival.

There are many further respects in which Liberal may be revised and expanded. For example, we could use the mental committee model to represent your *values* in addition to your credences. For instance, we could identify members of your mental state with combinations of precise credences, subjective risk functions, and value functions. Then your having incommensurable values might be represented by members of your mental state having distinct value functions. Conditional values might be represented by dependencies between the credences and values of your mental committee members.

In addition, we could endorse more general procedures for deriving normative facts from features of your mental state. For instance, it may be that your permissible actions are not restricted to actions sanctioned by some member of your mental state, but instead include actions sanctioned by reasonable aggregations of the preferences of those members. For example, suppose that some member of your mental state is certain that you will survive the Cabinet, and some member is certain that you will not survive. Then it may be permissible for you to act as if you are not confident of either claim, not because your mental state contains some third member with this moderate opinion, but because moderate actions are preferred by some reasonable aggregation of your immoderate preferences. The same goes for accepting the certain forty-five cents. That action may be permissible, not because your mental state contains some risk-averse members, but because it is preferred by some reasonable aggregation of your immoderate preferences.

The theories I have defended are inspired by extant theories of agents with incommensurable values. If an agent is torn between stringent values, then intuitively she may sometimes act according to a moderate compromise of those values. That may be because she takes some moderate value function to be an additional legitimate expression of her character, or it may be that someone who identifies only with stringent values may nevertheless act according to a reasonable aggregation of those values. Either way, we can say just the same thing about agents acting in the face of indeterminacy. The real

[18] Buchak 2013 develops and defends this permissive alternative to standard expected utility theory.

world is full of compromises. You may join the army while still visiting your mother every weekend. You may accept an investment from one broker while hedging your bets with another. When it comes to agents with incommensurable values, we have both intuitive and highly theorized judgments regarding mind changing, faultless disagreement, hedging, and compromise. These judgments provide us with fruitful resources for defending time-slice theories of action under indeterminacy.

REFERENCES

Arntzenius, Frank. 2003. "Some Problems for Conditionalization and Reflection." *Journal of Philosophy*, vol. 100 (7): 356–70.

Bayes, Thomas. 1763. "An Essay toward Solving a Problem in the Doctrine of Chances." *Philosophical Transactions of the Royal Society of London*, vol. 53: 370–418.

Bratman, Michael. 2012. "Time, Rationality, and Self-Governance." *Philosophical Issues*, vol. 22 (1): 73–88.

Briggs, Rachael. 2009. "Distorted Reflection." *Philosophical Review*, vol. 118 (1): 59–85.

Broome, John. 2000. "Incommensurable Values." In *Well-Being and Morality: Essays in Honour of James Griffin*, Roger Crisp, editor, 21–38. Clarendon, Oxford.

Buchak, Lara. 2013. *Risk and Rationality*. Oxford University Press, Oxford.

Burge, Tyler. 1993. "Content Preservation." *Philosophical Review*, vol. 102 (4): 457–88.

Chang, Ruth, editor. 1997a. *Incommensurability, Incomparability, and Practical Reason*. Harvard University Press, Cambridge, MA.

Chang, Ruth. 1997b. "Introduction." In Chang (1997a), 1–34.

Christensen, David. 1991. "Clever Bookies and Coherent Beliefs." *Philosophical Review*, vol. 100 (2): 229–47.

Dougherty, Tom. 2014. "Vague Value." *Philosophy and Phenomenological Research*, vol. 89 (2):352–72.

Elga, Adam. 2010. "Subjective Probabilities Should Be Sharp." *Philosophers' Imprint*, vol. 10 (5): 1–11.

Ferrero, Luca. 2012. "Diachronic Constraints of Practical Rationality." *Philosophical Issues*, vol. 22 (1): 144–64.

van Fraassen, Bas C. 1984. "Belief and the Will." *Journal of Philosophy*, vol. 81 (5): 235–56.

Hedden, Brian. 2013a. "Options and Diachronic Tragedy." *Philosophy and Phenomenological Research*. Article published online first. DOI: 10.1111/phpr.12048.

Hedden, Brian. 2013b. "Time-slice Rationality." MS, Dept. of Philosophy, Oxford University. Forthcoming in *Mind*.

Hinchman, Edward. 2003. "Trust and Diachronic Agency." *Noûs*, vol. 37 (1): 25–51.

Holton, Richard. 2009. *Willing, Wanting, Waiting*. Oxford University Press, Oxford.

van Inwagen, Peter. 1990. *Material Beings*. Cornell University Press, Ithaca.

Joyce, James. 2010. "A Defense of Imprecise Credences in Inference and Decision Making." *Philosophical Perspectives*, vol. 24 (1): 281–323.

Korsgaard, Christine. 2008. *The Constitution of Agency*. Oxford University Press, Oxford.

Lackey, Jennifer. 2008. *Learning from Words: Testimony as a Source of Knowledge*. Oxford University Press, Oxford.

Lewis, David K. 1988. "Desire as Belief." *Mind*, vol. 97 (418): 323–32.

Moss, Sarah. 2012. "Updating as Communication." *Philosophy and Phenomenological Research*, vol. 85 (2): 225–48.

Moss, Sarah. 2014. "Credal Dilemmas." MS., Department of Philosophy, University of Michigan.

Price, Huw. 1989. "Defending Desire-as-belief." *Mind*, vol. 98 (389): 119–27.

Raz, Joseph. 1997. "Incommensurability and Agency." In Chang (1997a), 110–28.

Rinard, Susanna. 2013. "A Decision Theory for Imprecise Credences." MS., Department of Philosophy, University of Missouri, Kansas City.

Sartre, Jean-Paul. 1946. "The Humanism of Existentialism." In *Jean-Paul Sartre: Essays in Existentialism*, Wade Baskin, editor, 31–62. Carol Publishing Group, New York.

Schervish, M. J., T. Seidenfeld, and J. B. Kadane. 2004. "Stopping to Reflect." *Journal of Philosophy*, vol. 101 (6): 315–22.

Schoenfield, Miriam. 2012. "Chilling out on Epistemic Rationality: A Defense of Imprecise Credences." *Philosophical Studies*, vol. 158 (2): 197–219.

Talbott, William. 1991. "Two Principles of Bayesian Epistemology." *Philosophical Studies*, vol. 62: 135–50.

Titelbaum, Michael. 2013. *Quitting Certainties: A Bayesian Framework for Modeling Degrees of Belief*. Oxford University Press, Oxford.

Weatherson, Brian. 2008. "Decision Making with Imprecise Probabilities." MS., Dept. of Philosophy, University of Michigan.

White, Roger. 2010. "Evidential Symmetry and Mushy Credence." In *Oxford Studies in Epistemology*, Tamar Szabó Gendler & John Hawthorne, editors, vol. 3, 161–86. Oxford University Press, Oxford.

Williams, Robbie. 2014. "Decision Making under Indeterminacy." *Philosophers' Imprint*, vol. 14 (4): 1–34.

Williamson, Timothy. 2000. *Knowledge and its Limits*. Oxford University Press, Oxford.

7. When Beauties Disagree

WHY HALFERS SHOULD AFFIRM ROBUST PERSPECTIVALISM

John Pittard

In this paper I present a variant of the "Sleeping Beauty" case that shows that the "halfer" approach to the original Sleeping Beauty problem is incompatible with an extremely plausible principle pertaining to cases of disagreement. This principle says that, in contexts where rationality is not "permissive," the weight I give to my view on *p* and to my disputant's view on *p* ought to be proportional to my estimation of the strength of our epistemic positions with respect to *p*. In requiring such proportionality, the principle denies the possibility of what I will call "robustly perspectival" contexts, contexts where two maximally rational disputants who are in perfect communication are rationally required to disagree despite knowing that their epistemic positions are equally strong. Given the plausibility and widespread acceptance of the proportionality principle, its incompatibility with the halfer approach to Sleeping Beauty gives us an apparently powerful new argument against the halfer position and for the alternative "thirder" view.

But I am a halfer, not a thirder. So I go on to argue that despite the principle's intuitive plausibility, there are good reasons for thinking that the case I present here does involve a robustly perspectival context and that the principle should therefore be rejected. I suggest that the lesson that we should draw from this case is *not* that we should accept the thirder view, but rather that rationality can be perspectival in a robust way that many may find quite surprising.

The paper will proceed as follows. In section 1, I present the key example, a variant of the Sleeping Beauty case that involves multiple subjects. I then show that an intuitive way of reasoning about the case leads to the conclusion that it involves a robustly perspectival context. In section 2, I show that this conclusion is in tension with the proportionality principle mentioned above, a principle that is often taken for granted in discussions of the epistemic significance of disagreement. In section 3, I argue that halfers and thirders are committed to different views on the key case: halfers should endorse the intuitive line of reasoning outlined in section 1 that leads to the acknowledgment of robustly perspectival contexts and the rejection of the proportionality principle; thirders, on the other hand, should deny that the case involves a robustly perspectival context and can therefore maintain a commitment to the proportionality principle. The counterintuitive perspectivalism implied by the halfer position could be seen as a significant and as yet unappreciated cost to the halfer view. But in section 4, I appeal to the rational significance of evidential selection procedures in order to defend the reasonability of the striking perspectivalism to which the halfer is committed. Finally, in section 5, I turn to another case that is arguably more difficult for the halfer to handle, a case where the halfer seems to be insufficiently deferential to an acknowledged expert. I point to two plausible lines of response available to the halfer.

It should be noted at the outset that I do not in this paper offer a positive argument for the halfer view. Rather, I show that the halfer view is committed to a counterintuitive form of perspectivalism and then attempt to defend the halfer view against the objection that its perspectivalist commitments are unreasonable.

1. DISAGREEING BEAUTIES

The original Sleeping Beauty case features Sleeping Beauty and some experimental philosophers who put her to sleep and then subject her to a certain number of awakenings and to memory tampering that prevents her from knowing which awakening she is presently experiencing (Elga 2000; more on this case later). The case that will be the primary focus of this chapter involves not one "Beauty" but four: Alvin, Brenda, Claire, and Dillon, who have each agreed to spend the night in the infamous Experimental Philosophy Laboratory.[1] Before they are put to sleep, the four subjects are brought together and informed of what will happen over the course of the evening. By means of random selection, one of the four subjects will be selected as the "victim." The experimenters will know the identity of the victim, but will not disclose this information to any of the subjects. At three different times during the night, the experimenters will simultaneously awaken the victim and exactly one of the other subjects. Each non-victim subject will be awakened only once during the night. The victim and the other awakened subject will converse for a time and will be asked to discuss their views about the probable identity of the victim. After each of these conversations, the victim and the other awakened subject will be given a drug that will put them back to sleep and that will also erase their memory of the awakening. So while the victim will be awakened three times, there will be no way for the victim to determine whether a given awakening is the first, second, or third awakening. And while the non-victim subjects will be awakened only once, during their awakenings they will have no way of knowing whether they are in their first and only awakening or whether they are the victim and possibly in a second or third awakening. Steps are taken to make sure that there are no differences between the waking experiences of the victim and those of the non-victims that might serve as clues as to the identity of the victim. (So it is not the case, for example, that one of the awakened subjects will feel especially tired

[1] So named by David Lewis (2001, 171).

or "drugged up" in comparison to the other awakened subject.) When the morning arrives, all four of the subjects will be woken at the same time and sent on their way, none of them possessing any memories of their awakening (or awakenings) during the preceding night. After being made certain of all of these features of the experiment, the subjects are put to sleep in the lab and the night begins.

Let us suppose that Alvin, Brenda, Claire, and Dillon are each ideally rational agents, and that they know this about one another. It is clear that at the moment after Claire has learned the setup of the experiment but before the subjects have been put to sleep for the night, her credence that she will be the victim will be 0.25. And when Claire and the other subjects are leaving the lab the next morning, her credence that she was the victim will again be 0.25. But suppose that sometime during the night, the experimenters wake Claire, and when Claire looks across the room, she sees that Dillon has also just been awakened. At this time, what should Claire's credence be for the proposition that she is the victim?

A very intuitive line of reasoning suggests that Claire's credence that she is the victim should not change during this awakening, but should remain 0.25. Consider first the moment after Claire has been awakened but before she has looked across the room to see who else is awake. At this moment, it seems that Claire has not gained any new evidence that is relevant to the question of whether she is the victim. For before she went to sleep, Claire knew that she would be awakened sometime during the night, whether or not she turned out to be the victim, and that the features of a victim's awakening would not be discernibly different from the features of a non-victim's awakening. Moreover, it does not seem that Claire has *lost* any evidence that she possessed just before the experiment began. And the following **Relevance Principle** seems to be a very basic principle of rationality: if I start off with a credence for p that is maximally rational given my evidence, and if I neither gain nor lose any evidence that is relevant to the likelihood of p, then I should not change my credence for p.[2] The Relevance Principle, together with the claim that Claire has neither gained nor lost relevant evidence, implies that after waking (but before identifying her "co-waker") Claire's credence that she is the victim should still be 0.25.

Next, it seems that Claire's credence that she is the victim should not change upon scanning the room and seeing that Dillon is also awake.[3] For suppose that after seeing that Dillon is awake, the value of Claire's credence that she is the victim should be c. Since Claire's sharing an awakening with Dillon

[2] In his defense of the halfer position on Sleeping Beauty, Hawley (2013) endorses the Relevance Principle under the label "Inertia." Note that the Relevance Principle is very different from the "Relevance-Limiting Thesis" (so named in Titelbaum 2008, 556) that is endorsed by some halfers, a thesis that holds that learning "self-locating" information should never lead to changes in credences for non-self-locating propositions.

[3] My statement of the argument in this paragraph benefited from the comments of an anonymous referee.

gives her no more or less reason to suspect that she is the victim than her sharing an awakening with Alvin or Brenda, Claire's credence should also be *c* if she were to look across the room and see Alvin or Brenda awake. And since before looking across the room, Claire knows that she will see one of these three subjects awake, she already knows that after seeing her co-waker her credence will be *c*. But if Claire knows that her credence will be *c* once she looks across the room, then she should already have a credence of *c*. Thus, after seeing that Dillon is awake, there should be no change in the value of Claire's credence. In looking across the room and identifying her co-waker, Claire does not learn anything that ought to raise or lower her credence for the proposition that she is the victim.

Taken together, the previous two paragraphs give us an argument for the conclusion that an ideally rational agent like Claire will, during her awakening or awakenings, continue to have a credence of 0.25 for the proposition that she is the victim. I will explore an objection to the first step of this argument later in the paper. But for the moment, let us assume that the reasoning just presented is correct and explore its implications. If Claire's credence that she is the victim ought to remain 0.25, then upon seeing that Dillon is awake, Claire's credence for the proposition that Dillon is the victim ought to go up from 0.25 to 0.75. For since Claire knows that in any given awakening, one of the two awakened subjects must be the victim, she now knows that the victim is either Dillon or herself. Given that she assigns a 0.25 credence to the proposition that she is the victim, probabilistic coherence requires that her credence that Dillon is the victim be 0.75. This increase in her credence for Dillon's being the victim does *not* seem to be in tension with the Relevance Principle. For while it seems that Claire has not learned anything that ought to increase her suspicion of herself, in seeing that Dillon is her co-waker it is clear that she *has* learned something that ought to increase her suspicion of Dillon.

Since Dillon is, like Claire, a perfectly rational agent, he will reason in an analogous way. This means that, upon being awakened and seeing that Claire is being awakened at the same time, Dillon will maintain his 0.25 credence for the proposition that he is the victim and raise his credence for the proposition that Claire is the victim to 0.75.

We have, then, a case of disagreement. And it is a most perplexing one. For starters, the disagreement is not the result of one party possessing better evidence than the other, or of one party processing their evidence more rationally. For the case is perfectly symmetrical: Claire and Dillon know that the other possesses equally strong evidence and that they are both perfectly rational in how they respond to that evidence. Second, it seems to make no difference if Dillon and Claire go to the effort of fully and perfectly disclosing all of their reasons for their view. For since Dillon and Claire share in common all of the knowledge and information that led them to their views, and since they both know that they are equally informed and equally rational, discussing their reasons with one another will not reveal anything about the other's reasoning and views beyond what could already be anticipated. Claire will already know that Dillon will employ reasoning that is exactly analogous to hers, and Dillon will likewise know that Claire will engage in reasoning just like his. So discussion and debate will not bring about any changes in their respective levels of confidence.

It would be natural to describe the disagreement as follows: Claire and Dillon are mutually acknowledged "epistemic peers" who have the exact same evidence and who are perfectly confident that the other person is reasoning impeccably, and yet they are nonetheless rational in remaining steadfast and dismissing the view of their disputant.[4] But one might reasonably question whether we should describe Claire and Dillon as having the same evidence. For while Claire and Dillon can affirm the same "third-person" facts about the way the world is, they obviously cannot make the same "first-person" affirmations about their own "location" within this world. Claire, for example, can assert that she opened her eyes and saw Dillon, while Dillon cannot assert this (though he can of course agree that *Claire* opened her eyes and saw Dillon). And if we suppose that Claire and Dillon will in fact disagree about the likelihood that Claire is the victim, then this seems to be a case where the credence that Claire should have for a hypothesis expressible in *third-person* terms is in part determined by Claire's *first-person* information about who she happens to be and thus what her vantage point on the world is. Since first-person information might possess such rational significance in this case, I do not want to simply assume that Claire's and Dillon's total relevant evidence is exhausted by their third-person information. Rather, I want to allow that the relevant evidence in this case could include information that is irreducibly "first person." Following others, I will use **uncentered information** to refer to purely "third-person" information about the world that in no way implies anything about my "location" in that world (e.g. who I am and where I am in space and time). **Centered information**, on the other hand, has at least some bearing on my location in the world. The extent to which two subjects can share their centered information is limited. For example, though Claire and Dillon can both affirm the centered proposition, "I am in the lab," Claire can truly affirm, "I am Claire," while Dillon cannot.

While Claire and Dillon cannot make all of the same first-person affirmations and thus do not share all of their centered information, this in no way should prevent them from acknowledging one another as epistemic peers, where (for present purposes) two people are epistemic peers with respect to p just in case they both possess the same quality and quantity of evidence bearing on p and are equally rational and reasonable in their assessment of that evidence.[5] For clearly neither Claire nor Dillon have any reason to think

[4] The term "epistemic peer," popular in the disagreement literature, was first introduced by Gutting (1982).

[5] I am using "peer" in a more fine-grained and demanding sense than is sometimes customary. It is important to keep this demanding sense of "peer" in mind in order to appreciate the apparent difficulty of maintaining confident and reasonable belief in a disagreement with a disputant who is an acknowledged peer.

that they possess some sort of epistemic advantage merely in virtue of being who they are. It would be ludicrous for Claire to think that, simply due to the fact she happens to be Claire and not Dillon, she is more likely than Dillon to have accurate views on the identity of the victim. Surely *being Claire* can be considered epistemically advantageous only if we think there is some *other* epistemically relevant factor (whether access to more evidence, greater intelligence, possession of a particular insight, or whatever) that distinguishes her and Dillon. But we are supposing that they know there to be perfect parity with respect to such matters. So neither Claire nor Dillon can take their centered information to confer some sort of epistemic advantage. We can say, then, that Claire and Dillon are mutually acknowledged epistemic peers who perfectly share their *uncentered* evidence and yet rationally disagree.

The mere fact of rational disagreement between mutually acknowledged epistemic peers would not be surprising if we suppose that there are "permissive contexts," contexts where, for one or more disagreeing subjects, the evidence and other rational factors bearing on p do not determine one credal attitude that is maximally rational for that subject to have towards p.[6] But it seems fairly clear that this is a *non*-permissive context where there is one maximally rational credence for Claire and one maximally rational credence for Dillon (even if there is, as we shall see, some debate as to what their credence assignments should be and whether or not they will differ). Even given that some context is non-permissive, it is clear that I can rationally affirm p while acknowledging that some epistemic peer denies p, as long as I am not in communication with this epistemic peer. For example, suppose I am one of ten subjects who each know that there is a 0.5 probability that an urn contains nine red balls and one black ball, and a 0.5 probability that it contains nine black balls and one red ball.[7] All of us are each allowed to draw one ball from the urn (without replacement), and initially we are not allowed to communicate with one another about the color of the ball we draw. If I draw a red ball, I am justified in being fairly confident that the urn originally contained nine red balls even though I know that there is at least one person who drew a black ball and who, on equally strong evidence, is justified in concluding that the urn probably contained only one red ball. The mere fact of peer disagreement need not pose an epistemic threat. But matters change

[6] Note that given the way I am using "permissive," denying that there are permissive contexts does not amount to accepting the "Uniqueness Thesis" endorsed by Feldman (2007) and others. To deny that there are permissive contexts is equivalent to affirming that, for any given person and proposition, there is exactly one maximally rational credal attitude for that person to have towards the proposition. But the Uniqueness Thesis goes further than this, holding that for any two subjects with the same evidence, the *very same* credal attitudes are maximally rational. So using the terminology I will employ here, we can say that the Uniqueness Thesis denies both the possibility of "permissive" contexts *and* the possibility of "perspectival" contexts.

[7] This example is adapted from Michael Titelbaum's "Mystery Bag" example (2012, 235–6), though Titelbaum's example involves additional components and is used to make a different point.

once I am in *communication* with a disagreeing peer, so that that peer becomes my *disputant*, where by 'disputant' I mean someone who not only disagrees with me, but who is also in communication with me about this disagreement. Suppose that on the next day I happen to cross paths with another one of the subjects who tells me that he drew a black ball. Earlier, my drawing a red ball gave me a reason for thinking that a subject who drew a black ball was not representative of the subjects as a whole. But now, my reason for thinking such a subject unrepresentative has been defeated by the fact that the one subject I happened to cross paths with drew a black ball. At this point, it seems as though I cannot rationally disagree with this subject on the likely number of red balls while acknowledging that he is my epistemic peer (in the very demanding sense of 'peer' stipulated above). Maintaining a disagreement requires thinking that I have some evidential advantage or that I am processing my evidence more rationally. So what *would* be surprising is if in a non-permissive context there could be a rational disagreement between two disputants who are mutually acknowledged epistemic peers. But if I am right that Claire and Dillon will rationally disagree about the likelihood that Claire is the victim, then the "multiple Beauties" example is precisely such a case.

Let us say that a context involving a dispute over some uncentered proposition p is **robustly perspectival** if and only if it is a non-permissive context where (i) there is full communication between two disputants and perfect sharing of their uncentered evidence; (ii) each disputant knows with full confidence that the other disputant possesses the same uncentered evidence and processes that evidence perfectly rationally; and (iii) the final credence for p that is rationally required of one disputant differs from the final credence for p that is rationally required of the other disputant. And let us use the label **Robust Perspectivalism** to refer to the view that there are robustly perspectival contexts. If it is true that Claire and Dillon will disagree upon awakening, then there are robustly perspectival contexts and Robust Perspectivalism is correct. As I will argue in the next section, Robust Perspectivalism is incompatible with an intuitively plausible principle that is often taken for granted in discussions of the epistemic significance of disagreement.

2. THE EPISTEMOLOGY OF DISAGREEMENT AND THE RULE OF PROPORTIONALITY

In this section, I characterize a highly intuitive epistemic principle that bears on disagreement and show that this principle is incompatible with the view that Claire ought to maintain her credence of 0.25 for the proposition that she is victim.

If I am highly confident that p and my disputant is highly confident that $\neg p$, consistency requires that I also believe that I am right about p and that my disputant is wrong about p. This is trivial and obvious. But can I rationally affirm that I am right and my disputant wrong while also affirming that my disputant is in at least as strong an epistemic position with respect to p as I am?

Arguably not, at least not if we understand the strength of one's epistemic position as taking into account *all* the dimensions of epistemic evaluation that bear on the likelihood of one's arriving at a reasonable position on *p* in the present circumstances, so that the strength of one's epistemic position takes into account such factors as the general reliability of one's cognitive faculties, the current level of functioning of those faculties, the presence or absence of any errors in reasoning, the presence or absence of bias, the quality and quantity of one's evidence, and the adequacy of one's overarching "epistemic framework." To be sure, there is nothing contradictory in affirming both that my disputant is in at least as strong an epistemic position with respect to *p* and that I am right and my disputant is wrong. For sometimes the epistemically disadvantaged can get lucky. An expert in probability theory may think it highly likely that in a particular series of coin tosses at least one coin has landed heads, and a toddler may firmly believe, without any evidence, that all of them have landed tails. It may turn out that, as luck would have it, the toddler is right. Nonetheless, the probability expert was clearly in a stronger epistemic position with respect to the question. But even if it is not contradictory to affirm both that I am right and that my disputant is in at least as strong an epistemic position, such an affirmation nonetheless seems epistemically problematic. For if I think that the superiority of my view on *p* is not due to any epistemic advantage I have over my disputant (such as greater insight or superior evidence), then I must think that my having the superior view is a matter of epistemic luck. And arguably, I cannot justifiably believe that I am lucky in this way without good evidence that I am lucky. Suppose I have such evidence. If my disputant does *not* have this evidence, then it would seem that this gives me reason for thinking that, in the present circumstances, my epistemic position with respect to *p* is superior. And if my disputant *does* have the evidence that I am the lucky one but does not give it proper weight, then this *also* gives me reason for thinking that my epistemic position is, on balance, superior (since I take my disputant to have improperly assessed a piece of evidence that ought to significantly shape her credence for *p*). So it seems that I cannot reasonably be confident that *p* while also thinking that my disputant's epistemic position with respect to *p* is at least as strong as my own.

The above line of reasoning suggests that the following **Rule of Proportionality** governs my confidence in *p* in the face of disagreement over the plausibility of *p*: the weight I give to my initial opinion regarding *p* and to my disputant's initial opinion regarding *p* ought to be proportional (in some sense) to my assessment of the strength of our epistemic positions with respect to *p*. To the extent that I think my epistemic position is stronger, I should weight my opinion regarding *p* more heavily, and to the extent that I think my disputant's epistemic position is stronger, I ought to weight her opinion regarding *p* more heavily.

It is important to see that the Rule of Proportionality does *not* require that I assess my and my disputant's epistemic credentials in a way that is independent of my views on the disputed matter. *That* requirement, often called the "independence" requirement, is quite controversial and leads fairly directly to a "conciliatory" position according to which disagreement has significant skeptical force (Christensen 2009, 758). The Rule of Proportionality is completely open to the possibility, defended by opponents of conciliatory views, that the reasoning behind my belief that *p* may serve to ground my judgment that those who dispute *p* suffer from some sort of epistemic disadvantage (e.g. that they are prone to a certain error in reasoning, have some sort of cognitive defect, lack some piece of evidence, etc.) and are for that reason in a weak epistemic position with respect to *p*. Advocates of conciliatory views as well as their opponents can both affirm that my being steadfast in the face of disagreement is reasonable only if I think that those on my side of the dispute are, on the whole, in a stronger epistemic position.

Most parties to the current debate over the epistemic significance of disagreement seem to take it for granted that the Rule of Proportionality is correct, or at least that it is correct in non-permissive contexts. If I am in a *permissive* context, and if I think that both my and my disputant's levels of confidence are in the range that is rationally permissible for me, then arguably I can remain steadfast in my level of confidence without thinking that my epistemic position is superior in some way.[8] But even if there are permissive contexts (which is controversial),[9] it seems that opposing sides in the disagreement debate are for the most part united in their acceptance of what we can call **Modest Proportionality**, which is the view that in non-permissive contexts, the Rule of Proportionality applies.[10]

There are, to be sure, epistemological positions that have adherents and that arguably conflict with Modest Proportionality. For example, Modest Proportionality might be incompatible with certain strong forms of "epistemic conservatism" that maintain that the fact that I believe *p* is a very resilient reason in favor of my continuing to believe *p*.[11] And certain forms of epistemic relativism that hold that there are no objective epistemic standards according to which the correctness of different "epistemic systems" can be measured might be in tension with the underlying motivations for Modest Proportionality.[12] But most philosophers currently engaged in the disagreement

8 Goldman (2010, 195–6) briefly argues for this possibility, as does Kelly (2010, 118–19).

9 For arguments against permissive contexts, see White (2005).

10 That proponents of conciliatory views affirm Modest Proportionality is rather obvious. See, e.g., Feldman (2006), Elga (2007), and Christensen (2011). Several opponents of conciliatory views also affirm something along the lines of Modest Proportionality. Lackey (2010, 277, 281–2), for instance, suggests that confidence in the face of disagreement requires that there be some "symmetry breaker," a reason for thinking that one's own epistemic situation is favorable to that of one's disputant. Along similar lines, Bergmann (2009, 342) thinks that reasonable steadfastness requires thinking that one is *probably* "internally" or "externally" more rational than one's disputant. van Inwagen (2010) argues in favor of several principles that are consonant with Modest Proportionality, though he stops short of firmly endorsing an anti-conciliatory view. Also see Enoch (2010, 975) and Fumerton (2010, 99).

11 For a critical discussion of different versions of epistemic conservatism, see Vahid (2004).

12 This characterization of epistemic relativism is based on Boghossian's discussion (2006, 73).

debate, even those who are not friendly to conciliatory views, have not taken such approaches. They have typically eschewed views on rationality that are explicitly perspectivalist, preferring instead views that are least moderately "objectivist." And those coming from such a perspective will be inclined to accept Modest Proportionality. What is interesting about the multiple Beauties case is that it challenges Modest Proportionality without relying on any premise (such as epistemic conservatism or relativism) that explicitly conflicts with a more objectivist understanding of rational norms. I will now make that challenge explicit.

Modest Proportionality implies that there cannot be a non-permissive context where two disputants rationally maintain different credences for *p* despite knowing with certainty that their epistemic positions with respect to *p* are equally strong. While Modest Proportionality is perfectly compatible with a subject's responding to a disagreement over *p* by concluding that her disputant is not in fact equally qualified to assess *p*, on the assumptions that each subject continues to know that the other subject is equally qualified and that the context is non-permissive, Modest Proportionality requires that both of the subjects give equal weight to the other's perspective. And if they both give equal weight to the other's perspective, then they will have the same final credence for *p*.[13] So any initial difference in confidence levels will disappear after full communication. Modest Proportionality thus implies that there are no robustly perspectival contexts. But, as already discussed, the view that Claire and Dillon will disagree about the likelihood that Claire is the victim implies that there are robustly perspectival contexts. So if we are to affirm that Claire and Dillon will disagree, then we must reject Modest Proportionality.

If I am correct in maintaining that Modest Proportionality is incompatible with the view that Claire and Dillon will rationally disagree, we have two options: deny that Claire and Dillon will disagree, or deny Modest Proportionality. I will now show that the first of these two options is available to those who advocate the "thirder" position on the original Sleeping Beauty case. Arguments for the thirder position, when adapted to the multiple Beauties case, support the conclusion that Claire and Dillon will agree on the likelihood that Claire is the victim. But those who advocate the alternative "halfer" position on Sleeping Beauty will be under rational pressure to conclude that Claire and Dillon will disagree, and thus that Modest Proportionality is false.

[13] This does not mean that their new credence for *p* will be halfway between their original credences. In some cases, both disputants might agree that when both perspectives are given equal weight, the correct compromise credence is closer to one of the original views than the other. For example, if I think Horse A won the race by a nose, and my friend thinks Horse B edged out Horse A at the finish line, we may agree that our final credence for a Horse B victory should be closer to my friend's pre-disagreement credence if we know that in close races between the two horses, Horse B wins the large majority of the time. In this case, the evidential significance of each judgment is canceled out by the other, leaving us back at our prior probability (which is closer to my friend's pre-disagreement credence than to mine). For another sort of example where equally weighting one another's views does not lead to middle-ground credences across all of the disputed propositions, see Thurow (2012).

3. AGREEING BEAUTIES? WHY THIRDERS AND HALFERS SHOULD DISAGREE ON WHETHER CLAIRE AND DILLON WILL DISAGREE

There are two basic positions on the original Sleeping Beauty problem: the "thirder" position and the "halfer" position. I will now argue that while "halfers" ought to affirm that Claire and Dillon will disagree, "thirders" ought to say that Claire and Dillon will not disagree but will both assign a credence of 0.5 to the proposition that Claire is the victim. Thus, the multiple Beauties case presents no problem for thirders, who can in fact use the case as a new argument for the thirder position, an argument that gains its strength from the apparent plausibility of Modest Proportionality. To show why halfers should maintain that Claire and Dillon will disagree and thirders should maintain that they will agree, I will describe the original Sleeping Beauty case, present the principal argument for the halfer position and one of the principal arguments for the thirder position, and show that analogous arguments in the multiple Beauties case lead halfers and thirders to disagree about whether Claire and Dillon will agree or disagree about the likelihood that Claire is the victim.

The original Sleeping Beauty case goes as follows. Sleeping Beauty is a perfectly rational agent who is about to be put to sleep on Sunday night by some experimental philosophers. The experimenters inform her that during the experiment she will either be awakened once (on Monday morning) or twice (on Monday and Tuesday morning), depending on the outcome of a coin toss. If the coin lands heads, Beauty will be awakened only on Monday. If the coin lands tails, Beauty will be awakened on Monday and Tuesday. Sleeping Beauty does not know the result of the coin toss, and thus does not know the number of awakenings she will experience. Moreover, the experimenters inform Beauty that she will have no way of distinguishing a Monday awakening from a Tuesday awakening. In order to make these waking experiences indiscernible, the experimenters will erase Beauty's memory of her Monday awakening before putting her back to sleep, so that if she awakens on Tuesday, she will not know whether this is her first or second awakening. When the experiment is over on Wednesday, the experimenters will awaken Beauty in a manner that ensures that Beauty will know it is Wednesday and not an awakening during the course of the experiment. (We can imagine that Beauty knows that on Wednesday she will be awakened by a song whose lyrics are, "It's Wednesday, the experiment is over!") The controversial question, which has given rise to a small cottage industry producing papers on the subject, is this: when Beauty is woken on Monday and experiences an awakening that is indistinguishable from a Tuesday awakening, what should her credence be for the proposition that the coin landed heads? Thirders argue that the rational credence is one third; halfers contend that it is one half.

Before presenting arguments for the thirder and halfer positions and showing how those arguments translate to the multiple Beauties case, an important clarifying note is in order. It is an unfortunate feature of the multiple Beauties case that arguments for the *thirder* position in the original Sleeping Beauty case support the conclusion that Claire's credence that she is the victim should, upon waking, be *one half*; and the principal argument for the *halfer* conclusion in the original Sleeping Beauty case supports the conclusion that Claire's credence in the relevant proposition should remain *one fourth*. So halfers support 1/2 in the original case and should (I will argue) support 1/4 in the multiple Beauties case, and thirders support 1/3 in the original case and should support 1/2 in the multiple Beauties case. Having warned of this potential source of confusion, I will now characterize the principal halfer and thirder arguments and show how they bear on the multiple Beauties case.

The principal consideration in favor of the halfer solution to the Sleeping Beauty Problem is that it seems that Sleeping Beauty has not learned anything of evidential significance when she is awakened on Monday. On Sunday, when Beauty's credence for the coin's landing heads was one half, Beauty *knew* that she would soon have at least one waking experience and that no qualitative features of a given awakening would provide clues as to whether it is a Monday awakening or a Tuesday awakening. So it seems that she does not acquire any new evidence relevant to the outcome of the coin toss when she learns that she is presently in the midst of a waking experience with such and such qualitative features. And if this is right, then application of the Relevance Principle leads us to conclude that her credence for heads ought to remain one half. Halfers endorse this "no new relevant evidence" reasoning and the application of the Relevance Principle, contending that Beauty's credence for the coin's landing heads should not change upon awakening.

The same "no new relevant evidence" reasoning that motivates the halfer view also supports the view that, *after* waking up but *before* identifying her co-waker, Claire's credence that she is the victim ought to still be 0.25. Once this has been granted, arriving at the conclusion that Claire and Dillon will disagree merely requires showing that Claire's credence that she is the victim should not change upon looking across the room and identifying her co-waker. The argument for this claim, rehearsed in section 1, is extremely strong. Since Claire knows that she will see a co-waker when she looks across the room, and since the particular identity of the co-waker doesn't make a difference to how much Claire suspects herself, in looking across the room and seeing that Dillon is awake, Claire does not learn anything that should lead her to change her credence that she is the victim. So it seems that halfers are committed to holding that Claire and Dillon will disagree, and thus that Robust Perspectivalism is true and that Modest Proportionality is false.

In saying that halfers are "committed" to the view that Claire and Dillon will disagree, I mean that any viable position that supports a halfer position in the Sleeping Beauty case will also imply that Claire and Dillon will disagree. There are, however, two reasons one might want to resist this claim. First, someone might argue that while halfers are rightly moved by the Relevance

Principle in the Sleeping Beauty case, the Relevance Principle is not absolute and is trumped in circumstances where it threatens to come into conflict with Modest Proportionality. I concede that such a position is at least superficially consistent. But it also strikes me as quite ad hoc. The Relevance Principle seems to be a more basic rational requirement than Modest Proportionality, so it is hard to see why the latter should override the former. And if the Relevance Principle does admit of exceptions so readily, then it is also questionable whether halfers are reasonable in thinking that it holds in the original Sleeping Beauty case, given the force of the thirder argument to be considered below.

But there is a more legitimate reason why one might want to resist the claim that halfers are committed to the view that Claire and Dillon will disagree. Some attempts to give a formalized generalization of the halfer approach (Halpern 2005; Meacham 2008; Briggs 2010) have associated the halfer approach with a policy that, when applied to the multiple Beauties case, does *not* support the conclusion that Claire and Dillon will disagree, but rather supports the conclusion that they will agree (a conclusion that I will associate with the thirder position). This policy, which Rachel Briggs calls the "Halfer Rule," requires "conditionalizing the *uncentered portion* of one's credence function on the *uncentered portion* of one's total evidence, and then within each [uncentered] world, dividing one's credence [for that world] among the doxastically possible centers [within that world]" (Briggs 2010, 9–10). Briggs thinks that halfers will be drawn to the view that only the *uncentered* component of my total evidence is of any relevance in determining the probability of some uncentered proposition, and the Halfer Rule requires that one form credences for uncentered propositions in a way that sets aside centered information as irrelevant. Since by hypothesis Claire and Dillon have the same uncentered evidence, clearly they will be in full agreement if they follow the Halfer Rule.

But even though the Halfer Rule does prescribe halfer credences in the original version of the Sleeping Beauty case, halfers have good reasons for rejecting the Halfer Rule. For in many cases, the Halfer Rule delivers prescriptions that are fundamentally at odds with the highly intuitive "no new relevant evidence" reasoning that drives people towards halfer conclusions in the first place (Titelbaum 2008, 591–7; Briggs 2010, 29). Consider, for instance, Titelbaum's "Technicolor Beauty" case (2008, 591–7).[14] In the original Sleeping Beauty case, it was stipulated that Beauty's Monday and Tuesday awakenings are qualitatively identical. The Technicolor Beauty example introduces qualitative differences that are evidentially irrelevant. In Technicolor Beauty, everything is just as it was in the original case except for the following addition. Beauty knows that at the same time the coin is to be tossed on Sunday night, a dice will also be rolled. If the dice roll comes out odd, a red piece of paper will be put in her room before she wakes up on Monday

[14] A similar case can also be found in Meacham (2008, 263).

and then, after she is put back to sleep, it will be replaced by a blue piece of paper that will be in the room throughout Tuesday. If the dice roll comes out even, the colors will be reversed: a blue piece of paper will be put in her room on Monday and a red piece of paper will be put in her room on Tuesday. Beauty knows all of this. Quite clearly, the fact that Beauty sees a red piece of paper upon awakening should make no difference to her credence for the proposition that the coin landed heads. For Beauty knew she would see either a red or blue piece of paper, and a piece of red paper is no more or less likely to be present on a Monday (or on a Tuesday) than a piece of blue paper. So if Beauty's credence for heads in the original case should be 0.5, then it should also be 0.5 in Technicolor Beauty: the "no new relevant evidence" reasoning is just as strong in both cases. But while the Halfer Rule delivers halfer prescriptions in the original case, it delivers *thirder* prescriptions in Technicolor Beauty. This "instability" (Briggs 2010, 27) of the Halfer Rule's prescriptions surely shows it to be mistaken. Those who wish to stick by the halfer view on the original case therefore ought to reject the Halfer Rule and seek a way of generalizing the halfer approach that is more closely aligned with the kind of "no new relevant evidence" reasoning that makes the halfer view so intuitive in the first place.[15] So the Halfer Rule does not, in my judgment, supply a plausible way of being a halfer while affirming that Claire and Dillon will agree.

I will now show that thirders ought to maintain that Claire and Dillon will find themselves in agreement during their awakening. While all of the arguments that I am aware of for the thirder position can be used to support the conclusion that Claire and Dillon will agree, I will focus here on an argument articulated in the piece by Adam Elga that introduced the Sleeping Beauty problem to the philosophical community. For purposes of presenting Elga's argument, I'll follow Elga in labeling the three possibilities open to Beauty upon awakening on Monday H1, T1, and T2; Table 7.1 shows which possibilities are designated by these labels:

	Heads	Tails
It's Monday	H1	T1
It's Tuesday	(Not possible)	T2

Table 7.1 *The possibilities open to Beauty upon awakening on Monday*

Let C be Beauty's (perfectly rational) credence function on Monday when she awakens. The thirder position follows from the following two premises:

(1) $C(T1) = C(T2)$

(2) $C(H1) = C(T1)$

[15] For a defense of the halfer view that rejects the Halfer Rule, see the recent work of Darren Bradley (2011a; 2012; and especially 2011b).

Since H1, T1, and T2 are mutually exclusive and exhaust all possibilities, we know that C(H1) + C(T1) + C(T2) = 1. From this and from (1) and (2), it follows that C(H1) = C(T1) = C(T2) = 1/3. And since Beauty's credence for Heads must be equal to her credence for H1, we can further conclude that her credence in Heads upon awakening on Monday will be 1/3. I'll now briefly rehearse the thirder's reasons for affirming (1) and (2).

Given the absence of any reason for thinking T1 more or less likely than T2, we have good reason for thinking that Beauty will assign the same credence to each of these two possibilities, and thus that (1) is correct. To be sure, some have contested the general "indifference principles" that have been offered as motivation for (1) (see, e.g., Weatherson 2005).[16] But even if such indifference principles are controversial, most everyone, halfers included, are inclined to accept that *in this particular case*, Beauty ought to assign equal credence to T1 and T2. The primary source of disagreement between halfers and thirders is (2).[17]

Elga's argument for (2) may be summarized as follows. First, since Beauty will be awakened on Monday irrespective of whether the coin comes up heads or tails, we can imagine that the experimenters do not toss the coin until *after* Beauty is awakened on Monday. And whether they do in fact toss the coin on Sunday night or after waking up Beauty on Monday should not, it seems, make any difference to the credences Beauty assigns to H1, T1, and T2. So let's imagine that Beauty knows that the coin is tossed on Monday, just after she is put back to sleep. Now, suppose that sometime after being woken up on Monday, Beauty is told that it is Monday. Upon learning this, Beauty learns that T2 is false, so that her credences for Tails and Heads will be identical to her updated credences for T1 and H1 (respectively). And what should her updated credences for H1 and T1 be? Well, her credence for H1 must be identical to her credence that a fair coin, *yet to be tossed*, will land heads, and her credence for T1 must be identical to her credence that a fair coin, *yet to be tossed*, will land tails. And surely, Elga contends, one's credence that a *future* toss of a fair coin will land heads should be 0.5. So upon learning that it is Monday, Beauty's credence for H1 ought to be 0.5.

From this, one more step is needed to arrive at the conclusion that (2) is correct. And it is a step that is frequently contested. Elga asserts the following:

(3) The credence that Beauty has for H1 *after* learning that it is Monday (and thus that H1 or T1) should be equal to the conditional credence C(H1 | H1 or T1) that she had *before* learning that it is Monday.[18]

[16] Briggs (2010, 12) offers a weaker indifference principle that avoids Weatherson's objections and motivates (1) on the condition that Beauty assigns both T1 and T2 some precise non-zero credence.

[17] One exception to this is Hawley (2013). Hawley is a halfer who accepts (2) but denies (1), arguing that it ought to be the case that C(T1) = 0.5 and C(T2) = 0.

[18] Since (3) simply asserts that Beauty ought to conditionalize on the information that it is Monday, it may seem that (3) should be accepted by anyone who is committed to Bayesian confirmation theory. But it is widely acknowledged that Bayesian conditionalization is not

	Claire is the victim	Dillon is the victim
It's the victim's 1st awakening	C1	D1
It's the victim's 2nd awakening	C2	D2
It's the victim's 3rd awakening	C3	D3

Table 7.2 *The possibilities open to Claire during her awakening with Dillon*

Given (3) and the fact that Beauty's credence for H1 after learning that it is Monday should be 0.5, it follows that *before* learning that it is Monday, it ought to be the case that C(H1 | H1 or T1) = 0.5. And from this, it follows that, before learning that it is Monday, it ought to be the case that C(H1) = C(T1), thus completing Elga's argument for (2) and delivering us the thirder result.[19]

We are now in a position to see how an argument that is analogous to the one Elga gives in support of the thirder position can be given in support of the conclusion that, upon waking up together, Claire and Dillon will agree that the probability that Claire is the victim is 0.5. When Claire is awakened and sees Dillon being awakened at the same time, she knows that one of six mutually exclusive possibilities obtain, which I'll label C1–C3 and D1–D3, in accordance with Table 7.2.

Let '**Agreement**' stand for the proposition that, upon being awakened during the night and seeing that Dillon has been awakened at the same time, Claire's credence for the proposition that she is the victim will be 0.5. (I'm calling this proposition 'Agreement' because the exact same argument would also show that Dillon would assign a credence of 0.5 to the proposition that Claire is the victim, thus resulting in his agreeing with Claire.) And let C be Claire's credence function after waking and seeing that Dillon is her co-waker. Agreement follows from the following two premises, which are analogous to (1) and (2) above:

(4) C(C1) = C(C2) = C(C3) and C(D1) = C(D2) = C(D3)

(5) C(C1) = C(D1)

applicable to certain types of cases where what is learned is self-locating information. Why? For starters, Bayesian conditionalization requires that certainties be preserved. But it is perfectly rational for an agent to go from being certain that it is morning to being certain at a later time that it is not morning. (For discussion, see, e.g., Meacham 2010; Bradley 2011b.) Whether or not conditionalization is appropriate when Beauty learns that it is Monday is contentious among halfers. In the original defense of the halfer view, Lewis (2001) affirmed that Beauty ought to conditionalize on this information, a view still held by some halfers (e.g. Jenkins 2005; Bradley 2011b). But most halfers today are "double halfers" who deny (3) and hold that Beauty's credence for Heads should remain $\frac{1}{2}$ even after learning that it is Monday (see, e.g., Bostrom 2007; Meacham 2008; Pust 2012).

[19] We can spell out this last step more explicitly. By Bayes' Theorem, C(H1 | H1 or T1) = C(H1 or T1 | H1) · C(H1) / C(H1 or T1). So on the assumption that Elga has shown that C(H1 | H1 or T1) = 0.5, we know that C(H1 or T1 | H1) · C(H1) / C(H1 or T1) = 0.5. Obviously, C(H1 or T1 | H1) is 1; so simplifying we get C(H1) / C(H1 or T1) = 0.5. Since H1 and T1 are mutually exclusive, C(H1 or T1) must be equal to C(H1) + C(T1). Thus, C(H1) / [C(H1) + C(T1)] = 0.5. From this, only algebra is needed to show that C(H1) = C(T1).

Since C1–C3 and D1–D3 are mutually exclusive and collectively exhaustive possibilities, we know that C(C1) + C(C2) + C(C3) + C(D1) + C(D2) + C(D3) = 1. From this and from (4) and (5), it follows that C(C1) = C(C2) = C(C3) = C(D1) = C(D2) = C(D3) = 1/6. And since Claire's credence that she is the victim must be equal to C(C1) + C(C2) + C(C3), it further follows that upon awakening, Claire's credence that she is the victim will be 1/2, giving us our conclusion that Agreement is correct.

The same "indifference" reasoning offered in support of (1) above supports premise (4). No matter who the victim is, Claire has no reason for thinking that it is more likely that this is the first (or second or third) awakening for the victim as opposed to either of the other two possibilities. It seems that rationality requires her to assign each of these possibilities equal credence, as (4) requires.

Premise (5), too, can be supported with an argument along the lines of the one given in support of (2). We can imagine that the experimenters select the victim in the following way: first, they randomly select two of the four subjects; then they wake both of these subjects during the first awakening of the night; and finally, *after* the two subjects have debated the probable identity of the victim and have been put back to sleep, they flip a fair coin in order to determine which of the two subjects just put back to sleep will be the victim. It seems that whether the experimenters use this method to select the victim or select the victim ahead of time should make no difference to Claire's credences for C1–C3 and D1–D3. So let's suppose that the experimenters use the two-stage approach to victim selection just described, and that Claire knows this. Now, suppose that sometime after being awakened at the same time as Dillon, Claire is told that this is the first awakening, and that the identity of the victim will be chosen via a coin toss after Claire and Dillon go back to sleep, with Claire being selected if the coin lands heads, and Dillon being selected if it lands tails. Upon learning this, Claire learns that either C1 or D1 is true, so that her credence for her being the victim and her credence for Dillon being the victim will be identical to her new credences for C1 and D1 (respectively). And since her new credence for C1 must be identical to her credence that a fair coin, *yet to be tossed*, will land heads, and her credence for D1 must be identical to her credence that a fair coin, *yet to be tossed*, will land tails, it seems that upon being told that this is the first awakening and the victim has yet to be selected, Claire's new credence for C1 (and for D1) ought to be 0.5.

Again, we need one more premise in order to conclude that (5) is correct:

(6) The credence that Claire has for C1 *after* learning that it is the first awakening (and thus that C1 or D1) should be equal to the conditional credence C(C1 | C1 or D1) that she had *before* learning this information.

Given (6) and the reasoning just rehearsed, it follows that *before* learning that it is the first awakening of the night, it ought to be the case that C(C1 | C1 or D1) = 0.5. And from this, it follows that, before learning that it is the first awakening, it ought to be the case that C(C1) = C(D1), completing the argument for (5) and thus for Agreement.

Given that thirders will think that Claire and Dillon will be in perfect agreement, thirders can continue to affirm Modest Proportionality. And given the intuitive appeal of Modest Proportionality, the fact that thirders can readily affirm it and halfers cannot (at least not without abandoning the principal motivation for their position) constitutes a new and not insignificant reason in favor of the thirder position. But however plausible Modest Proportionality may at first appear, I will argue that there are plausible reasons for thinking that the multiple Beauties case does involve a robustly perspectival context, and that Modest Proportionality is therefore false. If I am right, halfers should not be worried by the fact that the halfer approach can conflict with Modest Proportionality.

4. EVIDENTIAL SELECTION PROCEDURES AND PERSPECTIVAL RATIONALITY

We can summarize the challenge to the halfer position in the following way. If we consider the perspective of Claire, there is an intuitive line of reasoning that leads to the conclusion that there is a 0.75 chance that *Dillon* is the victim. And if we consider the perspective of Dillon, an exactly analogous line of reasoning supports the conclusion that there is a 0.75 chance that *Claire* is the victim. But clearly we, as third party observers of Claire's and Dillon's situation, have no reason for privileging either Claire's or Dillon's perspective over the other's. To prefer one side or the other would be completely arbitrary. We ought therefore to assign an equal probability to Claire's being the victim as to Dillon's being the victim. But (and here is the critical though misguided move) Claire has no more reason than we do for privileging her own perspective over Dillon's. Or at least she has no more *epistemic* reason than we do for privileging her perspective. For the mere fact that a perspective happens to be hers rather than Dillon's is no reason for thinking that that perspective will better serve the aim of true belief and accurate credences. Given that Claire knows that neither she nor Dillon possesses any epistemic advantage over the other, privileging her own perspective would amount to arbitrarily selecting one out of two perspectives that, from a disinterested point of view, are equally likely be the more reliable guide to the identity of the victim. Rationality, it seems, would require that such arbitrary selection be avoided and that perspectives with equal epistemic standing be given equal weight. Thus, the halfer view on the multiple Beauties case, and the "no new relevant evidence" reasoning that motivates that view, ought to be rejected as fallacious.

Against the above line of reasoning, I will claim that Claire's privileging her own perspective is not epistemically arbitrary, but is rationally required given the appropriate background assumptions about the process by which Claire has come to acquire the evidence she receives during her awakening with

Dillon. I am not the first to argue that the rationally required credences can vary for two people even in contexts where the uncentered information that grounds those credences is shared by both parties. Arnold Zuboff (2000) and Nick Bostrom (2000) have both offered examples in support of this perspectivalist claim. But the Robust Perspectivalism that I will argue for using the multiple Beauties case goes beyond the merely moderate perspectivalism implied by Zuboff's and Bostrom's examples. For as I argue in the notes, in the examples of Zuboff and Bostrom, the perspectivalist results depend on the inability of the multiple parties to share certain bits of uncentered information that, while not themselves evidentially relevant, cannot be shared without generating *new* evidence that *is* evidentially relevant and that would lead both parties to converge on the same credences.[20] So for these examples to yield

[20] Zuboff explicitly acknowledges that in his example communication between the disagreeing subjects must be disallowed in order for the example to yield a (moderately) perspectivalist result. While Bostrom's example yields a perspectivalist result only if we assume that communication is not possible, Bostrom does not explicitly stipulate that there is no communication. But Bostrom does argue that the perspectivalist results do not support the possibility of mutually agreeable bets between the disagreeing parties (2000, 105–6). And the same reasoning that shows why there are no mutually agreeable bets also shows that the differences in credences will not persist through communication and the full sharing of uncentered information. To see why the perspectivalist result is undermined by the full sharing of uncentered information, consider the following case, which is structurally just like Bostrom's case. Suppose that in an experiment some scientists leave me on a desert island in the middle of an undisclosed ocean. I know the following: first, if a particular coin flip conducted by the scientists came up heads, then one subject has been left on a desert island in the Indian Ocean and one subject has been left on a desert island in the Pacific Ocean; if it came up tails, then one subject has been left on a desert island in the Indian Ocean and *ten* subjects have been left on ten different desert islands in the Pacific Ocean; second, each subject has an electronic device that at 12:00 a.m. on January 1 will display which ocean they are located in; third, no subjects will have any evidence beyond this that can help them determine which ocean they are in or how many subjects are involved in the experiment. Now suppose that at 12:00 a.m. on January 1, my device informs me that I am in the Indian Ocean; at the same time, Fiona, another subject, gets her message indicating that she is in the Pacific Ocean. According to Bostrom, I should at this point be more confident than Fiona that the coin landed heads. For the chance of my being the one person in the Indian Ocean is much more likely given heads than it is given tails. But Fiona knows that *someone* has just learned that he (or she) is in the Indian Ocean. But since *she* is not that person, it should not have the same rational import (for her) as my learning that *I* am in the Indian Ocean. And of course I know that at least one person has just learned that his or her island is in the Pacific. But this knowledge does not have the same import (for me) as Fiona's learning that *her* island is in the Pacific. So centered information is the difference-maker here. Nonetheless, the perspectivalist result depends on imperfect sharing of uncentered information. For suppose that somehow Fiona and I were able to share all of our uncentered information. In this case, I would know not only that some subject is on an island in the Pacific, but also that *Fiona* is on an island in the Pacific. And Fiona would know that I know this. But now, Fiona has a new piece of evidence that is relevant to heads or tails: namely, that while I know that *Fiona* is on a Pacific island, I do not know the names of any other subjects on Pacific islands. And from Fiona's perspective, the chance of my knowing about *her* as opposed to some other subject is much more likely given heads (in which case, she is the only Pacific island subject I *can* know about) than it is given tails (in which case, there is only a one in ten chance that Fiona would be the only subject I know about). With this new evidence made possible by our sharing all of our uncentered information, Fiona's credence for heads will converge with my credence for heads. Essentially, communication often enables interlocutors to gain evidence about how representative they and their perspective are, and this evidence undermines any pre-communication perspectivalism.

results that could plausibly be called "perspectivalist," it must be assumed that the parties with opposed views cannot communicate. The examples therefore do not establish the possibility of *robustly* perspectival contexts. The kind of perspectivalism implied by these examples is thus less surprising, and less significant for the epistemology of disagreement, than the Robust Perspectivalism for which I will now argue. Still, the diagnosis of the multiple Beauties case that follows can be understood as an application and elaboration of some of the insights of Zuboff and Bostrom.

To understand why the credences that are rational for Claire and Dillon are perspective-dependent, it will be helpful to consider cases where certain facts about the evidential selection procedure give a third party reason for preferring either Claire's or Dillon's perspective. Suppose that the experimental philosophers have concluded their experiment with Alvin, Brenda, Claire, and Dillon, and after their results are recorded into a database, the experimenters allow you to make certain queries of the database and to see the answers the database returns. You have no knowledge concerning the identity of the victim other than what you learn in response to your queries. Suppose, first, that you instruct the database to first display the name of one randomly selected subject and then to display the name of someone who shared an awakening with that subject. If the first subject displayed is *not* the victim, then the next subject displayed will be the only person that this person shared an awakening with, i.e. the victim; and if the first subject displayed *is* the victim, then the database will randomly select one of the other subjects (each of whom shared an awakening with the victim). And suppose that in response to these instructions, the database program first displays the name "Claire" and then the name "Dillon." Let's call the procedure just described "Procedure 1." It is uncontroversial that if Procedure 1 is the only basis for your knowledge that Claire shared an awakening with Dillon, then you should have a 0.75 credence for the proposition that Dillon is the victim. For in response to the first query that is part of the procedure, three times out of four the program will display the name of someone who was *not* the victim, which means that the next query will display the name of the victim (since every non-victim shares an awakening only with the victim). Matters would have been very different if instead you had employed Procedure 2, where Procedure 2 consists in your directing the database program to randomly select one of the awakenings and then display the names of the two subjects involved in that awakening. If you had learned by Procedure 2 that Claire and Dillon shared an awakening, you would know that one of these two was the victim, but would have no basis for thinking either one of them is more likely to be the victim than the other.

Interestingly, Procedure 1 and Procedure 2 are both random procedures that are equally likely to result in your learning that Claire and Dillon shared an awakening; for Procedure 1 is equally likely to turn up information about the subjects in the first awakening as it is the second or third awakening. Nevertheless, if Procedure 1 is your method for arriving at the information that Claire and Dillon shared an awakening, it would be irrational for you to respond to this information as though all you had learned was that a randomly selected awakening involved Claire and Dillon. For you have learned something else that is evidentially relevant, namely that Dillon was a randomly selected "co-waker" of a randomly selected subject. This additional knowledge changes the rationally required response. The key point, one recently defended by Darren Bradley (2012), is that the process by which the evidence was selected is often itself a critical piece of evidence. And if this process is not known, background views and assumptions about the *likely* process will often play a critical role in determining the rational response to a piece of evidence.

Claire's epistemic situation upon awaking with Dillon is, I will argue, relevantly like that of someone who, employing Procedure 1, has randomly selected Claire from among the four subjects and then learned that Dillon is a co-waker of this randomly selected subject. So upon waking up and seeing Dillon, Claire is justified in adopting the same credences as someone who has employed Procedure 1. Of course it is true that Claire did not randomly select herself from among four subjects. Nor is it possible, given the constraints of the experiment, for Claire to randomly choose one of the subjects in order to learn the identity of one of that subject's co-wakers. Since Claire is not privy to information about any awakenings that do not involve herself, she cannot expect to learn the identity of any particular subject's co-wakers except for herself. So Claire's epistemic situation is relevantly like someone who has employed Procedure 1 only if she is justified in thinking of herself as a "randomly" selected subject. Claire can, I suggest, legitimately think of herself this way. Even though Claire is constrained in which of the subjects she is able to learn about, and even though this constraint biases Claire toward learning about one of her own co-wakers rather than another subject's co-wakers, this biasing constraint does not undermine the analogy between Claire's situation and Procedure 1. For the fact that constrains Claire to "select" herself (namely, the fact that subjects cannot gain information about any awakenings in which they are not involved) is probabilistically independent of whether or not Claire is the victim. And such probabilistic independence is all that the "randomness" of Procedure 1 was meant to achieve. Given the lack of probabilistic correlation between Claire's reason for selecting herself and the identity of the victim, Claire may legitimately think of herself as a "randomly" selected subject.

To help illustrate why such probabilistic independence is sufficient, imagine that instead of using the researchers' database to carry out Procedure 1, you are going to randomly select one of the four subjects and hypnotize that subject in order to retrieve the memory of one (randomly selected) forgotten awakening. To your dismay, it turns out that Alvin, Brenda, and Dillon are not susceptible to hypnosis. Claire, however, is able to be hypnotized. As long as you know that whether a subject is susceptible to hypnosis is probabilistically independent of whether that subject was the victim, then there is no problem in your thinking of Claire as a "randomly selected" subject. Upon hypnotizing her and learning the identity of a randomly selected co-waker, the rational implications will be the same as Procedure 1 as originally described. Similarly, Claire's reason for "selecting" Claire is a result of the epistemic constraints imposed by the experiment and the centered fact about her identity—facts that have no probabilistic correlation with Claire's being the victim. Thus, it seems that Procedure 1 is a fully adequate model for Claire's situation and that her credences should be identical to someone who has employed Procedure 1, randomly selected Claire, and then learned that Dillon is a co-waker of Claire's.

Of course Dillon is also justified in treating himself as a randomly selected subject and in adopting the credences of someone who, performing Procedure 1, selected Dillon randomly and then learned that Claire is a co-waker of Dillon's. And since Claire knows this, one might think that Claire's epistemic situation is best modeled by someone who has performed Procedure 1 twice, the first time randomly selecting Claire and then learning that Dillon is a randomly selected co-waker of this randomly selected subject, and the next time randomly selecting Dillon and learning that Claire is a randomly selected co-waker of this randomly selected subject. If this were the best model of Claire's situation, then it would indeed be the case that Claire ought to put equal credence in her being the victim and in Dillon's being the victim. But it would be a mistake for Claire to think that her situation is analogous to the situation of someone who has performed Procedure 1 twice. As we have seen, Procedure 1 is an adequate model only if the basis for the selection of the first subject is probabilistically independent of the identity of the victim. The fact that accounts for Claire's "selection" of Dillon from among the other subjects is the fact that she is currently sharing an awakening with Dillon. And this fact *is* probabilistically correlated with the identity of the victim, since having an awakening with Dillon is three times more likely if Dillon is the victim than if he is not the victim.

We are now in a position to appreciate why the rational credences are perspectival in this case, despite the perfect sharing of all uncentered evidence. The evidential significance of the information that Claire and Dillon share during their awakening depends on the process by which that information has been acquired. If the process yields information about a randomly selected co-waker of *Claire's* for reasons that are probabilistically independent of whether or not Claire is the victim, then the evidential significance of the information will be different than if the process yields information about a co-waker of *Dillon's* for reasons that are probabilistically independent of whether or not Dillon is the victim. Learning that Claire and Dillon share an awakening by the first kind of process (as in an instance of Procedure 1 where Claire is the randomly selected subject) can have no bearing on the likelihood of Claire's being the victim, and learning this information by the second kind of process (as in an instance of Procedure 1 where Dillon is the randomly selected subject) can have no bearing on the likelihood of Dillon's being the victim. But whether or not one has learned this evidence by a process of the first type or of the second type (or by some other type of process, like Procedure 2) depends on features of one's causal history that can vary from one subject to another. Thus, the rational significance of the evidence can depend on one's observational standpoint. Such seems to be the case in the multiple Beauties example. Claire and Dillon can share all of their uncentered information, but they cannot share their causal histories and thus cannot share the same observational standpoint. As a result, the evidential significance of their shared information differs for each of them, and different credences are called for.

One might worry that the Robust Perspectivalism I defend here stands in tension with Robert Aumann's agreement theorem, which says that two people who have the same prior credences will also have the same posterior credences for an event *A* if their posteriors are "common knowledge" (where persons 1 and 2 have common knowledge of event *E* if and only if "both know it, 1 knows that 2 knows it, 2 knows that 1 knows it, 1 knows that 2 knows that 1 knows it, and so on") (1976, 1236).[21] Essentially, Aumann shows that two ideal Bayesians with common priors will converge on the same posteriors if they fully disclose their credences to one another. On the supposition that ideally rational agents have the same priors, the agreement theorem would rule out Robust Perspectivalism, since according to the latter view, ideally rational agents can be required to disagree even in a context of perfect communication. But the view on the multiple Beauties case I have been defending here gives us reason to reject the supposition that maximally rational agents must have the same priors. I have argued that centered information like "I am Claire" can be evidentially relevant to an assessment of an uncentered proposition, since knowledge of who you are (and what your causal history is) may be needed to determine what evidential selection procedure you have used to acquire your evidence.[22] Thus, the prior credences that collectively determine Claire's views on the identity of the victim will include credences for certain centered propositions. Since Claire's prior for "I am Claire" is 1, and Dillon's prior for "I am Claire" is 0, their relevant prior credences are not the same and Aumann's agreement theorem does not apply. And in this case the divergent priors are clearly not due to any rational shortcoming.

5. A HARDER CASE?

In the previous section, I gave what I take to be a satisfactory justification for the striking perspectivalism that halfers seem committed to in the multiple

[21] Thanks to an anonymous referee for suggesting that I show how Robust Perspectivalism can be squared with Aumann's theorem.

[22] The claim that one's evidence for an uncentered proposition is not always exhausted by one's uncentered information can be supported without committing to any particular stance on Robust Perspectivalism. For example, see the "Mystery Bag" example in Titelbaum (2012, 235–6).

Beauties case. But I would now like to look briefly at another variant of the Sleeping Beauty case that is arguably more difficult for halfers to accommodate. Like the multiple Beauties case, this case also shows that halfers are committed to a surprising disconnect between how we assess someone's epistemic credentials and how much weight we give their views.

Let us suppose that alongside Sleeping Beauty during her time in the Experimental Philosophy Laboratory is a second subject named Informed Beauty. Informed Beauty will be put to sleep at the same time as Sleeping Beauty, and he will be awakened on Monday at the same time as Sleeping Beauty and then put to sleep at the same time after this awakening. But no matter what the outcome of the coin toss, Informed Beauty will be awakened on Tuesday, either along with Sleeping Beauty if her coin landed tails, or by himself if her coin landed heads. Another critical difference between Informed Beauty's situation and Sleeping Beauty's situation is this: Informed Beauty will not be subject to any memory tampering, and both he and Sleeping Beauty know this. So when Informed Beauty is awakened on Monday, he will know what day it is, and likewise when he is awakened on Tuesday. Unfortunately, the two Beauties are not allowed to communicate, so Sleeping Beauty cannot benefit from Informed Beauty's information.

Like Sleeping Beauty, Informed Beauty is known to be a paragon of rationality. Given that this is the case, it seems that when Sleeping Beauty awakens along with Informed Beauty on Monday morning, she ought to regard Informed Beauty as an "expert" (relative to her) with respect to the probability of Heads, where someone is an **expert** on p relative to S just in case his epistemic position with respect to p is at least as good as S's in every respect (including possession of evidence, rationality of judgment, and functioning of cognitive faculties) and is superior to S's in at least one respect. Sleeping Beauty ought to regard Informed Beauty as an expert because he has all of the evidence that Sleeping Beauty has, is just as rational as Sleeping Beauty, and possesses one piece of pertinent knowledge that Sleeping Beauty does *not* have: namely, knowledge of what day it is. Since knowledge of what day it is is relevant to assessing the likelihood of Heads (since if one knows that it is Tuesday and whether or not Sleeping Beauty is awake, one can confirm whether Heads or Tails is true), it seems that Sleeping Beauty ought to prefer Informed Beauty's credence to her own, and thus ought to mirror Informed Beauty's credence as best as she can. For it seems that rationality requires that our current credences "reflect" the credences of acknowledged experts. Specifically, the following seems to be a rational constraint:

> **Expert Reflection**: If I know that S is an expert on p relative to myself, then my credence for p conditional on S's credence for p being x should also be x. Formally, if C is my credence function and C_S is S's credence function, then it ought to be the case that $C(p \mid C_S(p) = x) = x$.

Like the Rule of Proportionality, Expert Reflection requires that my credences be calibrated with my views on the epistemic credentials of myself and others.

An upshot of Expert Reflection is that if S is an acknowledged expert on p, and if I have precise credences for all the possibilities for S's credence for p, then my current credence for p should be equal to the expected value of S's credence for p.

Expert Reflection has been endorsed by some philosophers, and it initially seems quite plausible.[23] And the claim that Sleeping Beauty ought to regard Informed Beauty as an expert is also very plausible. But halfers must reject one of these claims or embrace a highly implausible version of the halfer view. To see why, suppose that upon awakening, Sleeping Beauty's credences for H1, T1, and T2 are those endorsed by most halfers: 0.5, 0.25, and 0.25, respectively. What will Sleeping Beauty's expected value be for Informed Beauty's credence for Heads? Well, if it is Monday, then Informed Beauty will not know if the coin landed heads or tails and will surely have a credence for Heads of 0.5. And if it is Tuesday, then Informed Beauty will know this (since he'll remember the Monday awakening the day before), and, upon seeing that Sleeping Beauty is awake, will assign Heads a credence of 0. Since Sleeping Beauty believes with 0.75 confidence that it is Monday, her expected value for Informed Beauty's credence for Heads is $0.75 \cdot 0.5 + 0.25 \cdot 0 = 0.375$. Since this value differs from Sleeping Beauty's credence for Heads of 0.5, Sleeping Beauty will violate Expert Reflection if she takes Informed Beauty to be an expert relative to her with respect to Heads.

On the assumption that Sleeping Beauty should have precise credences for H1, T1, and T2, the only positions that allow Sleeping Beauty to reflect Informed Beauty's credences are those that maintain that her credence for H1 should be equal to her credence for T1.[24] So thirders have no problem accommodating Expert Reflection in this case. An "optimistic halfer" position that maintains that, upon awakening on Monday, Beauty ought to be certain that it is Monday (with a credence of 0.5 for H1 *and* a credence of 0.5 for T1) also has no problem accommodating the view that Sleeping Beauty ought to reflect Informed Beauty's credences. But while this view has at least one defender (Hawley 2013), I think most would judge it exceedingly implausible. If we set optimistic halfism aside, the halfer is left with two options: deny Expert Reflection, or deny that Sleeping Beauty ought to count Informed Beauty as an expert.

[23] While I have defined an "expert" as someone whose epistemic position is at least as strong as mine in every respect (including both information and judgment) and superior in at least one respect, and then offered Expert Reflection as an intuitive proposal about how we ought to reflect the credences of such an expert, Elga (2007, 479–80), adapting the "expert" terminology from Gaifman (1988), *defines* an expert as someone whose credences we reflect, and suggests that someone ought to be treated as an expert if she has all of the information I have and more, and if her judgment is at least as good as mine. So Elga does seem to endorse Expert Reflection. Titelbaum (2012, 147–8) also endorses an "interpersonal Generalized Reflection principle" that has Expert Reflection as a consequence.

[24] Let C be Sleeping Beauty's credence function; Sleeping Beauty will reflect Informed Beauty's credences if and only if $C(H1) = (C(H1) + C(T1))\ 0.5 + C(T2)\ 0$. Solving, we get $C(H1) = C(T1)$.

The latter option is a promising response for those halfers who affirm that Sleeping Beauty ought to conditionalize upon learning that it is Monday (e.g. Lewis 2001; Jenkins 2005; Bradley 2011b). If these halfers are correct in holding that Sleeping Beauty's credence for Heads should move to 2/3 after learning that it is Monday (even if the coin has yet to be tossed), then at this stage Sleeping Beauty must have some sort of evidence that is not possessed by Informed Beauty and that justifies a credence that departs from the objective probability of the coin toss. The fact that Sleeping Beauty alone possesses this evidence gives her reason for not taking Informed Beauty to be an expert. But what could this evidence be? Here again, the differences in evidence could be explained in terms of differences in evidential selection procedures. From Sleeping Beauty's perspective, the current awakening could have been a Monday or Tuesday awakening. She thus possesses the following piece of evidence: this awakening that had to be a Monday awakening conditional on Heads and that was equally likely to be a Monday or Tuesday awakening conditional on Tails turned out to be a Monday awakening. From Sleeping Beauty's perspective, a "randomly" selected awakening proved to be a Monday awakening.[25] This evidence, which serves to confirm Heads, is perspectival evidence since it cannot be possessed by Informed Beauty. While Informed Beauty can affirm that the present awakening could have been a Monday or Tuesday awakening *from Sleeping Beauty's perspective*, from his perspective the awakening was guaranteed to be a Monday awakening irrespective of whether Heads or Tails obtains, so that the fact of its being a Monday awakening has no evidential significance. In short, on this view we have another robustly perspectival context: because the day of the present awakening was initially uncertain for Sleeping Beauty but not for Informed Beauty, the fact that it is Monday has evidential significance for Sleeping Beauty that it cannot (and should not) have for Informed Beauty. And because on Monday there is evidence that is accessible to Sleeping Beauty and inaccessible to Informed Beauty, Sleeping Beauty should not count Informed Beauty as an expert, and her failing to reflect Informed Beauty's credences therefore does not violate Expert Reflection.

However plausible the above response may or may not be, it is not available to the majority of halfers who deny that Sleeping Beauty ought to conditionalize on the information that it is Monday. These "double halfers" (e.g. Bostrom 2007; Meacham 2008; Pust 2012) maintain that after learning that it is Monday, Sleeping Beauty's credence for Heads ought to remain at 1/2 (rather than moving to 2/3, which was her prior conditional credence for Heads conditional on its being Monday). Since double halfers deny that Sleeping Beauty has any epistemic advantage relative to Informed Beauty once Sleeping Beauty has learned what day it is, the claim that she has some epistemic advantage lacked by Informed Beauty *before* learning that it

[25] This response was proposed by Darren Bradley in personal correspondence. See also Jenkins (2005).

is Monday is unmotivated. Given that Sleeping Beauty knows that she will converge on Informed Beauty's credences once she acquires knowledge of what day it is, so that Informed Beauty's credences constitute an epistemic target for her, surely Sleeping Beauty must acknowledge that Informed Beauty is an expert. So the double halfer is stuck with having to reject Expert Reflection. Perhaps this result does not significantly intensify the already substantial worries about the double halfer position stemming from its rejection of conditionalization in this case. But since conditionalization is a *diachronic* constraint and Expert Reflection is a *synchronic* constraint, there is no guarantee that an explanation for the inapplicability of conditionalization will also provide the material for an explanation of the inapplicability of Expert Reflection.[26] So the apparent incompatibility of the double halfer position and Expert Reflection does provide additional grounds for worry.

While Expert Reflection is extremely plausible upon first inspection, I shall argue that there are cases where one should not reflect the credences of known experts, and further suggest that Informed Beauty is such a case. To see why Expert Reflection is false, consider the following case.[27] On Friday morning, Natasha announces to her online network of hundreds of friends that on Friday night she is going to perform a very important coin toss. She also informs her friends of the following: If the coin lands heads, Natasha will tell no one of the result. If the coin lands tails, Natasha will immediately share this result with exactly one person who has been randomly selected in advance from the large pool of Natasha's out-of-state friends. No one besides Natasha knows the identity of this person, and Natasha ensures that none of her friends will talk to others about whether or not Natasha has contacted them about the coin toss.

[26] Indeed, some ways that double halfers have attempted to argue for the inapplicability of conditionalization when Sleeping Beauty learns that it is Monday quite clearly do not help with explaining why Expert Reflection should also fail to apply. Consider, for example, Pust's recent (2012) attempt to defend the double halfer position. Pust first notes that conditionalizing on evidence *E* requires that one have a prior credence for *E*, and thus that *E* be in the domain of one's prior credence function. He then argues that when Sleeping Beauty acquires the temporally indexical evidence that it is Monday *now*, the proposition that expresses that evidence is one that, due to its temporally indexical nature, could not be grasped by Sleeping Beauty at any other time. The proposition is therefore not in the domain of any prior credence functions, rendering conditionalization completely inapplicable. Or, if we are working in a framework where propositions express only eternal truths, then according to Pust, whatever else we may substitute in the place of a proposition to serve as the object of Sleeping Beauty's knowledge that it is Monday *now* will also be outside of the domain of her prior credence function. I don't think this attempt to defend the double halfer position succeeds. But the important point here is that even if it *did* succeed, it would not explain why Expert Reflection should fail to apply to Sleeping Beauty. For unlike conditionalization (as applied to temporally indexical knowledge), Expert Reflection does not require that one respond to temporally indexical evidence in a way that is coordinated with one's credal attitude towards that very same (not yet known) evidence at some *prior* time. Rather, it merely requires that one's *current* credence for some proposition be coordinated with one's *current* views on what an expert *currently* thinks about that proposition.

[27] Weatherson (2009) offers a somewhat similar counterexample to Expert Reflection in an online weblog.

As it happens, I am an in-state friend of Natasha's, so there is no chance that I will learn the result of the coin toss. But on Saturday morning, I know that there is one privileged friend of Natasha's who, conditional on the coin coming up tails, knows the result. Call this person "Special Friend." Let's suppose that I know that all of Natasha's friends are supremely rational thinkers, and that all of the information I have that could possibly bear on the outcome of Natasha's coin toss is information that is also possessed by all of Natasha's friends. This enables me to conclude that, relative to me, Special Friend is an expert with respect to the outcome of Natasha's coin toss. For Special Friend has all of my evidence plus knowledge of whether he or she has been contacted by Natasha, knowledge that will either very slightly confirm heads if he or she has not been contacted (since an out-of-state friend's not being contacted is *guaranteed* conditional on heads and only highly likely conditional on tails) or, if he or she *has* been contacted, fully confirm tails. So whatever the outcome of the coin toss, I can expect that Special Friend's credences for heads and tails will be more accurate than my own.

While Special Friend is clearly an expert in the stipulated sense, this is a case where I should *not* conform to the dictates of Expert Reflection. Because Special Friend's "expertise" is strongly skewed towards a particular direction, so that Special Friend is highly privileged with respect to evidence for tails but only slightly privileged with respect to evidence for heads, my "reflecting" Special Friend's credence in accordance with Expert Reflection would inevitably skew my credence for heads downwards.[28] And once I lower my credence for heads in order to reflect Special Friend's credence, I will be even *more* confident that Special Friend has a credence of 0 for heads, and I will have to lower my credence again, prompting yet greater confidence that Special Friend's credence for heads is 0 and calling for yet another decrease in my credence for heads. This process will continue indefinitely: given the extreme way in which Special Friend's expertise is skewed, the only credence for heads that stably satisfies Expert Reflection in this case is 0.[29] But clearly, rationality does not in this case require me to be perfectly confident that Natasha's coin landed tails! Thus, Expert Reflection is false.

I suggest that there is a similar sort of skewing effect in the Informed Beauty case that gives us reason for thinking that the requirement posited by

[28] For Special Friend's expertise to be skewed in this way, Special Friend must be ignorant of the fact that he or she is Special Friend (unless he receives the call from Natasha). Otherwise, the absence of definitive evidence for tails would itself be definitive evidence for heads. But this ignorance does not disqualify Special Friend from being an expert relative to myself on the outcome of the coin toss. For I also am ignorant of the identity of Special Friend, and there is no doubt that Special Friend's epistemic position is superior to my own.

[29] To see this, let c be my credence that Natasha's coin landed heads and r be the value that I expect Special Friend's credence for heads to be if the coin in fact landed heads. In this case, Expert Reflection requires that $c = c \cdot r + (1 - c) \cdot 0$. Solving, we get $c = c \cdot r$. Since I know that r is not 1 (recall: given heads, Special Friend will not know that he or she is Special Friend, and will only know that he or she has not been contacted—evidence that will justify an r slightly above 0.5), the only way that this constraint can be satisfied is if $c = 0$.

Expert Reflection does not apply. In this case, however, what is skewed is not Informed Beauty's expertise, but rather his expertise *conditional on Sleeping Beauty being awake and able to reflect his credences*. For while Informed Beauty can confirm either Heads or Tails on Tuesday, when Sleeping Beauty is awake he can only confirm Tails. Thus, if Sleeping Beauty is awake and is considering the expected value of Informed Beauty's credence for Heads *right now*, the fact that he might at some point be certain that Heads can make no difference (since he is *not* certain of Heads right now given that Sleeping Beauty is awake), while the fact that he might currently be certain of Tails *will* make a difference to the expected value. Evaluated from Sleeping Beauty's vantage point, Informed Beauty is effectively a biased expert (in the manner of Special Friend), even though Informed Beauty is not biased when evaluated from the perspective of someone who is awake on both Monday and Tuesday. This gives us grounds for doubting whether Expert Reflection expresses a genuine rational requirement in this case.

Clearly, more must be said in order to articulate a corrected "expert reflection" principle and to determine whether this principle vindicates the halfer or thirder position. I think, though, that enough has been said to significantly blunt any worries that may have resulted from the realization that the majority halfer position is in tension with Expert Reflection.

6. CONCLUSION

I have argued that the highly intuitive reasoning behind the halfer solution to the Sleeping Beauty problem also leads to counterintuitive perspectivalist results. Halfers who continue to stand by this reasoning must affirm that the multiple Beauties case involves a failure of Modest Proportionality, and it seems that double halfers are committed to the view that the Informed Beauty case involves a failure of Expert Reflection. While these counterintuitive results may appear to constitute a significant challenge to the halfer position, I have attempted to show that there are plausible reasons for thinking that Modest Proportionality and Expert Reflection do in fact fail in the cases described. But whether or not one finds my diagnoses of the cases convincing, I hope to have at least demonstrated that there is a rather surprising connection between the debate concerning Sleeping Beauty and the apparently orthogonal debate concerning the epistemic significance of disagreement: halfers are committed to Robust Perspectivalism and therefore must deny Modest Proportionality, a principle that plays an important role in both conciliatory and non-conciliatory approaches to disagreement.[30]

[30] Thanks to Darren Bradley, Lara Buchak, David Christensen, Keith DeRose, Blake Roeber, and Alex Worsnip for discussions of the issues raised here and (in some cases) for comments on earlier versions of this paper. Two anonymous referees also provided very helpful comments.

REFERENCES

Aumann, Robert J. 1976. "Agreeing to Disagree." *The Annals of Statistics* 4 (6) (November): 1236–9.

Bergmann, Michael. 2009. "Rational Disagreement after Full Disclosure." *Episteme* 6 (3): 336–53.

Boghossian, Paul Artin. 2006. *Fear of Knowledge: Against Relativism and Constructivism*. New York: Oxford University Press.

Bostrom, Nick. 2000. "Observer-Relative Chances in Anthropic Reasoning?" *Erkenntnis* 52 (1): 93–108.

Bostrom, Nick. 2007. "Sleeping Beauty and Self-Location: A Hybrid Model." *Synthese* 157 (1): 59–78.

Bradley, Darren. 2011a. "Confirmation in a Branching World: The Everett Interpretation and Sleeping Beauty." *The British Journal for the Philosophy of Science* 62 (2): 323–42.

Bradley, Darren. 2011b. "Self-Location Is No Problem for Conditionalization." *Synthese* 182 (3): 393–411.

Bradley, Darren. 2012. "Four Problems about Self-Locating Belief." *Philosophical Review* 121 (2): 149–77.

Briggs, Rachael. 2010. "Putting a Value on Beauty." *Oxford Studies in Epistemology* 3: 3–34.

Christensen, David. 2009. "Disagreement as Evidence: The Epistemology of Controversy." *Philosophy Compass* 4 (5): 756–67.

Christensen, David. 2011. "Disagreement, Question-Begging and Epistemic Self-Criticism." *Philosophers' Imprint* 11 (6) (March): 1–22.

Elga, Adam. 2000. "Self-Locating Belief and the Sleeping Beauty Problem." *Analysis* 60 (266): 143–7.

Elga, Adam. 2007. "Reflection and Disagreement." *Noûs* 41 (3): 478–502.

Enoch, David. 2010. "Not just a Truthometer: Taking Oneself Seriously (but not too Seriously) in Cases of Peer Disagreement." *Mind* 119 (476): 953 –97.

Feldman, Richard. 2006. "Epistemological Puzzles about Disagreement." In *Epistemology Futures*, ed. Stephen Hetherington, 216–36. New York: Oxford University Press.

Feldman, Richard. 2007. "Reasonable Religious Disagreements." In *Philosophers without Gods: Meditations on Atheism and the Secular Life*, ed. Louise M. Antony, 194–214. New York: Oxford University Press.

Fumerton, Richard. 2010. "You Can't Trust a Philosopher." In *Disagreement*, ed. Richard Feldman and Ted Warfield, 91–110. New York: Oxford University Press.

Gaifman, Haim. 1988. "A Theory of Higher-order Probabilities." In *Causation, Chance and Credence*, ed. Brian Skyrms and William L. Harper, 191–219. Dordrecht: Kluwer Academic.

Goldman, Alvin. 2010. "Epistemic Relativism and Reasonable Disagreement." In *Disagreement*, ed. Richard Feldman and Ted Warfield, 187–215. New York: Oxford University Press.

Gutting, Gary. 1982. *Religious Belief and Religious Skepticism*. Notre Dame, IN: University of Notre Dame Press.

Halpern, Joseph. 2005. "Sleeping Beauty Reconsidered: Conditioning and Reflection in Asynchronous Systems." *Oxford Studies in Epistemology* 1: 111–42.

Hawley, Patrick. 2013. "Inertia, Optimism and Beauty." *Noûs* 47 (1): 85–103.

van Inwagen, Peter. 2010. "We're Right. They're Wrong." In *Disagreement*, ed. Richard Feldman and Ted Warfield, 10–28. New York: Oxford University Press.

Jenkins, C. S. 2005. "Sleeping Beauty: A Wake-up Call." *Philosophia Mathematica* 13 (2) (June 1): 194–201. DOI:10.1093/philmat/nki015.

Kelly, Thomas. 2010. "Peer Disagreement and Higher-order Evidence." In *Disagreement*, ed. Richard Feldman and Ted A. Warfield, 111–74. New York: Oxford University Press.

Lackey, Jennifer. 2010. "What Should We Do When We Disagree?" *Oxford Studies in Epistemology* 2: 274–93.

Lewis, David. 2001. "Sleeping Beauty: Reply to Elga." *Analysis* 61 (271): 171–6.

Meacham, Christopher J. G. 2008. "Sleeping Beauty and the Dynamics of De Se Beliefs." *Philosophical Studies* 138 (2): 245–69.

Meacham, Christopher J. G. 2010. "Unravelling the Tangled Web: Continuity, Internalism, Uniqueness and Self-Locating Belief." *Oxford Studies in Epistemology* 3: 86–125.

Pust, Joel. 2012. "Conditionalization and Essentially Indexical Credence." *The Journal of Philosophy* 109 (4): 295–315.

Thurow, Joshua C. 2012. "Does Religious Disagreement Actually Aid the Case for Theism?" In *Probability in the Philosophy of Religion*, 209–24. Oxford: Oxford University Press.

Titelbaum, Michael G. 2008. "The Relevance of Self-Locating Beliefs." *Philosophical Review* 117 (4): 555–606.

Titelbaum, Michael G. 2012. *Quitting Certainties: A Bayesian Framework Modeling Degrees of Belief*. Oxford: Oxford University Press.

Vahid, Hamid. 2004. "Varieties of Epistemic Conservatism." *Synthese* 141 (1): 97–122.

Weatherson, Brian. 2005. "Should We Respond to Evil with Indifference?" *Philosophy and Phenomenological Research* 70 (3): 613–35.

Weatherson, Brian. 2009. "Trusting Experts." *Thoughts Arguments and Rants*. <http://tar.weatherson.org/2009/05/06/trusting-experts/>.

White, Roger. 2005. "Epistemic Permissiveness." *Philosophical Perspectives* 19 (1): 445–59.

Zuboff, Arnold. 2000. "The Perspectival Nature of Probability and Inference." *Inquiry* 43 (3): 353–8.

8. Knowledge Is Belief for Sufficient (Objective and Subjective) Reason

Mark Schroeder

[W]hen the holding of a thing to be true is sufficient both subjectively and objectively, it is *knowledge*.

Immanuel Kant, *Critique of Pure Reason*, A822/B850

This paper outlines and lays the basis for the defense of a simple thesis: that knowledge is belief for reasons that are both objectively and subjectively sufficient. The analysis of knowledge, of course, is one of the most famous failed projects in analytic philosophy. Attempts to analyze knowledge can generally be categorized as (at least) one of 1) prone to counterexample, 2) too vague to make real predictions, or 3) so complex as to make it puzzling why knowledge is so important. But it is the thesis of this paper that with the right understanding of the chief difficulties encountered in the Gettier literature, and with the right perspective on the place of epistemology within normative inquiry more generally, we can see that many of the early approaches to the analysis of knowledge were essentially on the right track after all, even though they made natural mistakes of implementation along the way. The analysis that I'll offer is simple, free from at least the most significant standard sources of counterexamples, and makes sense of why knowledge is important and interesting.

In section 1, I'll set up the problem and define the space for its solution by explaining why knowledge must consist in a kind of *match* between objective and subjective conditions. Then in section 2, I'll introduce an old idea about how to analyze knowledge that is well motivated by the observations in section 1, and explain the chief difficulties this idea ran into, when it was originally introduced—the difficulties that eventually convinced so many that the analysis of knowledge was a failed project. In sections 3–5, I'll set up each of the three key moves that I will argue allow us to retain the key insights of this old approach while not only avoiding the problems it faces, but offering an explanation of where those problems came from, and delegating details, where appropriate, to more general problems for which we require solutions on independent grounds. Finally, I'll close in section 6 by summarizing what we've accomplished and how.

This paper does not constitute an exhaustive defense of the analysis of knowledge that I propose—on the contrary, it comes with distinct and non-trivial commitments, and at least on its simplest version, it faces further potential obstacles that I won't have the space to take up here. But what I do hope to accomplish, in this paper, is to remind us of how natural and well motivated the basic idea is, that knowledge is belief whose justification "stands up," in the right way, to the facts, and to show that the most famous problems with analyses of knowledge that fit this schema have been problems with implementation, rather than with the spirit of the idea. If I can make each of these claims plausible, then that will help make the case that the distinctive commitments of my analysis are worth exploring further and taking seriously, and that it is worth exploring such a theory's versatility to respond to further potential objections.

1. KNOWLEDGE AS MATCH

Our story begins with the idea that knowledge is a distinctive kind of *match* between objective (or worldly) and subjective (or psychological) conditions. In this section I want to emphasize three aspects of this matching character of knowledge. Each of these aspects will later be important.

1.1. *Aspect 1: Primeness*

That knowledge consists in *some* kind of match between objective and subjective conditions is demonstrated by Timothy Williamson's (2000) argument that knowledge is what he calls *prime*. What Williamson means by the thesis that knowledge is prime, is that there is no way of separating out knowledge into "internal" and "external" factors, in such a way that to know *p* is to satisfy both the internal and external components. Since belief and justification are internal, and since truth is external, the idea is that you can't add either any internal condition or any external condition, which, together with belief, truth, and justification, is what it is to know.

Williamson's argument for the primeness thesis is simple. What he does is to construct pairs of cases, A and B, of subjects who know something, such that an internal duplicate of A who is in B's external circumstances does not know it. If knowledge is just a conjunction of internal and external factors, then this should be impossible—for A must satisfy the internal factor of knowledge (since she knows), and B must satisfy the external factor of knowledge (since she knows), so C—who has A's internal makeup and B's external circumstances—must satisfy both the internal and external factors of knowledge, and hence must know as well.

One of the simple examples that Williamson considers is a subject, A, who sees water normally through her right eye, but whose left eye, by chance, is receiving light rays "emitted by a waterless device just in front of that eye," but a head injury prevents further processing of signals from her left eye. This subject processes the visual signals from her right eye, and believes that there is water in front of her on the visual evidence. Intuitively, she knows that there is water in front of her. Subject B is just like subject A, except that everything is reversed—the left eye sees water normally, the right eye is confronted by the "waterless device," and it is the signals from the right eye that are internally impaired. By parity of reasoning, subject B knows that there is water in front of her. To complete the argument, subject C is exactly like subject A internally—she is receiving water-like light rays in both eyes, but the signal from her left eye is prevented by a head injury—but is in exactly B's external circumstances—the real water is in front of her left eye, and the waterless device is in front of her right eye. Intuitively, subject C does not know that there is water in front of her—for the only thing leading her to believe this is her illusory perceptual experience as of water, through her right eye.

What Williamson's examples show is that knowledge can't consist merely in the conjunction of internal and external conditions. It must involve the right *match* between these conditions. It is this match between the internal and external that Williamson's examples exploit. Subject A's internal component matches her external component, and subject B's internal component matches her external component. Subject C fails to know, however, because for her, these components no longer match.

The idea that prime conditions can result from a *match* between internal and external components should not be surprising, and Williamson even explicitly acknowledges that an analysis of knowledge on which it requires such a match is not eliminated by his argument for primeness. Being prime is a very far cry from being unanalyzable; even the state of *believing the truth about p* is a prime condition. If A believes *p* and is in a situation in which *p* is true and B believes ~*p* and is in a situation in which ~*p* is true, each believes the truth about *p*—but C, who like A believes only *p* and like B is in a situation in which ~*p* is true, does not believe the truth about *p*. Why not? Her internal state no longer *matches* her external state. The thesis of knowledge as match explains the primeness of knowledge in exactly the same way.

1.2. Aspect 2: Defeater Pairing

So to begin with, we know that knowledge involves some kind of match between internal and external conditions. But in fact, we know more about what kind of match this must be. For there is independent evidence that knowledge involves such a match, deriving from the way in which *defeaters* for knowledge pair into objective and subjective versions. As I will use the term, a *defeater* for knowledge is just a further detail, which, when added to a case in which we presume that someone knows, makes it turn out, other things equal, that she doesn't know after all.

For example, suppose that Jones is driving through some scenic countryside, and looks over and sees a barn by the side of the road.[1] "Hey," he thinks to himself, "that's a cool-looking old barn." Intuitively, we would

[1] See Goldman (1976).

presume that in such a case, Jones knows that he is looking at a barn. However, if we add to the case the detail that the barn Jones is looking at is actually the only barn for miles around, and all of the other apparent barns that he has been driving past are really just barn façades set up by Hollywood filmmakers, that changes our judgment about the case. Now we conclude that Jones doesn't really know that he is looking at a barn after all. The fact that he is driving through fake barn country is a *defeater* for his knowledge, because it is the detail of the case that makes it the case that he doesn't know after all.

The fake barn country case is what we might call an *objective defeater* for knowledge, because it is a worldly condition that defeats his claim to knowledge. In addition to objective defeaters, however, there are also *subjective defeaters* for knowledge. Suppose, instead of adding to our story that Jones is actually driving through fake barn country, we instead added that he *believes* that he is driving through an area where all but one of the things that look like barns are really just façades set up by Hollywood filmmakers. Nevertheless, as Jones drives by this particular thing that looks like a barn, he still thinks that it is really a barn. In this case it doesn't seem like Jones knows, either. So just as the fact that he is really driving through fake barn country can defeat his knowledge, so can the fact that he believes that he is, whether or not that belief is true. Since this defeater is a condition of Jones's belief state, we may call it a *subjective* defeater for his knowledge.

Notice that in the fake barn cases, the objective defeater and the subjective defeater come paired. The very same proposition whose truth defeats Jones's knowledge in the objective defeater case is one such that Jones's mere belief in it defeats his justification in the subjective defeater case. This turns out to be no coincidence—it is an important and general fact that objective and subjective defeaters for knowledge *always* come paired. To see this, compare a different sort of case. The fake barn country case involves what has come to be known as an *undercutting* defeater. The fact that Jones is driving through fake barn country undercuts his justification for believing that he is looking at a barn, because it renders his visual evidence less useful. Undercutting defeaters are typically contrasted with *countervailing* defeaters, which involve contrary reasons.

For example, if you read in an academic article that a study has shown that axillary dissection is indicated for breast cancer, the fact that this study used an unrepresentative sample would be an undercutter for the conclusion that axillary dissection is so indicated, but the fact that there are several other studies that show no positive net effects for axillary dissection unless the sentinel lymph node tests positive for metastatic disease would be a countervailing defeater. Like undercutting defeaters, countervailing defeaters for knowledge come paired in matching objective and subjective varieties. If you form a belief that axillary dissection is indicated for breast cancer after reading only one article, then even if this is true, you don't really know it, if there is good research to the contrary. This is the objective defeater case. But if you *have* read the contrary literature and believe the first article anyway in spite of the evidence, then you don't know, either. That is the subjective defeater case.

The importance of the pairing of objective and subjective defeaters for knowledge is illustrated by the literature on *pragmatic encroachment* on knowledge. Some authors have argued—very controversially—that knowledge depends not only on evidence and other truth-related factors, but also on what is at stake over a question for the believer.[2] But importantly, advocates of such pragmatic encroachment hold that high stakes can make it harder to know in each of two different ways. It can be harder to know either because the stakes are actually high, regardless of whether the agent realizes that they are, or because the agent believes the stakes to be high, regardless of whether they really are. The former cases—called *ignorant high stakes* by Stanley (2005)—are putative objective defeaters for knowledge, and the latter are putative subjective defeaters for knowledge.

It is no surprise that pragmatic encroachers like Stanley will think that there are two different ways in which stakes can make it harder to know, because it follows from our general principle that defeaters for knowledge always come paired in this way. Consequently, whether or not you follow Stanley and the others in believing that there actually *is* pragmatic encroachment on knowledge, the fact that those who are tempted to think there is are naturally led to postulate two corresponding types of defeat is further evidence for the centrality of the phenomenon of defeater pairing.[3]

1.3. *Aspect 3: Explanatory Power*

The phenomenon of defeater pairing is not only another important piece of evidence that knowledge involves a kind of match between objective and subjective conditions; it also tells us something important about what *kind* of match this must be. It suggests that the relevant match must be between the relationship a belief bears to the agent's other beliefs, and the relationship it bears to the facts. It is because these two relationships must match, that analogous upsets in either suffice to defeat knowledge. Before going on, however, there is one more aspect of this match that it will be important for us to observe. It is from this third aspect, that we can learn something about just what the relationship between a belief and other beliefs must be, which the relationship between that belief and the facts must match.

For our third aspect of the matching character of knowledge, we return again to Williamson, who argues that knowledge has a distinctive explanatory power, over and above belief and justified belief. Williamson argues for this distinctive explanatory power for his own distinctive dialectical reasons, and for our purposes we will not need everything that Williamson means to get

[2] See especially Fantl and McGrath (2002), Hawthorne (2004), Stanley (2005), Fantl and McGrath (2009), Schroeder (2012a), and Ross and Schroeder (2014).

[3] For a similar observation about lottery cases, see Nelkin (2000).

out of this argument.[4] But what I do think is clearly correct about Williamson's point, is that there are at least some cases in which the fact that someone knows provides a better explanation of their action than the fact that they believe, or that they justifiably believe.

Williamson's leading example of an explanation in which knowledge plays a distinctive explanatory role is the case of a burglar who "spends all night ransacking a house, risking discovery by staying so long." Williamson's explanation of why the burglar stayed so long is that he knew that there was a diamond in the house. The burglar's behavior is not explained by the fact that he believed that there was a diamond in the house—because several hours of ransacking with no results to show for it would in most cases suffice to make it rational for someone who believes, but does not know, that there is a diamond in the house, to give up that belief. Similarly, the burglar's behavior is not explained by the fact that he justifiably believed that there was a diamond in the house—for even a very good justification to believe that there is a diamond in the house can be defeated by the accumulation of the kind of counter-evidence one is bound to come by in the course of eight or nine hours of searching for it with no luck. In contrast, Williamson claims, the burglar's searching all night *can* be explained by the fact that he *knew* that there was a diamond in the house.

The reason the burglar's knowledge can provide a better explanation for his behavior than the fact that he believed, or even that he justifiably believed, is that knowledge involves a match between the burglar's belief state and the facts. It is because it involves such a match that it can explain why the burglar *still believes*, and indeed still *justifiably* believes, that there is a diamond in the house, even after eight or nine hours of looking for it with no luck. Here is an intuitive gloss on how it does this: it does it because knowledge is belief whose justification *stands up to the facts*. The fact that the burglar knows explains why he is justified in not ceasing to believe, even once he has acquired a fair bit of new evidence that there is no diamond in the house after all, because it involves having a justification that stands up to—and hence is robust in the face of—such evidence.[5]

How does this relate to our idea that knowledge is a match between the relationship between a belief and an agent's other beliefs, and the relationship between that belief and the world? It tells us something about what that relationship is. A belief's justification depends on its relationship to the agent's

4 Compare, for example, Molyneux (2007).

5 Note that I am *not* claiming that knowledge requires justification that is robust in the face of *any* discovery. This is clearly not the case. Sometimes even though you know, you can learn something that defeats your knowledge, by making you no longer justified in your belief. See section 2.2. But in general, this happens only when there is some other fact that, if only you learned it as well, would restore both your justification and your knowledge. Such evidence is *misleading*, in the proper sense. The intuitive force of the paradox of dogmatism, introduced by Harman (1973) and attributed to Saul Kripke, trades on the ambiguity between evidence that is misleading in this sense, and *is* properly ignored, and evidence that merely supports something false, which is *not* properly ignored.

other beliefs. The idea of knowledge as match tells us that a belief must bear a similar relationship to the world as it bears to other beliefs, in virtue of which it is justified. It's because of this match between the facts and one's justification, that because the burglar knows, his justification is robust in the face of the facts. This third aspect of the match involved in knowledge is one of the important things we'll be able to explain once my account is on the table.

2. TRADITIONAL PROBLEMS FOR THE ANALYSIS OF KNOWLEDGE

In section 1 I've collected three important observations that motivate the idea that knowledge involves a kind of match between internal and external conditions. In fact, I argued, these observations motivate a much more specific idea about the kind of match that is involved. They motivate the idea that the justificatory status a belief has, in virtue of the agent's other beliefs, must be *good enough* that it "stands up to" the rest of the facts. This intuitive idea not only provides an intuitive explanation of Williamson's observations about the explanatory power of knowledge, it also explains the simplest sorts of Gettier cases—"false lemma" cases like Gettier's (1963) original *Brown in Barcelona*, undercutting defeater cases like Jones in fake barn country, and even countervailing defeater cases like the breast cancer research case. The appeal of this sort of idea should therefore be clear, and it is no surprise that many authors in the early decades of research into the Getter problem offered versions of the idea that knowledge is justified belief that stands up in some way to the facts.[6]

The analysis of knowledge did not, therefore, become the most famous failed project of analytic philosophy because it was mysterious how to get this far. What became notoriously difficult was nailing down the details. Two issues, in particular, turned out to pose repeated challenges, no matter how theorists tried to contort the details of their accounts.

2.1. *An Illustrative Account*

Many of the early accounts of knowledge presented in the late 1960s and throughout the 1970s had trouble with both of the main difficulties in which I will be interested. For concreteness, I'll illustrate them with a particularly simple and natural account due to Peter Klein (1971), which fully captures the spirit of our guiding idea that knowledge is belief whose justification "stands up to" the facts.

According to Klein (1971), *S* knows *p* (at time t_1) just in case at t_1 *S* truly believes *p*, *p* is evident to *S*, and "there is no true proposition such that if it became evident to S at t_1, *p* would no longer be evident to *S*" (1971, 475). This captures very well the idea that knowledge is belief whose justification stands

[6] See, in particular, Clark (1963), Sosa (1964), (1970), Lehrer (1965), (1970), (1974), Lehrer and Paxson (1969), Klein (1971), Annis (1973), Ackerman (1974), Swain (1974), Johnsen (1974), Unger (1975), Olin (1976), and Barker (1976) for some of the highlights of this tradition.

up to the facts. Klein captures the way in which belief must stand up to the facts by supposing that there must be no fact such that were it added to S's beliefs, S's justification for believing *p* would go away. This account explains why knowledge involves the kind of match that makes it prime. It explains why we observe defeater pairing, because the objective conditions that defeat knowledge are just the things that, were they to be justifiedly believed, would subjectively defeat knowledge by defeating justification. And it explains Williamson's thesis about explanatory power, because it explains why someone who knows will in general continue to be justified in her belief even when she discovers new evidence, as the burglar does after spending the entire night ransacking the house in search of diamonds.[7]

Klein's account explains all of our observations from section 1 because it makes good on the idea that knowledge is belief whose justification stands up to the facts. As we'll see in sections 2.2 and 2.3, it is subject to predictable counterexamples. But it is important to keep clear on whether these counterexamples reflect poorly on the core idea that knowledge is belief whose justification stands up to the facts, or they only reflect poorly on the way in which Klein tried to make this idea precise. It will be the thesis of this paper that the major problems besetting this account and others like it derive from mistakes in implementation, rather than from any failure of the core insight that knowledge is belief whose justification stands up to the facts in the right way.

2.2. *The Defeater Dialectic*

Suppose that you see Tom Grabit come out of the library, pull a book from under his shirt, cackle gleefully, and scurry off. In this case, absent further information, it looks like you know that Tom stole a book. But if Tom has an identical twin, Tim, from whom you could not distinguish him, then it seems that you don't really know after all. Tim would therefore be a defeater for your knowledge that Tom stole a book. Klein's account can capture this defeater for knowledge, since if you were to find out that Tom has an identical twin, then you would no longer be justified in believing that Tom stole a book solely on the basis of your visual evidence. So far, so good.

But unfortunately, just as knowledge can be defeated, defeaters for knowledge can also be defeated. Suppose, for example, that Tim's wedding was scheduled for today in another state. If this is the case, then it seems that you can know that Tom stole a book on the basis of your visual evidence alone, even though he has an identical twin. So the potential defeater for your knowledge is itself defeated. But Klein's account is too strong, and gets this wrong. According to Klein, since finding out that Tom has a twin (without also

[7] Note, however, that it does not explain exactly the right degree of robustness of knowledge in the face of new evidence. This will become clear in section 2.2.

finding out that Tim's wedding is scheduled to take place in another state) would make your justification go away, you don't know.[8]

In the early 1970s a great deal of published work on the analysis of knowledge went into trying to characterize the conditions on which a true proposition is a defeater that is not itself defeated. This turned out to be very difficult to do, in part because just as knowledge can be defeated and defeaters can be defeated, defeater-defeaters can also be defeated. For example, if Tim called off the wedding, then the fact that it was scheduled to be today in another state doesn't interfere with Tim's interfering with your visual evidence that Tom stole a book. And if the reason Tim called off the wedding was to elope to Bali instead, then it seems that you can know after all. But if all of the flights to Bali have been cancelled, then perhaps you don't. What cases like this show is that defeaters and defeater-defeaters can go on ad infinitum.[9]

This means that it is not enough for an analysis of knowledge to predict the ways in which knowledge can be defeated. It must also be able to predict the ways in which the defeaters for knowledge can themselves be defeated, so that the agent knows after all. An analysis that fails to allow for defeaters will be too expansive, allowing for knowledge that there is not. But an analysis that fails to allow for defeater-defeaters will be too narrow, failing to account for knowledge that there is. And one that fails to allow for defeater-defeater-defeaters will be too expansive again. Just talking about the phenomenon is a bit dizzying; it's easy to see why so many attempts to analyze knowledge ended up with epicycles—the phenomenon seems to cry out for them.

It turns out that the defeater dialectic is very familiar to moral philosophers, as pushed by proponents of moral particularism. Whereas the defeater dialectic for knowledge starts with the problem that knowledge can be defeated in a range of ways, and then adds that defeaters can be defeated, and even defeater-defeaters can be defeated, the particularist dialectic in moral philosophy starts with the facts that the wrongness of an action can be defeated in a variety of ways, and even those defeaters can themselves be defeated. Just as the defeater dialectic in epistemology poses a problem for the analysis of knowledge, particularists argue that the defeater dialectic in moral philosophy poses a problem for the possibility of posing any informative generalizations at all about—let alone any analyses of—moral wrongness.[10]

[8] In general, whenever there is a potential objective defeater for your knowledge that is itself defeated, learning the potential defeater without learning its defeater-defeater will undermine both justification and knowledge. Defeaters which are themselves defeated can thus provide *misleading* evidence, if you learn them without also learning of their defeater. This shows that the extra explanatory power offered by knowledge in burglar cases like Williamson's is limited.

[9] Compare especially Levy (1977) for discussion of this point. This case is a variation on a case introduced by Lehrer and Paxson (1969), variations on which are common in the literature cited in note 4.

[10] Compare especially Dancy (2004), Schroeder (2011a) for discussion.

The similarity between the defeater dialectic in epistemology and in moral philosophy will be important for my eventual solution.

2.3. *The Conditional Fallacy*

The other major difficulty faced by many early attempts to analyze knowledge is the conditional fallacy. An account commits the conditional fallacy by attempting to analyze a categorical property in conditional terms. For example, we saw that on Klein's account you know if you justifiedly believe the truth, and moreover there is no true proposition, such that were you to (justifiedly) believe it, you would cease to be justified in your belief. This conditional account attempts to capture the idea that knowledge is true belief whose justification is good enough to "stand up to" the facts, and uses the conditional in order to gloss what it is to be good enough to stand up to the facts. The idea is that it *is* good enough, if it *would still be* sufficient for justification, if the agent (justifiedly) had beliefs in those facts.

We've already seen, in section 2.2, that this account runs into trouble with the defeater dialectic. But it also runs into trouble with the conditional fallacy. To see why in the abstract, note that the conditional analysis has us evaluate whether some justification *is* good enough for knowledge by looking at whether it *would be* good enough, at the closest world at which the agent has the other belief. But in some cases, the closest world in which the agent has the relevant belief will also be a world where other things happen. Perhaps the agent *would* know, in such a world, even though she actually does not. Or perhaps she would not, even though actually she does. Either of these scenarios can create conditional fallacy trouble for such an account.

Once we understand how the conditional fallacy works, it is easy to imagine what such counterexamples must look like, but the one offered by Shope (1983) is simple: suppose that S knows that she is not justified in believing some proposition *r*. But suppose that *r* is true. Finally, add to the case that were S to justifiedly believe *r*, this justification would be transparent to her, so that she would no longer be justified in believing that she is not justified in believing it. According to Klein's account, S does not know.

If Shope's counterexample to Klein strikes you as a matter of detail, rather than spirit, then you and I are on the same page. The counterexample shows that Klein was wrong to try to cash out the way in which justification must "stand up to" the facts in terms of a subjunctive conditional. But the problem is a general one for the use of such subjunctive conditionals to try to analyze categorical properties. We would be remiss if we took it to cast aspersion on the core insight that knowledge is belief whose justification stands up to the facts. We will simply need to find a way of understanding this "standing up to" relation in categorical, non-conditional, terms.

The defeater dialectic and the conditional fallacy are the two main problems for the analyses of knowledge canvassed by Shope (1983), which is cited by pessimists like both Williamson (2000) and Kvanvig (2003) as an authoritative treatment of the persistent problems facing analyses of knowledge.

There is no guarantee that an analysis of knowledge that avoids these two problems without arcane twists and turns will also be free of other problems or objections, but if there is an independently motivated and natural account that is free of these two problems, that should at least make us question what grounds we have for inductive pessimism about the Gettierological project.

3. FIRST MOVE: TWO KINDS OF SUFFICIENCY

In this paper I will be defending the same sort of account as we have been describing so far—one on which knowledge consists in the right sort of match between one's justification and the facts, where that match involves the justification somehow being "good enough" to "stand up to" the facts. But unlike earlier attempts to cash out this idea, the account that I offer will be based on some broad principles that render it immune to the chief difficulties encountered by earlier accounts.

In the next three sections, I will be breaking up the steps required in order to explain how my favored version of this analysis of knowledge works, by dividing them into three principal moves. The first move, in the remainder of section 3, is to distinguish between two different ways in which the reasons for a belief can be sufficient, by distinguishing between two kinds of reasons: objective and subjective. This move establishes the key concepts employed by my account. The second move, in section 4, is to show how to defend a categorical account of the sufficiency of reasons. This move lets us avoid the conditional fallacy. And the third move, in section 5, is to defend an important thesis about the weight of reasons. This thesis will allow us to make the right predictions about the defeater dialectic.

What all three of these moves have in common is that they appeal to natural and independently motivated claims about *reasons*. I will offer no general argument here that knowledge needs to be understood in terms of reasons, besides trying to exhibit the explanatory power and resourcefulness of such an account. But one piece of circumstantial evidence that this is not a crazy idea is the pervasiveness of the idea, at least in moral philosophy, that reasons are the basis of normativity. If questions about knowledge and justification are normative questions, then it follows from this general idea that we should expect them to ultimately be questions about reasons. Of course, it could be that knowledge and justification are the cases that prove this general idea to be false. But given the broad appeal of this general idea, it is at least hardly ad hoc or strained to investigate what resources reasons provide us for the analysis of knowledge.

3.1. *Objective and Subjective Reasons*

So far I've been saying a few things: that knowledge involves a kind of *match*, and that this match involves one's justification being *good enough* to stand up to the facts, in some way. But we know, from the problems posed by the conditional fallacy, that it will not do to try to cash out this notion of being "good enough" in counterfactual terms. So we will need to understand it in terms of some categorical relationship between one's belief state and the facts. In this and the following subsection I will argue that we can do so by appeal to some independently important and well-motivated distinctions—distinctions that are important not only for the study of epistemology, but for the study of reasons more generally.

The most important distinction that we will need, is that between what are sometimes called *objective* and *subjective* reasons. The intuitive distinction goes like this: if Max is smiling, that is reason to believe that he is happy. But if no one realizes that Max is smiling, no one *has* that reason to believe that Max is happy. I'll call the sense in which the fact that Max is smiling is a reason to believe that he is happy, even if no one knows about it, the *objective* sense of "reason," and I'll call the sense in which in this case no one has a reason to believe that Max is happy the *subjective* sense of reason. Objective reasons, then, are facts or true propositions, and subjective reasons are propositions to which agents have some sort of epistemic access—the kind of access, whatever it is, that is lacked when no one has the reason to believe that Max is happy.

Some people believe that subjective reasons, so understood, must themselves be objective reasons. They believe that talk about whether someone *has* the reason to believe that Max is happy may be taken literally, as talk about some sort of possession relation that agents might bear to things that are reasons to believe that Max is happy.[11] I call this view the *Factoring Account*, and my view is that it is wrong. I believe that the "has" in "Caroline has a reason to believe that Max is happy" is pleonastic, as in "Caroline has a golf partner." The latter does not mean that there is someone who is a golf partner and whom, moreover, Caroline has; it just means that there is someone who is Caroline's golf partner. Similarly, on my view, talk about the (subjective) reasons that someone has is just talk about the things that are reasons for her (in the subjective sense of "reason"). I have argued at length against the Factoring Account elsewhere;[12] in this paper I will simply assume that this theory is wrong, and that subjective reasons need not themselves be objective reasons—someone may have a subjective reason without there being any corresponding objective reason.

I mention the Factoring Account and my view that it is false because it will be important in what follows that subjective reasons do not need to be based on *true* beliefs. Since objective reasons must be truths, the idea that subjective reasons are just objective reasons to which you stand in some possession relationship implies that subjective reasons *must* be based on true beliefs. That is why I think it is important to see that this view is false. At any rate, I will assume that subjective reasons can be based on false beliefs in what follows.[13]

11 In Schroeder (2008) I make the case that there is much circumstantial evidence that this thesis has been widely accepted by epistemologists. Errol Lord (2010) defends it explicitly, in responding to Schroeder (2008).

12 Schroeder (2008).

13 For argument, see Schroeder (2008).

Other people—well, some of them are the same people—believe that having a subjective reason requires having a justified belief in that proposition, or even knowing it.[14] I believe that this theory is also wrong; on my preferred view, having a subjective reason requires only having a belief—*or* a perceptual state with a propositional object (perhaps intellectual seemings will also do the trick). Having a reason does not require that the belief (or the perceptual state or intellectual seeming) be justified, though I do hold that when a belief is *un*justified, the subjective reason that the agent thereby has is guaranteed to be defeated, so it is not possible to "bootstrap" yourself into having *good* reasons to believe something, simply by having unjustified beliefs that support it. Again, I have argued for these views elsewhere at length.[15] Here I will simply assume that we do not need to appeal to the concepts of rationality, justification, or knowledge in understanding what it is for someone to have a subjective reason. This is important, because this allows us to *use* the concept of having a reason in order to analyze rationality, justification, and knowledge—which would not otherwise be possible, without circularity. For our purposes, we can see this as one of the important, non-trivial, commitments of the analysis of knowledge advocated in this paper.

I want to emphasize three things about the objective/subjective distinction. First, it is a natural and intuitive distinction. This is illustrated not only by the fact that it is easy to give an intuitive sense for what such talk is about, but by the facts that the same distinction applies to reasons for action as for reasons for belief, and that the same distinction can be made for evidence as for reasons—presumably because evidence matters in epistemology because it is a particularly important kind of reason for belief. Second, I want to emphasize that for my main claims in this paper I do not need any particular claims about the ontology of subjective reasons. You may hold, as I have in previous work,[16] that they are the contents of beliefs, or you may hold that they are the belief state itself. All that matters for my view about knowledge is that there is a way of mapping between subjective reasons and objective reasons.

And third, I will be assuming both that subjective reasons need not be factive, and that we can understand them independently of knowledge and justification. These are the assumptions that will be important for me in what follows, and though I have argued for each of them separately before, here the principal argument for these claims will be by reference to their fruits, as illustrated by the way in which they allow for subjective reasons to play a role in the analysis of knowledge.

[14] Compare, for example, Feldman (1988, 227): "If I believe, for no good reason, that P and I infer (correctly) from this that Q, I don't think we want to say that I 'have' P as evidence for Q. Only things that I believe (or could believe) rationally, or perhaps, with justification, count as part of the evidence that I have. It seems to me that this is a good reason to include an epistemic acceptability constraint on evidence possessed . . ." See also Williamson (2000) and Hawthorne and Stanley (2008).

[15] Schroeder (2011b).

[16] Schroeder (2007), (2008), (2011b).

3.2. *Rationality and Correctness*

The objective/subjective distinction among reasons corresponds to an important distinction between *rationality* and *correctness*. Leaving epistemology aside for a moment, it can be rational for Bernie, who believes that his glass contains gin and tonic, to take a sip, even though this is not the correct or advisable thing for him to do, since in fact his glass contains gasoline. Similarly, it can be correct for him to set his glass down without taking a sip, without that being a rational course of action for him, given that he's been looking forward to a drink all day and doesn't want to offend his host. In the theory of practical reason, it is natural to hold that subjective reasons are related to rationality in the same way that objective reasons are related to correctness. Though Bernie's subjective reasons are *sufficient*, or *good enough* to make taking a sip rational, his objective reasons to take a sip are *not* sufficient to make it correct—for there is a decisive objective reason for him not to take a sip—namely, that his glass is full of gasoline.

The distinction between rationality and correctness is also important for belief. A belief is generally held to be *correct* just in case it is true, but many false beliefs are rational, and many true beliefs are not rational. A false belief will be rational for someone who has sufficient evidence that it is true—that is, who has good enough *subjective* reasons to believe that it is true. Similarly, a true belief will fail to be rational, for someone who has conclusive reason *not* to believe the proposition in question. So it is natural to think that the rationality of beliefs is related to subjective reasons for belief in the same way that the rationality of action is related to subjective reasons for action.

However, many epistemologists believe that the *correctness* of beliefs has nothing to do with reasons.[17] Whereas the correctness of an action may depend on the objective reasons in favor of or against it, it is commonly observed that the correctness of a belief depends only on whether it is true. Consequently, many epistemologists assume that for belief, correctness just *is* truth. This assumption is premature and misguided. On the assumption that the fact that *p* is false is always a conclusive objective reason not to believe *p*, we can derive the fact that it is correct to believe *p* only if *p* is true from the generalization that, like action, belief is correct just in case there are no conclusive reasons against it. On the natural assumption that "correct" is univocal as applied to action and belief, this is a much better motivated way of accounting for this data.

So this leads us to a picture on which there are two important kinds of sufficiency, corresponding to the two important kinds of reason: when the objective reasons to believe *p* are sufficient, it is *correct* to believe *p*, and when the subjective reasons to believe *p* are sufficient, it is *rational* to believe *p*. This kind of rationality of believing *p* is what epistemologists would refer to as

[17] Compare Gert (2008).

propositional justification.[18] Whether it is rational *for* a subject to believe *p* in the sense of whether the subject has a propositional justification to believe *p* can depend solely on her subjective reasons, and whether they are good enough. But it is also often important to know not only whether *p* is a rational thing for a subject to believe, but whether she is rational *in* believing *p*. This is what epistemologists typically refer to as *doxastic* justification.[19]

It is straightforward to make sense of doxastic justification in our framework. Doing so just requires introducing a third important sense of "reason"—what moral philosophers refer to as *motivating* reasons. The motivating reason for which you do something is just the reason *for which* you do it. Similarly, the motivating reason for which you believe something is just the reason *for which* you believe it. By calling such things "motivating reasons" I do not mean to judge whether believing something for a reason deserves to be called "being motivated" in any robust sense of the term, merely to observe that there is an exact analogue in epistemology of what moral philosophers refer to as "motivating reasons." Epistemologists tend to prefer to talk about the *basing relation*, but I find this awkward and less clearly grounded in pretheoretically important talk. We know pretheoretically that people do and believe things for reasons; motivating reason talk is just talk about the reasons for which they do and believe these things.

Although motivating reason talk is I think perfectly pretheoretically sensible, that is not to say that it is easy to analyze. Importantly, motivating reasons figure in explanations of why an agent does or believes what she does, but not all explanations of why an agent does or believes what she does are reasons-explanations. It turns out that it is hard to say what this difference is, but rather than get distracted by this, I will simply rely on an intuitive understanding of talk about motivating reasons, since it is a distinction that I take it everyone needs to be able to make sense of, independently of their account of knowledge. I'll have more to say in the final section of the paper about the detrimental effects in the epistemological literature of premature attempts to analyze important concepts like this one.[20]

I will take it that at least in normal cases and possibly in all cases, the reasons for which someone believes something are themselves subjective reasons for her to believe it. I will further take it that in at least some cases, the reasons for which someone believes something are themselves objective reasons to believe it—this happens, I take it, when the reasons for which someone believes something are true.[21] My own view is that what makes it possible

[18] Here I equate justification with rationality. I take it from its use within epistemology that "justification" is a property of belief which is clearly intuitively necessary for knowledge, and present in Gettier cases. I think, but will not argue here, that the most natural such property of which we have an independent grasp is simply that of being rational.

[19] Kvanvig and Menzel (1990) also distinguish doxastic and *personal* justification, but I will not be concerned with this distinction in what follows.

[20] Compare Lehrer (1971), whose account of what it is to believe something for a reason was one of Shope's (1978) leading examples of the conditional fallacy.

[21] Compare Dancy (2000).

for someone's motivating reasons for belief to be both subjective reasons to believe and objective reasons to believe, is that the same kind of thing plays all three roles: propositions are objective reasons when true, subjective reasons when believed, and motivating reasons when, by being believed, they play a certain role in bringing about another belief or in maintaining that other belief. But nothing in what follows will turn on that—so long as your conceptual framework can make sense of what it means for the reason for which someone believes something to be a good objective reason to believe it or a good subjective reason to believe it, you will have allowed for everything that I need.

Before going on, it is important to emphasize that these concepts—of objective and subjective reasons for belief, and of the motivating reason for which someone believes something—are not special inventions created for the purpose of understanding knowledge, or even for the purpose of epistemology. The very same concepts are central in the study of practical reason—moral philosophers make the same distinctions between objective and subjective reasons for action and for attitudes other than belief, and distinguish both from the motivating reasons for which someone acts or for which they hold a certain attitude. Indeed, my own previously mentioned arguments against the Factoring Account are based on the practical case. Moreover, it should not be surprising that belief is subject to at least some of the same categories that we use in trying to understand action and other attitudes, nor that "reason" should turn out to be unambiguous in "reason for action," "reason for belief," and "reason for intending."[22]

With the concept of a motivating reason in hand, however, we may say that an agent is *doxastically rational* in believing *p* just in case the reasons *for which* she believes *p* are subjectively sufficient. This means simply that the reasons for which she believes *p* include among them subjective reasons for her to believe *p* that are sufficient to make it propositionally rational for her to believe *p*. Corresponding to the notion of *doxastic rationality* in the subjective domain is the notion of *well-groundedness* in the objective domain. A belief is

[22] It is sometimes said that "reason" cannot be unambiguous across "reason for action" and "reason for belief," or at least that reasons for action and reasons for belief are very different kinds of thing, because reasons for action must be believed, and reasons for action must only be true. The foregoing distinction between objective and subjective reasons should make clear, however, that this allegation compares subjective reasons for belief with objective reasons for action—which *do* behave differently, because they are on opposite sides of the objective/subjective distinction. It is perfectly understandable, moreover, why we are more interested in objective reasons in ethics and more interested in subjective reasons in epistemology—after all, in epistemology it is widely agreed that what we objectively ought to do is to believe the truth—the whole problem is how to accomplish that. So we know what the objective reasons support—they support the truth; the problem is getting there. Whereas in moral philosophy, one of the main issues at stake is what action is supported by the facts, even if we know what those facts are—and consequently we do, in fact, spend more time discussing objective reasons in moral philosophy. The fact that discussions in epistemology focus mostly on subjective reasons and discussions in ethics focus mostly on objective reasons is therefore easy to understand, and should not confuse us into thinking that reasons for action and reasons for belief are fundamentally different topics.

well grounded just in case the reasons for which it is held are *objectively* sufficient. This means simply that the reasons for which it is held include among them objective reasons sufficient to make it correct for it to be held.

Knowledge, I claim, at least at a first pass, is just belief that is both doxastically rational and well grounded.[23] That is, as Kant says, it is belief for reasons that are both objectively and subjectively sufficient.[24] Now we just need to know how to make sense of this talk about sufficiency.[25]

4. SECOND MOVE: SUFFICIENCY AS BALANCE

In section 3 I laid out the key concepts and elements of my approach—enough to see what my final analysis of knowledge will look like. This account makes good on all of the structural features that underlay our original motivating idea that knowledge is belief whose justification stands up to the facts. At its core is a "match" between objective and subjective conditions, validating the primeness of knowledge. The relevant match is between the structure of one's justification and the facts, as suggested by the phenomenon of defeater pairing, and the account explains why we should predict the phenomenon of defeater pairing, because the same motivating reasons for belief that could fail to be subjectively sufficient because of some further belief could fail to be objectively sufficient because of a corresponding further fact. And it explains the explanatory power of knowledge in cases like Williamson's, because it imposes the requirement that a knower's reasons for belief need to be sufficient not only in the face of her other subjective reasons, but also in the face of the facts.

However, in order to make good on this account, we will need to see how it can avoid the main problems for similar accounts of knowledge. In particular, we will need to see how it can avoid the conditional fallacy, and how it will be able to make the right predictions about the defeater dialectic. In the remainder of section 4, I will introduce a categorical account of *sufficiency*, in order to address the first issue. Then in section 5, I'll show how to make the right predictions about the defeater dialectic.

[23] Note that well-foundedness ensures truth given our assumption that the fact that ~*p* is always a conclusive reason not to believe *p*, assuming bivalence. So given this assumption we need not separate truth condition in our analysis.

[24] In a pair of fascinating discussions, Chignell (2007a) and especially (2007b) explores what Kant means by this remark, and I think makes the case that Kant does in fact mean by it roughly what I do, although by Chignell's account, Kant has somewhat different views about objective reasons and subjective reasons than I would accept.

[25] Like Kant, on Chignell's (2007b) reading, Robert Fogelin (1994) defends an account that bears strong similarities to the one outlined in this paper, but with some differences in detail. According to Fogelin, "s knows that P iff S justifiably came to believe that P on grounds that establish the truth of P" (1994, 28). This is essentially my account, with "establish the truth of P" as a gloss on what makes grounds objectively sufficient. I'm not very clear, however, what it means for grounds to "establish the truth of P," if that is to be weak enough to make room for ordinary claims to knowledge. And Fogelin does not help on this score, using his account to argue for Pyrrhonian skepticism. My account requires only that the reasons for which you believe must outweigh the reasons against belief.

4.1. Balance as a Categorical Account of Sufficiency

On the picture I described in section 3.2, doxastic rationality and well-groundedness are strictly analogous properties of belief. One holds in virtue of the relationship between the reasons for which one holds the belief and the rest of one's *subjective* reasons, and the other holds in virtue of a strictly analogous relationship between the reasons for which one holds the belief and the rest of one's *objective* reasons. It is this strict analogy that makes good on our idea from section 1.2 that the kind of match that knowledge involves is a match between the way one's belief is related to one's other beliefs, and the way it is related to the world. Our account takes reasons as its primitive, and uses them to explain both justification and knowledge.

But Klein's account, introduced in section 2.1, takes justification, rather than reasons, as primitive, and does not have something more detailed to say about the relationship between reasons and justification. So when Klein is looking for something to say about the way in which one's justification must "stand up to" the facts, in order to constitute knowledge, he is limited to saying things that he can say using only the concept of justification. This is an important part of what pushes him to employ a counterfactual test (besides the fact that the test seems to do okay in a range of intuitive cases).

But another way of reading Klein's account is as offering an implicit picture of what makes the reasons for which one believes objectively sufficient. We can get this picture by adding our view of reasons and justification to Klein's view, and seeing what it implies about what it is for reasons to be objectively sufficient. On this picture, the reasons for which one believes are objectively sufficient if there is no truth such that if it were added to one's beliefs, then the reasons for which one believes would fail to be subjectively sufficient. This picture assumes an account of subjective sufficiency, and tries to piggyback an account of objective sufficiency by means of Klein's counterfactual test. The conditional fallacy suggests that this is a bad way to go. Rather than understanding the way in which the reasons for which the agent believes must "stand up to" the facts in terms of *justification*, we should understand it as a direct, categorical relationship between those motivating reasons and the facts. And it should be the exact same relationship on both the objective and subjective sides.

Moral philosophers have a simple idea about the relationship between reasons and justification. According to this idea, an action is rational just in case the agent's subjective reasons to do it are at least as good as her subjective reasons not to do it. Similarly, an action is correct just in case the agent's objective reasons to do it are at least as good as her objective reasons against doing it.[26] This is what I call the idea of *sufficiency as balance*.[27] According to the idea of sufficiency as balance, reasons determine what it is rational to

[26] Compare Parfit (2011), who takes this idea to be so natural that he stipulates that he will talk about "ought in the sense of most reason."

[27] For a fuller defense of *sufficiency as balance*, see Schroeder (forthcoming).

do by competing against one another. When the (subjective) reasons to do something are at least as good as their competitors,[28] that is a rational thing to do. Similarly, when the (objective) reasons to do something are at least as good as their competitors, it is a correct thing to do. When an agent's reasons to do something are at least as good as her reasons against doing it, we may say that they are *sufficient*, and the same definition works, whether we are talking about objective or subjective reasons. So on this picture, sufficiency is a categorical relationship between reasons, determined wholly by their relative *weights*—by how "good" of reasons they are, or how significant a role they play in the competition between reasons.

It is easy to extend this categorical account of sufficiency to smaller sets of reasons. Just as we may say that the set of *all* of an agent's subjective reasons to do something is sufficient just in case together they are at least as good as the set of all of the agent's subjective reasons against doing it, similarly we may say that some *arbitrary* set of an agent's subjective reasons to do something are sufficient just in case together they are at least as good as the set of all of the agent's subjective reasons against doing it. And similarly for objective reasons. This extension of the concept of sufficiency from applying to *total* sets of reasons to arbitrary sets is what allows us to apply it to the reasons *for which* an agent believes, which often do not include all of her subjective reasons to believe and always do not include all of the objective reasons to believe. Because a subset of an agent's subjective reasons can be sufficient only if the set of all of her reasons is sufficient, the sufficiency of even the subset of her reasons for which she actually believes is enough to guarantee the rationality of the belief. Similarly, because a subset of the objective reasons can be sufficient only if the set of all objective reasons is sufficient, the sufficiency of even the subset of objective reasons for which an agent believes is enough to guarantee the belief's correctness.

Because sufficiency as balance relies on a categorical relationship between reasons—their comparative weight—it doesn't introduce any liability to conditional fallacy-type problems. From the wider, more inclusive perspective that includes work in moral philosophy, appealing to something like sufficiency as balance looks like a no-brainer.[29] But in the next subsection I'll explain why this has not seemed like such an obvious thing to say about the rationality of *belief*.

4.2. *Harman's Challenge*

As we saw in the previous section, the idea of sufficiency as balance is familiar and important from moral philosophy. However, it has not always seemed like such an obvious idea in epistemology. To see why, let's start by being a bit more careful in our talk about reasons for and against belief. As illustrated by Pascal's Wager, not just anything that "counts in favor" of a belief helps

[28] Or at least: *not outweighed by* their competitors.

[29] Compare especially Ross (1930).

to make it rational in the way required for knowledge. Rather than getting caught up in whether there is *some* sense of "rational" in which becoming convinced by Pascal's argument can make it "rational" for you to believe in God, let us define *epistemic rationality* as the strongest kind of rationality that is entailed by knowledge. Because Pascalian considerations do not affect knowledge, they clearly do not affect epistemic rationality. So whether or not Pascalian considerations count as reasons for belief for some purposes, they are not the "right kind" of reason for belief to be important for the study of epistemic rationality or knowledge. They are not, as we may say, *epistemic* reasons.

What does affect epistemic rationality, it seems, is *evidence*. Without evidence of some sort, belief is not rational. This thought leads to the view that when we are talking about reasons for belief in the context of knowledge—that is, when we are talking about *epistemic* reasons for belief—what we are really talking about is evidence.[30] But unfortunately for sufficiency as balance, it is not at all plausible that your evidence is sufficient to make believing *p* rational just in case it is at least as good as the competing evidence. If your evidence for *p* is merely *as* good as the competing evidence, then it is generally *irrational* for you to believe *p*; instead, you should remain agnostic. Gilbert Harman (2004) argues that this means that though sufficiency as balance works just fine for the rationality of action, it is inadequate for the rationality of belief, and argues on these grounds that rationality of action and rationality of belief are disanalogous.

It is therefore plausible that thinking of epistemic reasons as evidence has played an important role in dissuading epistemologists from appealing to sufficiency as balance. But this reasoning goes too quickly. The putative problem for sufficiency as balance arises because in addition to believing *p* and believing ~*p*, there is an important third option—*believing neither*. But all of the evidence either supports *p* or supports ~*p*, and so all of the evidence is either reason to believe *p* or reason to believe ~*p*; there isn't any evidence left over to be reason to believe neither. So if epistemic reasons—the reasons that bear on epistemic rationality—are exhausted by the evidence, then sufficiency as balance can't be the right account of what makes them sufficient.

But it was a hasty overgeneralization from Pascal's case to conclude that just because Pascalian considerations are not epistemic reasons, there can't be any reasons that bear on epistemic rationality that are not evidence. Rather than concluding on the basis of Harman's observation that the rationality of belief and the rationality of action are deeply disanalogous, a perfectly good alternative would have been to conclude that not all epistemic reasons are evidence—and in particular, that there are epistemic reasons *against* belief that are not evidence.[31]

[30] Compare BonJour (1985), as well as Parfit (2001), Piller (2001), and Hieronymi (2005).

[31] See Schroeder (2012b).

This should not be a surprising claim. As I've introduced the term, "epistemic reason" isn't just a shorthand for "evidence"; it's a term for those reasons, whatever they are, that bear on epistemic rationality, which is the strongest kind of rationality entailed by knowledge. And there do seem to be non-evidential factors that bear on whether belief is rational. For example, suppose that both Sophia and Zoe have significantly better evidence for p than for $\sim p$, but their situations differ in the following way: although Sophia is in a position to be confident that no further evidence that might bear on the matter is forthcoming, Zoe is waiting on the results of an experiment that has the potential to provide more conclusive evidence than any of her evidence collected so far. Zoe's expectation of further evidence is not evidence against either conclusion, but it does seem to raise the bar for how conclusive her existing evidence must be, in order to make it rational for her to believe. Here is a natural explanation of why: it's because the expectation of further evidence is an epistemic reason *not* to believe.

A full evaluation of whether there are indeed epistemic reasons against belief that are not evidence would take us substantially astray.[32] The point I want to make in this section is that there is a *prima facie* obstacle, which is enough of an obstacle to explain why epistemologists have generally not appealed to sufficiency as balance, but easily enough overcome that it should not dissuade us from realizing the virtues of having a categorical account of sufficiency that is consistent with a uniform picture of the sufficiency of reasons for action and reasons for belief. With sufficiency as balance in hand, we can avoid any risk of conditional fallacy-type problems. So all that remains is to get an understanding of the more complicated features of the defeater dialectic. To that we turn in section 5.

5. THIRD AND FINAL MOVE: WEIGHING REASONS

So far, I've introduced the familiar distinction between objective and subjective reasons, characterized knowledge as belief for reasons that are both objectively and subjectively sufficient, shown that this account fits with the way in which we already observed that knowledge involves a "match" between objective and subjective factors, and showed that by appealing to familiar and general ideas from moral philosophy, we can characterize sufficiency in categorical terms, avoiding the need to fall into the conditional fallacy. What remains is to see why the resulting picture should lead us to predict, rather than be frustrated by, the defeater dialectic. In section 5.1 I'll explain on general grounds why the defeater dialectic is exactly what we should expect given general and independently motivated observations about the weight of reasons, and in section 5.2 I'll isolate a simple conjecture that, if true, would explain why this would be true.

[32] See especially Schroeder (2012a) and (2012b).

5.1. *The Structure of Defeaters*

To see why the picture that I've already described should lead us to anticipate the defeater dialectic, start by observing that one of the distinctive virtues of *sufficiency as balance*, as an account of the sufficiency of reasons, is that it readily explains the difference between undercutting and countervailing defeaters. A set of reasons fails to be sufficient only if the competing reasons are better. And when we add a further detail to a case, that can make this happen in exactly one of two ways. The further detail we add might itself be one of the competing reasons, or might reveal that the competing reasons are better than we otherwise would have presumed. In that case, it is a countervailing defeater. Or it might instead reveal that the reasons we are interested in are not, after all, as good as we would have presumed. In that case, it is an undercutting defeater. Because a set of reasons can fail to be sufficient either by being reduced in weight or by facing even stronger competition, this yields a natural and important distinction between these two kinds of defeat.

The fact that *sufficiency as balance* explains the naturalness and importance of an intuitive distinction that has widely been taken to be important is evidence in its favor. But it also points us in the direction of a general reason to expect the defeater dialectic: we should expect defeater-defeaters, defeater-defeater-defeaters, and so on, precisely if this is what we find for the weight of reasons in general. But in fact, this *is* what we find for reasons in general. In normal cases, the fact that telling Kenny that *p* would be a lie is a weighty reason not to do so. But if you are playing the game Diplomacy, this is not such a weighty reason—for lying is a normal and expected part of the game. What this shows is that ordinary cases of reasons for action can have their weight lowered by further facts, making these further facts defeaters. But these defeaters can also be defeated. For example, if Kenny is your husband and you ended up in a bitter fight the last time you lied to him during a game of Diplomacy, the fact that telling Kenny that *p* would be a lie may be a weighty reason not to do so after all. So though normally the fact that you are playing Diplomacy lowers the weight of this reason, under this circumstance it does not—and hence we have an example of a defeater-defeater.

We can make the same observations in the case of reasons for belief, merely by focusing on plausible judgments about objective reasons. If Jones sees something that looks like a barn in broad daylight, under ordinary circumstances we would take that to be an excellent objective reason to believe that it is a barn. But if he is in fake barn country, visual evidence of a barn is not such a great reason to believe he is seeing a barn after all. However, if he is in real barn state within fake barn country, that defeater is defeated, and his visual evidence seems like a good reason to believe that he is seeing a barn after all. What this case shows is that features of how the cases lead to defeaters for knowledge line up precisely with plausible judgments about the force of objective reasons for belief.

The literature on particularism in ethics is full of examples like this, which mirror the structure of the defeater dialectic in epistemology.[33] What this shows, I believe, is that the fact that the weight of reasons can be lowered or eliminated by further considerations, whose relevance can itself be eliminated by yet further considerations, and so on, is a general and independently motivated observation about the weights of reasons. But it is precisely the observation about the weight of reasons that would need to be true, in order for us to expect the defeater dialectic in epistemology, given my analysis of knowledge. And so I conclude that given what we independently know to be true about the weight of reasons, my analysis of knowledge predicts the defeater dialectic, rather than being frustrated by it. It does so not by explaining it in itself, but by delegating that explanation to independently observable facts about how the weight of reasons can be affected by further facts.

5.2. *Reasons for Reasons*

Still, we might want more of an explanation of why it is that the defeater dialectic works in this way: why it is true not only that the weight of reasons can be defeated, and defeaters for the weight of reasons can be defeated, but this phenomenon seems to go "all of the way up"—admitting of no a priori limit. In other work, I've advocated a simple conjecture that would explain *why* this is what we should expect. In its most general form, this conjecture is that the considerations that reduce the weight of reasons are themselves reasons. If reasons are lowered in weight by other reasons, then it is only to be expected that those reasons could also, at least in principle, be lowered in weight—by further reasons. And that is precisely the structure that would lead us to expect no a priori limit to when things might end.

Of course, if the weight of reasons is itself affected by other reasons, that raises many important questions about exactly how this works. We will not ultimately have any complete *explanation* for the defeater dialectic until we know exactly how this is.[34] However, the right standard for determining whether we are on the right track for the analysis of knowledge is not whether we have in hand a complete analysis of everything to which we appeal. That erroneous standard is what led earlier accounts astray, with premature attempts to analyze general concepts. What we need, in order to be confident that we are on the right track, is only that the features of the unanalyzed concepts to which we appeal, in order to get plausible predictions about knowledge, are just the ones that we have independent reason to expect out of any acceptable analysis of those subsidiary concepts. And that is what I have been arguing is true about the weight of reasons. We know on independent grounds that it is a general fact—true about reasons

[33] See especially Dancy (2004).

[34] See Shackel (forthcoming) for criticisms of the particular way that I tried to do this in chapter 7 of *Slaves of the Passions*.

for action as well as reasons for belief—that reasons can be lowered in weight by further considerations, and that yet further considerations can interfere with this weight-lowering, and so on. That is all that we need in order to be confident that our account has the right structure to expect and ultimately to explain, rather than to be frustrated by, the defeater dialectic.

6. OVERVIEW

In this paper I've been defending the idea that by appeal to general and independently motivated claims about reasons, we can make good on the natural idea that knowledge is belief that "stands up to" the facts, without falling into the familiar traps set by the conditional fallacy and the defeater dialectic, which were responsible for so many of the arcane twists and turns of the Gettier literature in the 1970s. The main lesson that I hope to draw from this is that the failures of attempts to make good on this general idea do not reflect poorly on the idea itself, so much as they reflect on the tools that were used to implement it. And this, I believe, should undermine what grounds this history of failures provides for inductive pessimism about the project of analyzing knowledge.

Moreover, if epistemology is just one branch of normative inquiry more generally—the branch concerned with the assessment of our cognitive capacities—then it should not be a surprise that it helps to take a broader perspective inspired by paying attention to normative concepts outside of epistemology, when focusing on the right way to implement this general, attractive, and well-motivated idea about the nature of knowledge. Paying attention to how our claims about reasons generalize to fit with cases outside of epistemology is neither ad hoc nor imperialist, on this view, but rather just the right kind of constraint to keep us from pursuing dead ends.

If I've been successful so far, then the virtues of the analysis of knowledge that I've been describing should be clear: it is motivated by its fit with our three observations about knowledge as match, by the natural solutions it provides to the most famous sorts of problems with the analysis of knowledge, and by the fact that these solutions appeal only to the sort of resources that we would expect, if we take seriously the idea that knowledge is a normative notion and the normative is to be explained in terms of reasons. It is also simple, and natural—it is simply the conjunction of two closely related properties: the property of being doxastically rational, and the property of being well grounded. Knowledge behaves in complex ways in interesting cases not because knowledge is complicated or ad hoc, but because of the complex behavior of reasons. And it is an important achievement for the same reasons that moral theorists from Aristotle through Kant and beyond have valued not only doing the right thing, but doing it for the right reasons. In short, knowledge is in the realm of belief what virtuous action is in the realm of action.

I don't claim that the account described here is free from all problems; as I noted earlier, we need to take on substantive and highly non-trivial commitments about the priority of reasons, justification, and knowledge even in order to get the project off of the ground, and along the way I've appealed to surprising claims about other things—particularly including the idea that there are epistemic reasons against belief that are not evidence. The account may also require refinement in order to deal with different kinds of cases. The main thing that I claim for it is not that it should be the final word in the analysis of knowledge, but that it offers a prima facie very promising space for an account, in a part of logical space about which there has been so much inductive pessimism, and does so in a way that is motivated by general principles. These things make me suspect that it is worthy of further attention.[35]

REFERENCES

Ackerman, Terrence (1974). "Defeasibility Modified." *Philosophical Studies* 26(5–6): 431–5.

Annis, David (1973). "Knowledge and Defeasibility." *Philosophical Studies* 24(3): 199–203.

Barker, John A. (1976). "What You Don't Know Won't Hurt You?" *American Philosophical Quarterly* 13(4): 303–8.

BonJour, Laurence (1985). *The Structure of Empirical Knowledge*. Cambridge, MA: Harvard University Press.

Chignell, Andrew (2007a). "Belief in Kant." *Philosophical Review* 116(3): 323–60.

Chignell, Andrew (2007b). "Kant's Concepts of Justification." *Noûs* 41(1): 33–63.

Clark, Michael (1963). "Knowledge and Common Grounds: A Comment on Mr. Gettier's Paper." *Analysis* 24(1): 46–8.

Dancy, Jonathan (2000). *Practical Reality*. Oxford: Oxford University Press.

Dancy, Jonathan (2004). *Ethics Without Principles*. Oxford: Oxford University Press.

Fantl, Jeremy, and Matthew McGrath (2002). "Evidence, Pragmatics, and Justification." *Philosophical Review* 111(1): 67–94.

Fantl, Jeremy, and Matthew McGrath (2009). *Knowledge in an Uncertain World*. Oxford: Oxford University Press.

Feldman, Richard (1988). "Having Evidence." In D. F. Austen, ed., *Philosophical Analysis: A Defense by Example* (Dordrecht: Springer), 83–104.

Fogelin, Robert (1994). *Pyrrhonian Reflections on Knowledge and Justification*. New York: Oxford University Press.

Gert, Joshua (2008). "Putting Particularism in its Place." *Pacific Philosophical Quarterly* 89(3): 312–24.

[35] I hope to be able to fill in more of this promise in future work. Special thanks to Barry Lam, Jake Ross, Stew Cohen, Ram Neta, Houston Smit, Juan Comesaña, Justin Lillge, two anonymous referees, and to audiences at Simon Fraser University and the University of Arizona.

Gettier, Edmund (1963). "Is Justified True Belief Knowledge?" *Analysis* 1963: 121–3.
Goldman, Alvin (1976). "Discrimination and Perceptual Knowledge." *Journal of Philosophy* 73(20): 771–91.
Harman, Gilbert (1973). *Thought*. Princeton: Princeton University Press.
Harman, Gilbert (2004). "Practical Aspects of Theoretical Reasoning." In Al Mele and Piers Rawling, eds., *The Oxford Handbook of Rationality*. Oxford: Oxford University Press, 45–56.
Hawthorne, John (2004). *Knowledge and Lotteries*. Oxford: Oxford University Press.
Hawthorne, John, and Jason Stanley (2008). "Knowledge and Action." *Journal of Philosophy* 105(10): 571–90.
Hieronymi, Pamela (2005). "The Wrong Kind of Reason." *Journal of Philosophy* 102(9): 437–57.
Johnsen, Bredo (1974). "Knowledge." *Philosophical Studies* 25(4): 273–82.
Klein, Peter (1971). "A Proposed Definition of Propositional Knowledge." *Journal of Philosophy* 68(16): 471–82.
Kvanvig, Jonathan (2003). *The Value of Knowledge and the Pursuit of Understanding*. Cambridge: Cambridge University Press.
Kvanvig, Jonathan, and Christopher Menzel (1990). "The Basic Notion of Justification." *Philosophical Studies* 59(3): 235–61.
Lehrer, Keith (1965). "Knowledge, Truth, and Evidence." *Analysis* 25(5): 168–75.
Lehrer, Keith (1970). "The Fourth Condition of Knowledge: A Defense." *Review of Metaphysics* 24(1): 122–8.
Lehrer, Keith (1971). "How Reasons Give us Knowledge, or, The Case of the Gypsy Lawyer." *Journal of Philosophy* 68(10): 311–13.
Lehrer, Keith (1974). *Knowledge*. Oxford: Oxford University Press.
Lehrer, Keith, and Thomas Paxson (1969). "Knowledge: Undefeated Justified True Belief." *Journal of Philosophy* 66: 225–37.
Levy, Stephen (1977). "Defeasibility Theories of Knowledge." *Canadian Journal of Philosophy* 7(1): 115–23.
Lord, Errol (2010). "Having Reasons and the Factoring Account." *Philosophical Studies* 149(3): 283–96.
Molyneux, Bernard (2007). "Primeness, Internalism, and Explanatory Generality." *Philosophical Studies* 135(2): 255–77.
Nelkin, Dana (2000). "The Lottery Paradox, Knowledge, and Rationality." *Philosophical Review* 109(3): 373–409.
Olin, Doris (1976). "Knowledge and Defeasible Justification." *Philosophical Studies* 30(2): 129–36.
Parfit, Derek (2001). "Rationality and Reasons." In Dan Egonsson, Jonas Jesefsson, Björn Petersson, and Toni Rønnow-Rasmussen, eds., *Exploring Practical Philosophy*. Burlington, VT: Ashgate, 17–39.
Parfit, Derek (2011). *On What Matters*, volumes 1 and 2. Oxford: Oxford University Press.
Piller, Christian (2001). "Normative Practical Reasoning." *Proceedings of the Aristotelian Society* 25, suppl. volume: 195–216.

Ross, Jacob, and Mark Schroeder (2014). "Belief, Credence, and Pragmatic Encroachment." *Philosophy and Phenomenological Research* 88(2): 259–88.

Ross, W. D. (1930). *The Right and the Good*. Oxford: Clarendon Press.

Schroeder, Mark (2007). *Slaves of the Passions*. Oxford: Oxford University Press.

Schroeder, Mark (2008). "Having Reasons." *Philosophical Studies* 139(1): 57–71.

Schroeder, Mark (2011a). "Holism, Weight, and Undercutting." *Noûs* 454(2): 328–44.

Schroeder, Mark (2011b). "What Does it Take to 'Have' a Reason?" In Andrew Reisner and Asbjørn Steglich-Petersen, eds., *Reasons for Belief*. Cambridge: Cambridge University Press, 201–22.

Schroeder, Mark (2012a). "Stakes, Withholding, and Pragmatic Encroachment on Knowledge." *Philosophical Studies* 160(2): 265–85.

Schroeder, Mark (2012b). "The Ubiquity of State-given Reasons." *Ethics* 122(3): 457–88.

Schroeder, Mark (forthcoming). "What Makes Reasons Sufficient?" Forthcoming in *American Philosophical Quarterly*.

Shackel, Nicholas (forthcoming). "Still Waiting for a Plausible Humean Theory of Reasons." Forthcoming in *Philosophical Studies*.

Shope, Robert (1978). "The Conditional Fallacy in Contemporary Philosophy." *Journal of Philosophy* 75(8): 397–413.

Shope, Robert (1983). *The Analysis of Knowing: A Decade of Research*. Princeton: Princeton University Press.

Sosa, Ernest (1964). "The Analysis of 'Knowledge that P.'" *Analysis* 25(1): 1–8.

Sosa, Ernest (1970). "Two Concepts of Knowledge." *Journal of Philosophy* 67(1): 59–66.

Stanley, Jason (2005). *Knowledge and Practical Interests*. Oxford: Oxford University Press.

Swain, Marshall (1974). "Epistemic Defeasibility." *American Philosophical Quarterly* 11(1): 15–25.

Unger, Peter (1975). *Ignorance*. Oxford: Oxford University Press.

Williamson, Timothy (2000). *Knowledge and its Limits*. Oxford: Oxford University Press.

9. Rationality's Fixed Point (or: In Defense of Right Reason)

Michael G. Titelbaum

Rational requirements have a special status in the theory of rationality. This is obvious in one sense: they supply the *content* of that theory. But I want to suggest that rational requirements have another special status—as *objects* of the theory of rationality. In slogan form, my thesis is:

> **Fixed Point Thesis** Mistakes *about* the requirements of rationality are mistakes *of* rationality.

The key claim in the Fixed Point Thesis is that the mistakes in question are *rational* mistakes. If I incorrectly believe that something is a rational requirement, I clearly have made a mistake in some sense, in that I have a false belief. But in many cases possession of a false belief does not indicate a *rational* mistake; when evidence is misleading, one can rationally believe a falsehood. According to the Fixed Point Thesis, this cannot happen with beliefs about the requirements of rationality—any false belief about the requirements of rationality involves a mistake not only in the sense of believing something *false* but also in a distinctly *rational* sense. While the Fixed Point Thesis is a claim about theoretical rationality (it concerns what we are rationally permitted to *believe*), it applies both to mistakes about the requirements of theoretical rationality and to mistakes about requirements of practical rationality.

Like any good philosophical slogan, the Fixed Point Thesis requires qualification. Suppose I falsely believe that what Frank just wrote on a napkin is a requirement of rationality, because I am misled about what exactly Frank wrote. In some sense my false belief is about the requirements of rationality, but I need not have made a rational mistake. This suggests that the Fixed Point Thesis should be restricted to mistakes involving a priori rational-requirement truths. (We'll see more reasons for this restriction below.) So from now on when I discuss beliefs about rational requirements I will be considering only beliefs in a priori truths or falsehoods.[1] It may be that the set of beliefs about rational requirements targeted by the Fixed Point Thesis should be restricted farther than that. As I build my case for the thesis, we'll see how far we can make it extend.

[1] By an "a priori truth" I mean something that can be known a priori, and by an "a priori falsehood" I mean the negation of an a priori truth.

Even restricted to a priori rational-requirement beliefs (or a subset thereof), the Fixed Point Thesis is surprising—if not downright incredible. As I understand it, rationality concerns constraints on practical and theoretical reasoning arising from consistency requirements among an agent's attitudes, evidence, and whatever else reasoning takes into account.[2] One does not expect such consistency requirements to specify particular contents it is irrational to believe. While there have long been those (most famously, Kant) who argue that practical rationality places specific, substantive requirements on our intentions and/or actions, one rarely sees arguments for substantive rational requirements on belief.[3] Moreover, the Fixed Point Thesis has the surprising consequence (as I'll explain later) that one can never have all-things-considered misleading total evidence about rational requirements.

Finally, the Fixed Point Thesis has implications for how one's higher-order beliefs (beliefs about what's rational in one's situation) should interact with one's first-order beliefs. Thus it has consequences for the peer disagreement debate in epistemology. Most philosophers think that in the face of disagreement with an equally rational, equally informed peer an agent should conciliate her opinions. Yet the Fixed Point Thesis implies that whichever peer originally evaluated the shared evidence correctly should stick to her guns.

Despite both its initial implausibility and its unexpected consequences, we can argue to the Fixed Point Thesis from a premise most of us accept already: that *akrasia* is irrational. After connecting the Fixed Point Thesis to logical omniscience requirements in formal epistemology, I will argue for the thesis in two ways from the premise that *akrasia* is irrational. I will then apply the Fixed Point Thesis to higher-order reasoning and peer disagreement, and defend the thesis from arguments against it.

1. LOGICAL OMNISCIENCE

I first became interested in the Fixed Point Thesis while thinking about logical omniscience requirements in formal theories of rationality. The best-known such requirement comes from Bayesian epistemology, which takes Kolmogorov's probability axioms to represent rational requirements on agents' degrees of belief. One of those axioms (usually called Normality) assigns a value of 1 to every logical truth. In Bayesian epistemology this entails something like a rational requirement that agents assign certainty to all logical truths. Logical omniscience in some form is also a requirement of such formal epistemologies as ranking theory and AGM theory.

[2] While some may want to use the word "rationality" in a more externalist way, I take it most of us recognize at least *some* normative notion meeting the description just provided (whatever word we use to describe that notion). That is the notion I intend to discuss in this essay, and will use the word "rationality" to designate. Later on I'll consider whether the Fixed Point Thesis would be true if framed in terms of other normative notions (justification, reasons, etc.).

[3] The main exception I can think of is Descartes' (1988) *cogito* argument, which (with some major reinterpretation of Descartes' original presentation) could be read as an argument that it's irrational for an agent to believe she doesn't exist.

Logical omniscience requirements provoke four major objections:

- There are infinitely many logical truths. An agent can't adopt attitudes toward infinitely many propositions, much less assign certainty to all of them. (Call this the Cognitive Capacity objection.)
- Some logical truths are so complex or obscure that it isn't a rational failure not to recognize them as such and assign the required certainty. (Call this the Cognitive Reach objection.)
- Rational requirements are requirements of consistency *among* attitudes toward propositions. They do not dictate particular attitudes toward *single* propositions, as logical omniscience suggests.[4]
- Logical truths play no different role in the theory of rationality than any other truths, and rationality does not require certainty in all truths. Garber (1983: p. 105) writes, "Asymmetry in the treatment of logical and empirical knowledge is, on the face of it, absurd. It should be no more *irrational* to fail to know the least prime number greater than one million than it is to fail to know the number of volumes in the Library of Congress."

The last two objections seem the most challenging to me. (In fact, much of this essay can be read as a response to these two objections when applied to attitudes toward rational requirements instead of attitudes toward logical truths.) The first two objections are rather straightforwardly met. For Cognitive Capacity, one need only interpret the relevant logical omniscience requirements as taking the form "*If* one takes an attitude toward a logical truth, *then* one should assign certainty to it." Logical omniscience then does not require that attitudes be taken toward any particular propositions (or every member of any infinite set of propositions) at all.

To respond to the Cognitive Reach concern, we can restrict logical omniscience so that it requires certainty only in logical truths that are sufficiently obvious or accessible to the agent. Notice that even if we respond to the Cognitive Capacity and Cognitive Reach objections as I've just suggested, the other two objections remain: Why should a theory of rationality be in the business of dictating particular attitudes toward particular propositions (that is, if attitudes toward those propositions are taken at all), and why should the class of logical truths (even when restricted to the class of *obvious* logical truths) have a special status in the theory of rationality? Of course, filling out a plausible obviousness/accessibility restriction on the logical omniscience requirement is no trivial matter. One has to specify what one means by "obviousness," "accessibility," or whatever, and then one has to give some account of which truths meet that criterion in which situations. But since it

[4] Alan Hájek first brought this objection to my attention; I have heard it from a number of people since then. There are echoes here of Hegel's famous complaint against Kant's categorical imperative that one cannot generate substantive restrictions from purely formal constraints. (See e.g. Hegel (1975: pp. 75ff.).)

was the objector who introduced the notion of obviousness or accessibility as a constraint on what can be rationally required, the objector is just as much on the hook for an account of this notion as the defender of logical omniscience.

Various writers have tried to flesh out reasonable boundaries on cognitive reach (Cherniak (1986), for instance), and formal theories of rationality can be amended so as not to require full logical omniscience. Garber (1983) and Eells (1985), for example, constructed Bayesian formalisms that allow agents to be less than certain of first-order logical truths. Yet it is an underappreciated fact that while one can weaken the logical omniscience requirements of the formal epistemologies I've mentioned, one cannot eliminate them entirely. The theories of Garber and Eells, for example, still require agents to be omniscient about the truths of sentential logic.[5]

Those wary of formal theorizing might suspect that this inability to entirely rid ourselves of logical omniscience is an artifact of formalization. But one can obtain logical omniscience requirements from informal epistemic principles as well. Consider:

> **Confidence** Rationality requires an agent to be at least as confident of a proposition y as she is of any proposition x that entails it.

This principle is appealing if one thinks of an agent as spreading her confidence over possible worlds; since every world in proposition x is also contained in proposition y, the agent should be at least as confident of y as x. But even without possible worlds, Confidence is bolstered by the thought that it would be exceedingly odd for an agent to be more confident that the Yankees will *win* this year's World Series than she is that the Yankees will *participate* in that series.

Given classical logic (which I will assume for the rest of this essay) it follows immediately from Confidence that rationality requires an agent to be equally confident of all logical truths and at least as confident of a logical truth as she is of any other proposition. This is because any proposition entails a logical truth and logical truths entail each other. One can add caveats to Confidence to address Cognitive Capacity and Reach concerns, but one will still have the result that if an agent assigns any attitude to a sufficiently obvious logical truth her confidence in it must be maximal.[6]

So special requirements on attitudes toward logical truths are not the sole province of *formal* epistemologies. Still, we can learn about such requirements by observing what happens to formal theories when the requirements are lifted. Formal theories don't require logical omniscience because formal theorists like the requirement; logical omniscience is a side-effect of systems

[5] Gaifman (2004) takes a different approach to limiting Bayesian logical omniscience, on which the dividing line between what's required and what's not is not so tidy as sentential versus first-order. Still, there remains a class of logical truths to which a given agent is required to assign certainty on Gaifman's approach.

[6] Notice that this argument makes no assumption that the agent's levels of confidence are numerically representable.

capturing the rational requirements theorists are after. Take the Bayesian case. Bayesian systems are designed to capture relations of rational consistency among attitudes and relations of confirmation among propositions. As I already mentioned, one can construct a Bayesian system that does not fault agents for failing to be certain of first-order logical truths. For example, one can have a Bayesian model in which an agent assigns credence less than 1 to $(\forall x)Mx \supset Ms$. Applied to a sample consisting entirely of humans, this model allows an agent to be less than certain that if all humans are mortal then the human Socrates is as well. But in that model it may also be the case that *Ms* no longer confirms $(\forall x)Mx$, one of the basic confirmation relations we build Bayesian systems to capture.[7] Similarly, in the imagined model the agent may no longer assign at least as great a credence to *Ms* as $(\forall x)Mx$; it will be possible for the agent to be less confident that the human Socrates is mortal than she is that all humans are mortal.[8]

This is but one example of a second underappreciated fact: You cannot give up logical omniscience requirements without also giving up rational requirements on consistency and inference.[9] What is often viewed as a bug of formal epistemologies is necessary for their best features. This second underappreciated fact explains the first; if one removed all the logical omniscience requirements from a formal theory, that theory would no longer place constraints on consistency and inference, and so would be vitiated entirely.

What does logical omniscience have to do with this essay's main topic—attitudes toward truths about rational requirements? In general, a rational requirement on consistency or inference often stands or falls with a requirement on attitudes toward a particular proposition.[10] I call such a proposition a "dual" of the requirement on consistency or inference.[11] Logical omniscience

[7] Here's how that works: Suppose that, following Garber, our model assigns credences over a formal language with an atomic sentence *A* representing $(\forall x)Mx$ and an atomic sentence *S* representing *Ms*. If our model has a basic Regularity requirement and we stipulate that $P(A \supset S) = 1$, we get the result that $P(S \mid A) > P(S \mid \sim A)$, so *S* confirms *A*. But if $P(A \supset S)$ is allowed to be less than 1, this result is no longer guaranteed.

[8] Taking the Garberian model from note 7, if $P(A \supset S) = 1 - c$ then $P(A)$ can exceed $P(S)$ by as much as *c*.

[9] As Max Cresswell has been arguing for decades (see, for example, Cresswell (1975)), a version of this problem besets theories that model logical non-omniscience using logically impossible worlds. Such theories cannot make good sense of logical connectives—if we can have a possible world in which *p* and *q* are both true but *p* & *q* is not, what exactly does "&" *mean*?—and so lose the ability to represent the very sentences they were meant to model reasoning about. (For more recent work on the difficulties of using impossible worlds to model logical non-omniscience, see Bjerring (2013).)

[10] Balcerak Jackson and Balcerak Jackson (2013) offer another nice example of this phenomenon. In classical logic an agent who can rationally infer *y* from *x* can also complete a conditional proof demonstrating $x \supset y$. Going in the other direction, if the agent rationally believes $x \supset y$ a quick logical move makes it rational to infer *y* from *x*. So the rational permission to infer *y* from *x* stands or falls with a rational permission to believe the proposition $x \supset y$. (See also Brandom's (1994: Ch. 2) position that material conditionals just *say* that particular material inferences are permitted.)

[11] Note that the duality relation need not be one-to-one: a given rational requirement may have multiple dual propositions, and a given proposition may serve as a dual for multiple rational requirements.

requirements reveal logical truths to be duals of rational requirements—if an agent is not required to take a special attitude toward a particular logical truth, other potential requirements on her reasoning fall away as well. The Fixed Point Thesis affirms that each rational requirement also has a dual in the proposition expressing that requirement.[12] If rationality permits an agent to disbelieve an a priori proposition describing a putative rational requirement, the putative requirement is not a genuine one.[13]

Of course I need to *argue* for this thesis, and I will begin to do so soon. But first I should clarify my commitments coming out of this phase of the discussion. In what follows I will be agnostic about whether Cognitive Capacity and Cognitive Reach are good objections to theories of rationality. The arguments and theses advanced will be capable of accommodating these concerns, but will not be committed to their having probative force. The Fixed Point Thesis, for example, requires agents *not* to have *false* beliefs about rational requirements (instead of requiring agents *to* have *true* beliefs) so that no infinite belief set is required. Similarly, each argument to come will be *consistent* with limiting rational requirements on an agent's beliefs to what is sufficiently obvious or accessible to her. But those arguments will not *require* such limitations, either.

2. THE AKRATIC PRINCIPLE

Before I can argue for the Fixed Point Thesis, I need to define some terms and clarify the kinds of normative claims I will be making. We will be discussing both an agent's doxastic attitudes (for simplicity's sake we'll stick to just belief, disbelief, and suspension of judgment) and her intentions. I will group both doxastic attitudes and intentions under the general term "attitudes." Because some of the rational rules we'll be discussing impugn combinations of attitudes without necessarily indicting individual attitudes within those combinations, I will not be evaluating attitudes in isolation. Instead I will examine rational evaluations of an agent's "overall state," which includes all the attitudes she assigns at a given time.

Evaluations of *theoretical* rationality concern only the doxastic attitudes in an agent's overall state. Evaluations of *practical* rationality may involve both beliefs and intentions. For example, there might be a (wide-scope) requirement of instrumental rationality that negatively evaluates any overall state

[12] I say "also," but on some understandings of logical truth the Fixed Point Thesis entails a logical omniscience requirement. Kant (1974) took logical truths to express the rules of rational inference. So for Kant, a requirement that one be maximally confident in logical truths just is a requirement that one remain confident in truths about rational requirements.

[13] I am *not* suggesting here that every time an agent makes an inference error she also has a mistaken belief about the requirements of rationality; plenty of poor inferrers have never even thought about the requirements of rationality. However we can *generate* plenty of cases in which an agent has explicit higher-level views, and then argue that in such cases the requirements at different levels match.

that includes an intention to ϕ, a belief that ψ-ing is necessary for ϕ-ing, and an intention not to ψ.[14]

Rules of rationality require or permit certain kinds of overall states. But which states are permitted for a particular agent at a particular time may depend on various aspects of that agent's circumstances. Different philosophical views take different positions here. An evidentialist might hold that which doxastic attitudes figure in the overall states permitted an agent depends only on that agent's evidence. One natural development of this view would then be that the list of rationally permitted overall states for the agent (including both beliefs and intentions) varies only with the agent's evidence and her desires. But we might think instead that which intentions appear in permitted states depends on an agent's reasons, not on her desires. Or if we want to deny evidentialism, we might suggest that an agent's beliefs in the past influence which doxastic attitudes appear in the overall states permitted to her in the present.[15]

To remain neutral on these points I will assume only that whatever the true theory of rationality is, it may specify certain aspects of an agent's circumstances as relevant to determining which overall states are rationally permitted to her. Taken together, these relevant aspects comprise what I'll call the agent's "situation." An agent's situation at a given time probably includes features of her condition at that time, but it might also include facts about her past or other kinds of facts.

Given an agent's current situation and overall state, we can evaluate her state against her situation to see if the state contains any rational flaws. That is, we can ask whether from a rational point of view there is anything negative to say about the agent's possessing that overall state in that situation. This is meant to be an *evaluative* excercise, which need not immediately lead to prescriptions—I am not suggesting a rational rule that agents ought only adopt rationally flawless states. In Section 7 I will assess the significance of such evaluations of rational flawlessness.

But in the meantime we have a more pressing problem. I want to be able to say that in a given situation some particular overall states are rationally without flaw, and even to say sometimes that a particular overall state is the only flawless state available in a situation. But English offers no concise, elegant way to say things like that, especially when we want to put them in verb phrases and the like. So I will repurpose a terminology already to hand for describing states that satisfy all the principles of a kind and states that uniquely satisfy the principles of that kind: I will describe an overall state

[14] For the "wide-scope" terminology see Broome (1999). One might think that some requirements of practical rationality involve not just an agent's intentions but also her actions. In that case one would have to include actions the agent is in the process of performing at a given time in her overall state along with her attitudes. For simplicity's sake I'm going to focus just on rational evaluations involving beliefs and intentions.

[15] Various versions of conservatism and coherentism in epistemology take this position.

with no rational flaws as "rationally permissible." A state that is not rationally permissible will be "rationally forbidden." And if only one overall state is flawless in a given situation, I will call that state "rationally required."[16]

I will also apply this terminology to individual attitudes. If an agent's current situation permits at least one overall state containing a particular attitude, I will say that that attitude is "rationally permissible" in that situation. If *no* permitted states contain a particular attitude, I will say that attitude is "rationally forbidden" in the current situation. If *all* permitted states contain an attitude I will say that attitude is "rationally required." Notice, however, that while talking about attitudes this way is a convenient shorthand, it is a shorthand for evaluations of *entire* states; at no point am I actually evaluating attitudes in isolation.

I realize that the "permitted" and "required" terminology I've repurposed here usually carries prescriptive connotations—we'll simply have to remind ourselves periodically that we are engaged in a purely evaluative project.[17] I also want to emphasize that I am evaluating states, not agents, and I certainly don't want to get into assignations of praise or blame. At the same time the states being evaluated are states of real agents, not states of mythical idealized agents. Even if you're convinced that a real agent could never achieve a rationally flawless set of attitudes, it can be worthwhile to consider what kinds of rational flaws may arise in a real agent's attitude set. Finally, my rational evaluations are all-things-considered evaluations. I will be asking whether, given an agent's current situation and taking into account *every* aspect of that situation pointing in whatever direction, it is all-things-considered rationally permissible for her to adopt a particular combination of attitudes.

Working with situations and overall states, we can characterize a variety of theses about rationality. There might, for instance, be a rational rule about perceptual evidence that if an agent's situation includes a perception that x, all the overall states rationally permissible for her include a belief that x. Such a rule relates an agent's beliefs to her evidence; other rational rules might embody consistency requirements strictly among an agent's beliefs.[18] Perhaps no situation rationally permits an overall state containing logically contradictory beliefs, or perhaps there's an instrumental ϕ/ψ rationality requirement of the sort described earlier. On the other hand, there may be no general rules of rationality at all. But even a particularist will admit that certain overall states are rationally required or permitted in particular

[16] Situations that allow for no rationally flawless overall states are rational dilemmas.

[17] When we get around to making *arguments* about what's required and permitted, it may appear that I'm assuming a substantive deontic logic (in particular something like Standard Deontic Logic) to make those arguments go through. But that will be a false appearance due to my idiosyncratic use of typically prescriptive language. Given the definitions I'm using for "required," "permitted," etc. all of my arguments will go through using only classical first-order logic (with no special deontic axioms or inference rules).

[18] The contrast is meant simply to be illustrative; I am not making any assumption going forward that an agent's evidence is not a subset of her beliefs.

situations; he just won't think any general, systematic characterizations of such constraints are available.

Using this terminology, the Fixed Point Thesis becomes:

> **Fixed Point Thesis** No situation rationally permits an a priori false belief about which overall states are rationally permitted in which situations.

I will argue to this thesis from a premise we can state as follows:

> **Akratic Principle** No situation rationally permits any overall state containing both an attitude *A* and the belief that *A* is rationally forbidden in one's current situation.

The Akratic Principle says that any akratic overall state is rationally flawed in some respect. It applies both to cases in which an agent has an intention *A* while believing that intention is rationally forbidden, and to cases in which the agent has a belief *A* while believing that belief is forbidden in her situation.[19] The principle does not come down on whether the rational flaw is in the agent's intention (say), in her belief about the intention's rational status, or somehow in the combination of the two. It simply says that if an agent has such a combination in her overall state, that state is rationally flawed. So the Akratic Principle is a wide-scope norm; it does *not* say that whenever an agent believes *A* is forbidden in her situation that agent is in fact forbidden to assign *A*.[20]

The irrationality of practical *akrasia* has been discussed for centuries (if not millennia), and I take it the overwhelming current consensus endorses the Akratic Principle for the practical case. Discussions of the theoretical case (in which *A* is a belief) tend to be more recent and rare. Feldman (2005) discusses a requirement on beliefs he calls "Respect Your Evidence," and for anyone who doubts the principle's application to the belief case it is well worth reading Feldman's defense.[21] (Requirements like Respect Your Evidence are also discussed in Adler (2002), Bergmann (2005), Gibbons (2006), and Christensen (2010).) Among other things, Feldman points out that an agent who violated the Akratic Principle for beliefs could after a quick logical step find herself

[19] I also take the Akratic Principle to apply to cases in which *A* is a combination of attitudes rather than a single particular attitude. While I will continue to talk about *A* as a single attitude for the sake of concreteness, any "*A*" in the arguments that follow can be read either as a single attitude or as a combination of attitudes.

[20] Arpaly (2000) argues (*contra* Michael Smith and others) that in some cases in which an agent has made an irrational mistake about which attitude rationality requires, it can still be rationally better for him to adopt the rationally required attitude than the one he thinks is required. In this case the Akratic Principle indicates that if the agent adopts the rationally required attitude then his overall state is rationally flawed. That is consistent with Arpaly's position, since she has granted that the agent's belief in this case about what's rationally required already creates a rational flaw in his overall state. Arpaly (2000, p. 491) explicitly concedes the presence of that rational flaw.

[21] Since Feldman is an evidentialist, he takes an agent's situation (for belief-evaluation purposes) to consist solely of that agent's evidence. His principle also concerns justification rather than rationality.

with a Moore-paradoxical belief of the form "*x*, but it's irrational for me to believe *x*."[22]

Still, objections to the Akratic Principle (in both its theoretical and practical applications) are available.[23] One important set of objections focuses on the fact that an agent might be mistaken about aspects of her current situation, or about aspects of her current overall state. Given my neutrality about the contents of situations, I cannot assume that all aspects of situations are luminous to the agents in those situations. So we might for instance have a case in which an agent believes that *p*, believes that it's rationally forbidden to believe something the negation of which one has believed in the past, actually did believe $\sim p$ in the past, but does not remember that fact now. I also do not want to assume that an agent is always aware of every element in her overall state.[24] So we might have a case in which an agent believes attitude *A* is rationally forbidden, possesses attitude *A*, but does not realize that she does. Or she might believe that attitudes meeting a particular description are forbidden, yet not realize of an attitude she has that it meets that description.

Rational evaluations in such cases are subtle and complex. The Akratic Principle might seem to indict the agent's overall state in all these cases, and I don't want to be committed to that. I have tried to formulate the principle carefully so as to apply only when an agent has the belief that her current situation, *described as her current situation*, rationally forbids a particular attitude. But that formulation may not handle all complications involving multiple descriptions of situations, and it certainly doesn't handle failures of state luminosity. Frankly, the best response to these objections is that while they are important, they are tangential to our main concerns here. For every case I will construct and every debate about such cases I will consider, that debate would remain even if we stipulated that the agent is aware of all the relevant situational features and of all her own attitudes (under whatever descriptions are required). So I will consider such stipulations to be in place going forward.[25]

[22] See Smithies (2012) for further discussion of such paradoxical statements.

[23] One objection we can immediately set aside is that of Audi (1990). Audi objects to the claim that if an agent judges better alternatives to an action to be available, then it's irrational for her to perform that action. But this is a narrow-scope claim, not our wide-scope Akratic Principle. We can see this in the fact that Audi's objection focuses exclusively on evaluating the *action*, and turns on a case in which the agent's judgment is mistaken. Audi does not investigate negative rational evaluations of that judgment, much less broader negative evaluations of the overall state containing both the judgment and the intention to perform the action, or of the agent herself. That his objection does not apply to such broader evaluations comes out when Audi writes, "I ... grant that incontinence counts against the rationality of *the agent*: one is not fully rational at a time at which one acts incontinently" (1990: p. 80, emphasis in original). (For further analysis of Audi's objection see Brunero (2013: Section 1).)

[24] For the sorts of reasons familiar from Williamson (2000).

[25] Williamson (2011) uses an example involving an unmarked clock to argue that "It can be rational for one to believe a proposition even though it is almost certain on one's evidence that it is not rational for one to believe that proposition." While that is not quite a direct counterexample to the Akratic Principle, it can easily be worked up into one. (See, for instance, Horowitz (2013, Section 6).) However Williamson's example is explicitly set up to make it

Before moving on, however, I should note that these objections to the Akratic Principle bring out further reasons why we need the a priori rider in the Fixed Point Thesis. An agent might have a false belief about what's required in her situation because she mistakes the content of that situation. She might also falsely believe that her current state is rationally permitted in her current situation because she is incorrect about what her overall state contains. But neither of these false beliefs necessarily reveals a rational mistake on the agent's part. Each of them is really a mistake about an a posteriori fact—the contents of her situation or of her overall state.

So what kind of false belief *is* rationally forbidden by the Fixed Point Thesis? One way to see the answer is to think of rational requirements as describing a function $\mathcal{R}$. Reading an overall state as just a set of attitudes, we can think of $\mathcal{R}$ as taking each situation S to the set $\mathcal{R}(S)$ of overall states that would be rationally flawless for an agent to hold in S.[26] The Fixed Point Thesis would then hold that there do not exist a situation S', a situation S, and an overall state $O \in \mathcal{R}(S)$ such that O contains a false belief about the values of $\mathcal{R}(S')$. In other words, no situation permits an agent to have false beliefs about which overall states $\mathcal{R}$ permits in various situations. This formulation isolates out issues about whether an agent can tell that her current situation is S and her current overall state is O. Further, I take it that facts about the values of $\mathcal{R}$ are a priori facts.[27] So this formulation clarifies why a priori false beliefs figure in the Fixed Point Thesis.[28]

These complications aside, there's a much more intuitive objection to the Akratic Principle available. Weatherson (ms) presents this objection—and its underlying commitments—in a particularly clear fashion.[29] He begins with an example:

unclear to the agent what her evidence is. So I read Williamson's case as a failure of situational luminosity, and hence will set it aside.

[26] Notice that overall states can be "partial," in the sense that they don't contain a doxastic attitude toward every proposition or an intention concerning every possible action. This reflects my earlier response to Cognitive Capacity that rationality need not require agents to take attitudes toward everything.

[27] Even if situations can include empirical facts not accessible to the agent (such as facts about her beliefs in the past), there will still be a priori truths about which situations rationally permit which overall states. They will take the form "*if* the empirical facts are such-and-such, *then* rationality requires so-and-so."

[28] There are still some complications lurking about states and situations under disparate descriptions. For instance, we might think that the sentence "my current overall state is in $\mathcal{R}(S)$" (for some particular S) expresses a "fact about the values of $\mathcal{R}(S)$" that an agent could get wrong because she misunderstands her current state. I'm really picturing $\mathcal{R}$ taking as inputs situations described in some canonical absolute form (no indexicals, no *de re* locutions, etc.) and yielding as outputs sets of states described in a similar canonical form. The Fixed Point Thesis bans mistakes about which canonically described situations permit which canonically described states, without addressing mistakes about which non-canonically described situations/states are identical to the canonically described ones. While the details here are complex, I hope the rough idea is clear.

[29] Weatherson ultimately wants to deny a version of the Akratic Principle for the theoretical case. But he gets there by first arguing against a version for the practical case, and then drawing an analogy between the practical and the theoretical. (For another example similar to Weatherson's Kantians, see the Holmes/Watson case in Coates (2012).)

> **Kantians:** Frances believes that lying is morally permissible when the purpose of the lie is to prevent the recipient of the lie performing a seriously immoral act. In fact she's correct; if you know that someone will commit a seriously immoral act unless you lie, then you should lie. Unfortunately, this belief of Frances's is subsequently undermined when she goes to university and takes courses from brilliant Kantian professors. Frances knows that the reasons her professors advance for the immorality of lying are much stronger than the reasons she can advance for her earlier moral beliefs. After one particularly brilliant lecture, Frances is at home when a man comes to the door with a large axe. He says he is looking for Frances's flatmate, and plans to kill him, and asks Frances where her flatmate is. If Frances says, "He's at the police station across the road," the axeman will head over there, and be arrested. But that would be a lie. Saying anything else, or saying nothing at all, will put her flatmate at great risk, since in fact he's hiding under a desk six feet behind Frances. What should she do?

Weatherson responds to this example as follows:

> That's an easy one! The text says that if someone will commit a seriously immoral act unless you lie, you should lie. So Frances should lie. The trickier question is what she should believe. I think she should believe that she'd be doing the wrong thing if she lies. After all, she has excellent evidence for that, from the testimony of ethical experts, and she doesn't have compelling defeaters for that testimony. So she should do something that she believes, and should believe, is wrong....
>
> For her to be as she should, she must do something she believes is wrong. That is, she should do something even though she should believe that she should not do it. So I conclude that it is possible that sometimes what we should do is the opposite of what we should believe we should do. (p. 12)

There are a number of differences between our Akratic Principle and the principle Weatherson is attacking. First, we are considering *intentions* while Weatherson considers what *actions* Frances should perform. So let's suppose Weatherson also takes this example to establish that sometimes what intention we should form is the opposite of what intention we should believe we should form. Second, Weatherson is considering what attitudes Frances *shouldn't* have, while we're considering what combinations of attitudes would be *rationally flawed* for Frances to have. Can Weatherson's Kantians example be used to argue against our Akratic Principle, concerning rationally flawed overall states?

When we try to use Kantians to build such an argument, the case's description immediately becomes tendentious. Transposed into rationality-talk, the second sentence of the Kantians description would become, "If you know that someone will commit a seriously immoral act unless you lie, you are rationally required to lie." This blanket statement rules out the possibility that what an agent is rationally required to do in the face of someone about to commit a seriously immoral act might depend on what evidence that agent has about the truth of various ethical theories. We might insist that if Frances has enough reason to believe that Kantian ethics is true, then Frances is rationally forbidden to lie to the axeman at the door. (And thus is not required to form an intention she believes is rationally forbidden.) Or, going in the other direction, we might refuse to concede Weatherson's claim that Frances "doesn't have compelling defeaters for" the testimony of her professors. If rationality truly requires intending to lie to the axeman, whatever reasons make that the case will also count as defeaters for the professors' claims. While these two responses move in opposite directions,[30] each denies that the case Weatherson has described (as transposed into rationality-talk) is possible.

These responses also bring out something odd about Weatherson's reading of the Kantians case. Imagine you are talking to Frances, and she is wondering whether she is rationally required to believe what her professor says. To convince her that she is, there are various considerations you might cite—the professor knows a lot about ethics, he has thought about the case deeply and at great length, he has been correct on many occasions before, etc.—and presumably Frances would find some of these considerations convincing.[31] Now suppose that instead of wondering whether she is required to *believe* what her professor says, Frances comes to you and asks whether she is required to *intend* as her professor prescribes. It seems like the points you made in the other case—the professor knows a lot about how one ought to behave, he has thought about her kind of situation deeply and at great length, he has prescribed the correct behavior on many occasions before, etc.—apply equally well here. That is, any consideration in favor of believing what the professor says is also a consideration in favor of behaving as the professor suggests, and vice versa.

Weatherson cannot just stipulate in the Kantians case what Frances is required to do, then go on to describe what her professor says and claim that she is bound by that as well. The professor's testimony may give Frances reasons to behave differently than she would otherwise, or the moral considerations involved may give Frances reason not to believe the testimony. So I don't think Kantians provides a convincing counterexample to the Akratic Principle.[32]

There is another kind of case in which what an agent should do might diverge from what she should believe she should do. I suggested above that when testimony offers normative advice, any reason to believe that testimony can also be a reason to obey it, and vice versa. Yet we can have cases in

[30] In Section 5 we will give the positions that engender these responses names. In the terminology of that section, the first response would be popular with "top-down" theorists while the second belongs to a "bottom-up" view.

[31] I am not taking a position here on whether testimony is a "fundamental" source of justification. Even if testimonial justification is fundamental, one can still adduce considerations to an audience that will make accepting testimony seem appealing. Fundamentalism about testimonial justification is not meant to choke off all discussion of whether believing testimony is epistemically desirable.

[32] Note that Kantians could be a rational dilemma—a situation in which no overall state is rationally permitted. In that case Kantians would not be a counterexample to the Akratic Principle because it would not constitute a situation in which an overall state is permitted containing both an attitude *A* and the belief that that attitude is forbidden. We will return to rational dilemmas in Section 7.

which certain reasons bear on behavior but not on belief. To see this possibility, consider Bernard Williams's famous example (1981) of the agent faced with a glass full of petrol who thinks it's filled with gin. For Williams, what an agent has reason to do is determined in part by what that agent would be disposed to do were she *fully* informed. Thus the fact that the glass contains petrol gives the agent reason not to drink what's in it. But this fact does not give the agent reason to believe that the glass contains petrol, and so does not give the agent any reason to believe she shouldn't drink its contents. For Williams, any true fact may provide an agent with reason to behave in particular ways if that fact is appropriately related to her desires.[33] Yet we tend to think that an agent's reasons to believe include only *cognitively local* facts. A position on which an agent has reason to believe only what she would believe were she *fully* informed makes all falsehoods impermissible to believe (and makes all-things-considered misleading evidence impossible in every case).

If we accept this difference between the dependence bases of practical and theoretical reasons, it's reasonable to hold that an agent can have most reason to act (or intend) in one direction while having most reason to believe she should act in another. What the agent has reason to *believe* about whether to drink the liquid in front of her is determined by cognitively local information; what she has reason to *do* may be influenced by nonlocal facts.[34] And if we think that what an agent *should* do or believe supervenes on what she has *most reason* to do or believe, we might be able to generate cases in which an agent should do one thing while believing that she should do another.

Yet here we return to potential distinctions between what an agent should do, what she has most reason to do, and what she is rationally required to do.[35] It's implausible that in Williams's example the agent is rationally required to believe the glass contains gin but rationally forbidden to drink what's in it. What one is *rationally* required to do or believe depends only

[33] Williams (1981: p. 103) writes, "[Agent] *A* may be ignorant of some fact such that if he did know it he would, in virtue of some element in [his subjective motivational set] *S*, be disposed to ϕ: we can say that he has a reason to ϕ, though he does not know it. For it to be the case that he actually has such a reason, however, it seems that the relevance of the unknown fact to his actions has to be fairly close and immediate; otherwise one merely says that *A* would have a reason to ϕ if he knew the fact." Notice that whether the unknown fact counts as a reason for the agent depends on how relevant that fact is to the agent's actions given his motivations, *not* how cognitively local the fact is to the agent.

[34] It's interesting to consider whether one could get a similar split between what an agent has reason to believe and what she has reason to believe *about* what she has reason to believe. If there is some boundary specifying how cognitively local a fact has to be for it to count as a reason for belief, then the dependency bases for an agent's reasons for first-order beliefs and her reasons for higher-order beliefs would be identical. In that case, it seems difficult to generate a Williams-style case in which an agent has reason to believe one thing but reason to believe that she has reason to believe another, because we don't have the excuse that the former can draw on sets of facts not available to the latter. In the end, this might make it even more difficult to deny versions of the Akratic Principle for the theoretical case (in which *A* is a doxastic attitude) than for the practical case (in which *A* is an intention).

[35] Not to mention the distinction between an agent's "subjective" and "objective" reasons. (See Schroeder (2008) for a careful examination of the intersection of that distinction with the issues considered here.)

on what's cognitively local—that's what made Cognitive Reach a plausible objection. As long as the normative notion featured in the Akratic Principle is rational requirement, Williams-style cases don't generate counterexamples to the principle.

Once more this discussion of potential counterexamples to the Akratic Principle reveals something important about the Fixed Point Thesis and the arguments for it I will soon provide. While I have framed the Fixed Point Thesis in terms of rational requirements, one might wonder whether it applies equally to other normative notions. (Could one be *justified* in a mistake about *justification*? Could one have *most reason* for a false belief about what *reasons* there are?) I am going to argue for the Fixed Point Thesis on the basis of the Akratic Principle, which concerns rational requirements. As we've just seen, that principle may be less plausible for other normative notions; for instance, Williams-style cases might undermine an Akratic Principle for reasons. But for any normative notion for which an analogue of the Akratic Principle holds, I believe I could run my arguments for a version of the Fixed Point Thesis featuring that normative notion. For normative notions for which a version of that principle is not plausible, I do not know if a Fixed Point analogue holds.

3. NO WAY OUT

I will now offer two arguments for a restricted version of the Fixed Point Thesis:

> **Special Case Thesis** There do not exist an attitude *A* and a situation such that:
> - *A* is rationally required in the situation, yet
> - it is rationally permissible in that situation to believe that *A* is rationally forbidden.

As a special case of the Fixed Point Thesis (concerning a particular *kind* of mistake about the rational requirements that an agent could make) the Special Case Thesis is logically weaker than the Fixed Point Thesis. Yet the Special Case Thesis is a good place to start, as many people inclined to deny the Fixed Point Thesis will be inclined to deny its application to this special case as well.[36]

While the Special Case Thesis may look a lot like the Akratic Principle, they are distinct. The Akratic Principle concerns the rational permissibility of an agent's assigning two attitudes at once. The Special Case Thesis concerns an agent's assigning a particular attitude when a particular rational requirement is in place. Yet despite this difference one can argue quickly from the principle to the thesis, and do so in multiple ways. I call my first argument from one

[36] For example, someone with Weatherson's inclinations might read Kantians as a case in which intending to lie is required, yet Frances is permitted to believe intending to lie is forbidden. If that reading were correct, Kantians would be a counterexample to the Special Case Thesis.

to the other No Way Out; it is a *reductio*. Begin by supposing (contrary to the Special Case Thesis) that we have a case in which an agent's situation rationally requires the attitude *A* yet also rationally permits an overall state containing the belief that *A* is rationally forbidden to her. Now consider that permitted overall state, and ask whether *A* appears in it or not. If the permitted overall state does not contain *A*, we have a contradiction with our supposition that the agent's situation requires *A*. (That supposition says that every overall state rationally permissible in the situation contains *A*.) So now suppose that the permitted overall state includes *A*. Then the state includes both *A* and the belief that *A* is forbidden in the current situation. By the Akratic Principle this state is not rationally permissible, contrary to supposition once more. This completes our *reductio*. The Akratic Principle entails the Special Case Thesis.

It's surprising that the Special Case Thesis is so straightforwardly derivable from the Akratic Principle. Part of the surprise comes from deriving something controversial (if not downright counterintuitive) from something that the large majority of philosophers believe. But I think another part of the surprise comes from deriving a *substantive* conclusion from a *structural* premise. Here I am borrowing terminology from Scanlon (2003), though not using it exactly as he does.[37] Structural constraints concern the way an agent's attitudes hang together, while substantive constraints explain which particular attitudes an agent's situation requires of her. In epistemology, structural norms of coherence and consistency among an agent's beliefs are often contrasted with substantive norms about how her beliefs should be driven by her evidence.

If one accepts this division, the Akratic Principle certainly looks like a structural rationality claim. The Special Case Thesis, meanwhile, says that when a particular fact is *true* in an agent's situation she is forbidden from disbelieving it in a certain way. The No Way Out argument moves from a premise about the general consistency of an agent's attitudes to a conclusion about what the specific content of those attitudes must be.[38]

[37] Scanlon distinguishes structural *normative claims* from substantive *normative claims*. Scanlon works in terms of reasons, and has a particular view about how the structural claims are to be understood, so he distinguishes structural from substantive normative claims by saying that the former "involve claims about what a person must, if she is not irrational, treat as a reason, but they make no claims about whether this actually *is* a reason" (2003: p. 13, emphasis in original). There's also the issue that in his earlier writings (such as Scanlon (1998)) Scanlon claimed only structural claims have to do with *rationality*, but by Scanlon (2003) he ceased to rely on that assumption.

[38] A similar move from structural to substantive occurred in my earlier argument from Confidence to the conclusion that logical truths require maximal credence. One might object that the No Way Out argument does not move solely from structural premises to a substantive conclusion, because that argument begins by assuming that there is at least one situation in which an attitude *A* is rationally required (which seems to involve a presupposed substantive constraint). I think that objection is harder to make out for the Confidence argument, but even with No Way Out a response is available. As I suggested in note 19, we can read "*A*" throughout the argument either as an individual attitude or as a combination of attitudes. Since structural constraints are requirements on combinations of attitudes, we can therefore run No Way Out for a case built strictly around structural assumptions. For a thorough presentation

That conclusion—the Special Case Thesis—may seem to run afoul of our earlier Cognitive Reach concerns. The thesis forbids believing that *A* is rationally forbidden whenever it's simply *true* that *A* is required; no mention is made of whether *A*'s being required is sufficiently accessible or obvious to the agent. This makes Special Case seem like an externalist thesis (in epistemologists' sense of "externalist"), which is worrying because many epistemologists consider rationality an internalist notion.[39] But this appearance is incorrect. Suppose you hold that in order for an attitude to be rationally required (or forbidden) of an agent in a situation, the relevant relation between the situation and that attitude must be sufficiently accessible or obvious to the agent. Under this view, whenever it's true that attitude *A* is required of an agent in a situation it's also true that *A*'s relation to the situation is sufficiently accessible or obvious to the agent. So whenever the Special Case Thesis applies to an agent, that agent has sufficiently obvious and accessible materials available to determine that it applies. The moment an internalist grants that *any* attitudes are required, he's also granted that there are propositions about rationality agents are forbidden to believe.

No Way Out has no consequences for the dispute between internalists and externalists in epistemology. But it does have consequences for the notion of evidential support. I said earlier that the evaluations discussed in our arguments are all-things-considered appraisals of rational permissibility. Most people hold that if an agent's total evidence supports a particular conclusion, it is at least rationally permissible for her to believe that conclusion. Yet the Special Case Thesis says there is never a case in which an attitude *A* is rationally required but it is rationally permissible to believe that attitude is forbidden. This means an agent's total evidence can never all-things-considered support the conclusion that an attitude is forbidden when that attitude is in fact required. Put another way, a particular type of all-things-considered misleading total evidence about rational requirements is impossible. The No Way Out argument moves from a premise about consistency requirements among an agent's attitudes (the Akratic Principle) to a strong conclusion about what can be substantively supported by an agent's evidence.

The Special Case Thesis is not the full Fixed Point Thesis. No Way Out concerns cases in which an agent makes a mistake about what's required by *her own* situation, and in which the agent takes an attitude that's *required* to be *forbidden*. To reach the full Fixed Point Thesis, we would have to generalize the Special Case Thesis in two ways:

(1) to mistakes besides believing that something required is forbidden; and

of such cases and an explicit derivation of the substantive from the structural, see Titelbaum (2014).

[39] Of course, Cognitive Reach concerns need not be exclusive to (epistemological) internalists. While accessibility is an internalist concern, externalists who reject accessibility as a necessary requirement for various positive epistemic evaluations may nevertheless hold that a relation must be sufficiently *obvious* to an agent for it to rationally require something of her.

(2) to mistakes about what's rationally required by situations other than the agent's current situation.

As an example of the first generalization, we would for example have to treat cases in which an attitude is rationally *forbidden* for an agent but the agent believes that attitude is *required*. This generalization is fairly easy to argue for, on the grounds that any well-motivated, general epistemological view that rationally permitted agents to have a belief at odds with the true requirements of rationality in this direction would permit agents to make mistakes in the other direction as well. (Any view that allowed one to believe something *forbidden* is *required* would also allow one to believe something *required* is *forbidden*.) Yet we already know from the Special Case Thesis that believing of a required attitude that it's forbidden is rationally impermissible. This rules out such epistemological views.[40]

The second generalization, however, is more difficult to establish. I'll argue for it by first presenting another route to the Special Case Thesis.

4. SELF-UNDERMINING

One strong source of resistance to the Fixed Point Thesis is the intuition that if an agent has the right kind of evidence—testimony, cultural indoctrination, etc.—that evidence can rationally permit her to mistakenly believe that a particular belief is forbidden. No Way Out combats the intuition that evidence might authorize false beliefs about the requirements of rationality by showing that an agent who formed such beliefs would land in a rationally untenable position. But that doesn't explain where the intuition goes wrong; it doesn't illuminate *why* evidence can't all-things-considered support such false beliefs. My next argument, the Self-Undermining Argument, focuses on what the requirements of rationality themselves would have to be like for these false beliefs to be rationally permissible.

Suppose, for example, that the following were a rule of rationality:

> **Testimony** If an agent's situation includes testimony that x, the agent is rationally permitted and required to believe that x.

By saying that the agent is both permitted and required to believe that x, I mean that the agent's situation permits at least one overall state and all permitted overall states contain a belief that x. The permission part is important, because I'm imagining an interlocutor who thinks that an agent's receiving testimony that x makes it acceptable to believe that x even if x is false or epistemically undesirable in some other respect. Of course Testimony is drastically oversimplified in other ways, and in any case testimony is not the only source from which an agent could receive evidence about what's

[40] In Section 6 I'll argue for another instance of the first generalization, one in which the mistake made about what's rational is less extreme than thinking what's required is forbidden (or vice versa).

rationally required. But after presenting the Self-Undermining Argument I'll suggest that removing the simplifications in Testimony or focusing on another kind of evidence would leave my main point intact.[41]

The Self-Undermining Argument shows by *reductio* that Testimony cannot express a true general rule of rationality. Begin by supposing Testimony is true, then suppose that an agent receives testimony containing the following proposition (which I'll call "*t*"):

> If an agent's situation includes testimony that *x*, the agent is rationally forbidden to believe that *x*.

By Testimony, the agent in this situation is permitted an overall state in which she believes *t*. So suppose the agent is in that rationally permitted state. Since the agent believes *t*, she believes that it's rationally impermissible to believe testimony. She learned *t* from testimony, so she believes that belief in *t* is rationally forbidden in her situation. But now her overall state includes both a belief in *t* and a belief that believing *t* is rationally forbidden. By the Akratic Principle, the agent's state is rationally impermissible, and we have a contradiction. The Akratic Principle entails that Testimony is not a true rule of rationality.

A moment ago I admitted that Testimony is drastically oversimplified as a putative rational rule, and one might think that adding in more realistic complications would allow Testimony to avoid Self-Undermining. For example, an agent isn't required and permitted to believe just *any* testimony she hears; that testimony must come from a particular kind of source. Instead of investigating exactly what criteria a source must meet for its testimony to be rationally convincing, I'll just suppose that such criteria have been identified and call any source meeting them an "authority." The Testimony rule would then say that an agent is required and permitted to believe testimony from an authority. And the thought would be that when the agent in the Self-Undermining Argument hears her source say *t*, she should stop viewing that source as an authority. (Anyone who says something as crazy as *t* certainly shouldn't be regarded as an authority!) The source's testimony therefore doesn't generate any rational requirements or permissions for the agent, the argument can't get going, and there is no problem for the (suitably modified) Testimony rule.

Whatever the criteria are for being an authority, they cannot render the Testimony norm vacuous. That is, a source can't qualify as an authority by virtue of agents' being rationally required and permitted to believe what he says. Usually a source qualifies as an authority by virtue of being reliable, having a track-record of speaking the truth, being trusted, or some such.

[41] As stated, Testimony applies only to an agent's beliefs. Yet following on some of the points I made in response to Weatherson's Kantians argument in Section 2, we could create a general testimony norm to the effect that whenever testimony recommends a particular attitude (belief or intention), rationality permits and requires adopting that attitude. The arguments to follow would apply to this generalized norm as well.

Whatever those criteria are, we can stipulate that the source providing testimony that *t* in the Self-Undermining Argument has met those criteria. Then the claim that the agent should stop treating her source as an authority the moment that source says *t* really becomes a flat denial of the Testimony rule (even restricted to testimony from authorities). The position is no longer that all testimony from an authority permits and requires belief; the position is that authorities should be believed unless they say things like *t*.

This point about the "authorities" restriction generalizes. Whatever restrictions we build into the Testimony rule, it will be possible to construct a case in which the agent receives a piece of testimony satisfying those restrictions that nevertheless contradicts the rule. That is, it will be possible unless those restrictions include a de facto exclusion of just such testimony. At that point, it's simpler just to modify the Testimony rule as follows:

> **Restricted Testimony** If an agent's situation includes testimony that *x*, the agent is rationally permitted and required to believe that *x*—unless *x* contradicts this rule.

Restricted Testimony performs exactly like Testimony in the everyday cases that lend Testimony intuitive plausibility. But the added restriction inoculates the rule against Self-Undermining; it stops that argument at its very first step, in which the agent's receiving testimony that *t* makes it permissible for her to believe *t*. *t* contradicts Restricted Testimony by virtue of providing an opposite rational judgment from Restricted Testimony on all *x*s received via testimony that don't contradict the rule.[42] Thus the restriction in Restricted Testimony keeps testimony that *t* from rationally permitting or requiring the agent to believe *t*.[43]

There's nothing special about Testimony as a rational rule here—we're going to want similar restrictions on other rational rules to prevent Self-Undermining. For example, we might have the following:

[42] If we read both *t* and Restricted Testimony as material conditionals universally quantified over a domain of possible cases, then as it stands there is no direct logical contradiction between them—both conditionals could be satisfied if neither antecedent is ever made true. But if we assume as part of our background that the domain of possible cases includes some instances of testimony that don't contradict the rule, then relative to that assumption *t* and Restricted Testimony contradict each other.

[43] One might think that the move from Testimony to Restricted Testimony is unnecessary, because a realistic version of the Testimony rule would exempt testimony from permitting belief when defeaters for that testimony are present. If *t*—or any other proposition that similarly contradicts the Testimony rule—counts as a defeater for any testimony that conveys it, then a Testimony rule with a no-defeaters clause will not be susceptible to Self-Undermining. Yet if one could successfully establish that such propositions always count as defeaters, then the no-defeaters Testimony rule would come to the same thing as the Restricted Testimony rule (or perhaps a Restricted Testimony rule with a no-defeaters clause of its own). And no-defeaters Testimony would still be susceptible to the looping problem I'm about to describe for Restricted rational rules.

Restricted Perceptual Warrant If an agent's situation includes a perception that x, the agent is rationally required to believe that x—unless x contradicts this rule.

Restricted Closure In any situation, any rationally permitted overall state containing beliefs that jointly entail x also contains a belief that x—unless x contradicts this rule.

The restriction may be unnecessary for some rules because it is vacuous. (It's hard to imagine a situation in which an agent *perceives* a proposition that directly contradicts a rational rule.) But even for those rules, it does no harm to have the restriction in place.

While these Restricted principles may seem odd or ad hoc, they have been seriously proposed, assessed, and defended in the epistemology literature—see Weiner (2007), Elga (2010), Weatherson (2013), and Christensen (2013).[44] But that literature hasn't noticed that restricting rules from *self*-undermining doesn't solve the problem. Rational rules must not only include exceptions to avoid undermining themselves; they must also include exceptions to avoid undermining each other. To see why, suppose for *reductio* that the three restricted rules just described are true. Now consider an unfortunate agent who both perceives that she has hands and receives testimony of the disjunction that either t is true or she has no hands (where t is as before). By Restricted Testimony, there is a state rationally permitted in that agent's situation in which she believes that either t is true or she has no hands. (Notice that this disjunctive belief does not logically contradict Restricted Testimony, and so does not invoke that rule's restriction.) By Restricted Perceptual Warrant, that permitted overall state also includes a belief that the agent has hands (which clearly doesn't contradict the Restricted Perceptual Warrant rule). By Restricted Closure, that permitted state also contains a belief in t (which, while it contradicts Restricted Testimony, does not contradict Restricted Closure). But t indicates that the agent is rationally forbidden to believe that either t is true or she has no hands, and we can complete our argument as before by the Akratic Principle.

At no point in this argument does one of our restricted rational rules dictate that a belief is required or permitted that logically contradicts *that rule*. Instead we have constructed a loop in which no rule undermines itself but together

[44] This discussion takes off from a real-life problem encountered by Elga. Having advocated a conciliatory "Split the Difference" position on peer disagreement like the one we'll discuss in Section 6, Elga found that many of his peers disagreed with that position. It then seemed that by his own lights Elga should give up his staunch adherence to Split the Difference. Elga's response is to argue that Split the Difference requires being conciliatory about all propositions except itself. More real-life self-undermining: The author Donald Westlake once joked that when faced with a t-shirt reading "Question Authority," he thought to himself "Who says?" And then there's this exchange from the ballroom scene in *The Muppet Show* (Season 1, Episode 11): "I find that most people don't believe what other people tell them." "I don't think that's true."

the rules wind up undermining each other.[45] Clearly we could expand this kind of loop to bring in other rational rules if we liked. And the loop could be constructed even if we added various complications to our perceptual warrant and closure rules to make them independently more plausible. For example, clauses added to Restricted Closure in response to Cognitive Capacity and Cognitive Reach concerns could be accommodated by stipulating that our unfortunate agent entertains all the propositions in question and recognizes all the entailments involved.

The way to avoid such loops is to move not from Testimony to Restricted Testimony but instead to:

> **Properly Restricted Testimony** If an agent's situation includes testimony that *x*, the agent is rationally permitted and required to believe *x*—unless *x* contradicts an a priori truth about what rationality requires.

and likewise for the other rational rules.

These proper restrictions on rational rules explain the points about evidence that puzzled us before. Rational rules tell us what various situations permit or require.[46] Rational rules concerning belief reveal what conclusions are *supported* by various bodies of evidence. In typical, run-of-the-mill cases a body of evidence containing testimony all-things-considered supports the conclusions that testimony contains, as will be reflected in most applications of Properly Restricted Testimony. But an agent may receive testimony that contradicts an (a priori) truth about the rational rules. Generalizing from typical cases, we intuitively thought that even when this happens, the evidence supports what the testimony conveys. And so we thought it could be rationally permissible—or even rationally required—to form beliefs at odds with the truth about what rationality requires. More generally, it seemed like agents could receive evidence that permitted them to have rational, false beliefs about the requirements of rationality.

But self-undermining cases are importantly different from typical cases, and they show that the generalization from typical cases fails. Rational rules need to be properly restricted so as not to undermine themselves or each other. The result of those restrictions is that testimony contradicting the rational rules does not make it rationally permissible to believe falsehoods about the rules. Generally, an agent's total evidence will never all-things-considered support an a priori falsehood about the rational rules, because the rational rules are structured such that no situation permits or requires a belief that contradicts them. There may be pieces of evidence that provide *some* reason to believe a falsehood about the rational rules, or evidence may provide prima facie

[45] There may have even been one of these loops in our original Self-Undermining Argument, if you think that the move from *t* and "my situation contains testimony that *t*" to "I am rationally forbidden to believe *t*" requires a Closure-type step.

[46] Put in our earlier functional terms, they describe general features of the function $\mathcal{R}$.

support for such false beliefs. But the properly restricted rules will never make such false beliefs all-things-considered rational.

Now it may seem that what I've called the "proper" restrictions on rational rules are an overreaction. For example, we could adopt the following narrower restriction on Testimony:

> **Current-Situation Testimony** If an agent's situation includes testimony that *x*, the agent is rationally permitted and required to believe that *x*—unless *x* contradicts an a priori truth about what rationality requires *in the agent's current situation.*

Current-Situation Testimony is restricted less than Properly Restricted Testimony because it prevents testimony from permitting belief only when that testimony misconstrues what the agent's *current situation* requires. Yet current-situation restrictions are still strong enough to prevent *akrasia* in the loop case. (Because *t* contradicts a fact about requirements in the agent's current situation, Current-Situation Closure would not require the agent to believe that *t*.) Current-Situation Testimony is also of interest because it would be the rule endorsed by someone who accepted the Special Case Thesis but refused to accept its second generalization—the generalization that goes beyond mistakes about what's required in one's current situation to mistakes about what's required in other situations.[47]

With that said, I don't find Current-Situation Testimony at all plausible—it's an egregiously ad hoc response to the problems under discussion. Yet by investigating *in exactly what way* Current-Situation Testimony is ad hoc we can connect the rational rules we've been considering to such familiar epistemological notions as justification, evidence, and reasons.

I keep saying that the evaluations involved in our rational rules are all-things-considered evaluations. If the Akratic Principle is true, the correct rational rules will be restricted in *some* way to keep an agent who receives testimony that *t* from being all-things-considered permitted to believe *t*. Plausibly, this means that the agent won't be all-things-considered justified in believing *t*, and that her total evidence won't all-things-considered support *t*. But that doesn't mean that *none* of her evidence will provide *any* support for *t*. And if we're going to grant that testimony can provide *pro tanto* or *prima facie* justification for believing *t*, we need to tell a story about what outweighs or defeats that justification, creating an all-things-considered verdict consistent with the Akratic Principle.

Similarly, if we respond to the loop cases by moving to Current-Situation Testimony (without going all the way to Properly Restricted Testimony), we still need to explain what offsets the incremental justification testimony provides for false claims concerning what's required in one's current situation. And if we accept the Special Case Thesis, we need to explain what justificatory arrangement makes it impermissible to believe that a rationally required

[47] I am grateful to Shyam Nair for discussion of Current-Situation rules.

attitude is forbidden. Certainly if attitude *A* is required in an agent's situation, the agent will have support for *A*. But that's different from having support for the proposition that *A* is required, or counter-support for the proposition that *A* is forbidden.

Ultimately, we need a story that squares the Akratic Principle with standard principles about belief support and justification. How is the justificatory map arranged such that one is never all-things-considered justified in both an attitude *A* and the belief that *A* is rationally forbidden in one's current situation? The most obvious answer is that every agent possesses a priori, propositional justification for true beliefs about the requirements of rationality in her current situation.[48] An agent can reflect on her situation and come to recognize facts about what that situation rationally requires. Not only can this reflection justify her in believing those facts; the resulting justification is also empirically indefeasible.[49]

I said this is the most obvious way to tell the kind of story we need; it is not the only way. But every plausible story I've been able to come up with is *generalizable*: it applies just as well to an agent's conclusions about what's rationally required in situations other than her own as it does to conclusions about what's required in her current situation. For example, take the universal-propositional-justification story I've just described. However it is that one reflects on a situation to determine what it rationally requires, that process is available whether the situation is one's current situation or not. The fact that a particular situation is currently yours doesn't yield irreproducible insight into its a priori rational relations to various potential attitudes. So agents will not only have a priori propositional justification for truths about the rational requirements in their own situations; they will have a priori justification for true conclusions about what's required in any situation.[50]

The generalizability of such stories makes it clear why the restriction in Current-Situation Testimony is ad hoc. Whatever keeps testimony from all-things-considered permitting false beliefs about one's *own* situation will also keep testimony from permitting false beliefs about *other* situations. This moves us from Current-Situation Testimony's weak restriction to Properly Restricted Testimony's general restriction on false rational-requirement beliefs. Properly Restricted Testimony and the other Properly Restricted rules

[48] For discussion of positions similar to this one and further references, see Field (2005) and Ichikawa and Jarvis (2013: Chapter 7).

[49] Let me be clear what I mean, because "indefeasible" is used in many ways. The story I'm imagining might allow that a priori propositional justification for truths about rational requirements could be opposed by empirical evidence pointing in the other direction, empirical evidence that has some weight. But that propositional justification is ultimately indefeasible in the sense that the empirical considerations will never outweigh it, making it all-things-considered rational for the agent to form false beliefs about what her situation requires.

[50] Another available backstory holds that everything I've just said about empirically indefeasible propositional justification is true for all a priori truths—there's nothing special about a priori truths concerning rational requirements. Clearly *that* story is generalizable, but assessing it is far beyond the scope of this essay.

then give us our second generalization of the Special Case Thesis. Properly Restricted Testimony keeps testimony from providing rational permission to believe anything that contradicts an a priori rational-requirement truth—whether that truth concerns one's current situation or not. Parallel proper restrictions on other rational rules prevent any rational permission to believe an attitude is forbidden that is in fact is required. This holds whether or not the situation under consideration is one's own. And that's the second generalization of the Special Case Thesis.

5. THREE POSITIONS

My argument from the Akratic Principle to the (full) Fixed-Point Thesis is now complete. It remains to consider applications of the thesis and objections to it. To understand the thesis's consequences for higher-order reasoning, we'll begin with an example.

Suppose Jane tells us (for some particular propositions p and q) that she believes it's not the case that either the negation of p or the negation of q is true. Then suppose Jane tells us she also believes the negation of q. $\sim(\sim p \vee \sim q)$ is logically equivalent to $p \& q$, so Jane's beliefs are inconsistent. If this is all we know about Jane's beliefs, we will suspect that her overall state is rationally flawed.

Let me quickly forestall one objection to the setup of this example. One might object that if we heard Jane describe her beliefs that way—especially if she described them immediately one after the other, so she was plainly aware of their potential juxtaposition—we would have to conclude that she uses words like "negation" and "or" to mean something other than our standard truth-functions. Now I would share such a concern about connective meaning if, say, Jane had directly reported believing both "p and q" and "not-q." But we cannot assume that whenever someone has what looks to us like logically inconsistent beliefs it is because she assigns different meanings to logical terms.[51] To do so would be to eliminate the possibility of logical errors, and therefore to eliminate the possibility of a normative theory of (deductive) rational consistency.

There is a delicate tradeoff here. At one extreme, if an apparent logical error is too straightforward and obvious, we look for an explanation in alternate meanings of the connectives. At the other extreme, if what is admittedly a logical inconsistency among beliefs is too nonobvious or obscure, Cognitive Reach concerns may make us hesitant to ascribe rational error. But if we are to have a normative theory of logical consistency at all, there must be some middle zone in which an inconsistency is not so obvious as to impugn connective interpretation while still being obvious enough to count as rationally mistaken. I have chosen a pair of beliefs for Jane that strikes me as falling

[51] At the beginning of my elementary logic course I find students willing to make all sorts of logical mistakes, but I do not interpret them as speaking a different logical language than I.

within that zone. While you may disagree with me about this particular example, as long as you admit the existence of the sweet spot in question I am happy to substitute an alternate example that you think falls within it.

Given what we know of Jane so far, we are apt to return a negative rational evaluation of her overall state. But now suppose we learn that Jane has been taught that this combination of beliefs is rationally acceptable. Jane says to us, "I understand full well that those beliefs are related. I believe that when I have a belief of the form $\sim(\sim x \vee \sim y)$, the only attitude toward y it is rationally permissible for me to adopt while maintaining that belief is a belief in $\sim y$." Perhaps Jane has been led to this belief about rational consistency by a particularly persuasive (though misguided) logic teacher, or perhaps her views about rational consistency are the result of cultural influences on her.[52]

We now have two questions: First, is there any way to fill in the background circumstances such that it's rationally permissible for Jane to have this belief about what's rationally permissible? Second, is there any way to fill in the background circumstances such that Jane's combination of p/q beliefs actually is rationally permissible—such that it's rationally okay for her overall state to contain both a belief in $\sim(\sim p \vee \sim q)$ and a belief in $\sim q$?

I will distinguish three different positions on Jane's case, divided by their "yes" or "no" answers to these two questions.[53] Begin with what I call the "top-down" position, which answers both questions in the affirmative. On this view Jane's training can make it rationally permissible for her to maintain the logically inconsistent beliefs, and also for her to believe that it is rationally acceptable for her to do so. According to the top-down view, Jane's authoritative evidence makes it rationally permissible for her to believe certain belief combinations are acceptable, then that permission "trickles down" to make the combinations themselves permissible as well. One might motivate this position by thinking about the fact that rational requirements are consistency requirements, then concluding that it is the consistency between an agent's attitudes and her beliefs about the rationality of those attitudes that is most important by rationality's lights. On this reading Jane's state need not exhibit any rational flaws.

I will read the top-down position as holding that no matter what particular combination of attitudes an agent possesses, we can always add more to the story (concerning the agent's training, her beliefs about what's rational, etc.) to make her overall state rationally permissible. One could imagine a less extreme top-down position on which certain obvious, straightforward

[52] Again, however we tell our background story about Jane we have to ensure that the connective words coming out of her mouth still mean our standard truth-functions. Perhaps Jane's attitude comes from an authoritative logic professor who taught her the standard truth-functional lore but accidentally wrote the wrong thing on the board one day—a mistake that Jane has unfortunately failed to recognize as such and so has taken to heart.

[53] Technically there are four possible yes-no combinations here, but the view that answers our first question "no" and our second question "yes" is unappealing and I don't know of anyone who defends it. So I'll set it aside going forward.

mistakes are rationally forbidden no matter one's background, then the rational latitude granted by training or testimony grows as mistakes become more difficult to see. To simplify matters I will stick to discussing the pure top-down view, but what I have to say about it will ultimately apply to compromise positions as well. On the pure view no evidence is indefeasible and no combination of attitudes is forbidden absolutely, because an agent could always have higher-order beliefs and evidence that make what looks wrong to us all right.

The opposition to top-down splits into two camps. Both answer our second question in the negative; they split on the answer to the first. What I call the "bottom-up" position holds that it is always rationally forbidden for Jane to believe both $\sim(\sim p \vee \sim q)$ and $\sim q$, and it is also always forbidden for her to believe that that combination is rationally permissible. According to this view, when a particular inference or combination of attitudes is rationally forbidden, there is no way to make it rationally permissible by altering the agent's attitudes about what's rational. What's forbidden is forbidden, an agent's beliefs about what's rational are required to get that correct, and no amount of testimony, training, or putative evidence about what's rational can change what is rationally permitted or what the agent is rationally permitted to believe about it.[54]

Between top-down and bottom-up is a third position, which I call the "mismatch" view. The mismatch view answers our second question "no" but our first question "yes"; it holds that while Jane's education may make it rationally acceptable to believe that her beliefs are permissible, that does not make those beliefs themselves permissible. The mismatch position agrees with bottom-up that Jane's attitudes directly involving p and q are rationally forbidden. But while bottom-up holds that Jane also makes a rational mistake in getting this fact about rationality wrong, mismatch allows that certain circumstances could make Jane's false belief about the rational rationally okay. (For our purposes we need not specify more precisely what kinds of circumstances those are—I'll simply assume that if they exist then Jane's case involves them.) Mismatch differs from top-down by denying that circumstances that rationally permit Jane's believing that her attitudes are acceptable make those attitudes themselves okay.[55]

[54] To be clear, the bottom-up position does not deny the possibility of defeaters in general. For example, if a statistical sample rationally necessitates a particular conclusion it will still be possible for additional, undercutting evidence to reveal that the sample was biased and so change what can be rationally inferred from it. The dispute between top-down and bottom-up views concerns additional evidence that is explicitly about a priori rational requirement truths, and whether such evidence may change both the agent's higher-order beliefs and what's permissible for her at the first order.

[55] Given my insistence on evaluating only overall states—in their entirety—how can we make sense of this talk about the mismatch view's permitting some components of Jane's state while forbidding others? The best way is to think about what overall states in the vicinity the mismatch view takes to be rationally permissible. For example, the mismatch position makes rationally permissible an overall state containing a belief that $\sim(\sim p \vee \sim q)$, a belief that q, and a belief like Jane's that the only attitude toward q permissible in combination with $\sim(\sim p \vee \sim q)$ is

How do the Akratic Principle, the Fixed Point Thesis, and our arguments apply to these positions? Hopefully it's obvious that the mismatch position contradicts the Fixed Point Thesis. On the mismatch reading, it's rationally *impermissible* for Jane to combine a belief in $\sim(\sim p \vee \sim q)$ with a belief in $\sim q$—yet it's *permissible* for Jane to believe this combination of beliefs is okay. Thus the mismatch view would rationally permit Jane to have a false belief about which belief combinations are rationally permissible. As we've seen, the Fixed Point Thesis can be grounded in the Akratic Principle, and the mismatch position is in tension with that principle as well. Mismatch holds that in order for Jane to square herself with all the rational requirements on her, she would have to honor her testimonial evidence by maintaining her beliefs about what's rationally permissible, while at the same time adopting some combination of p/q attitudes like $\sim(\sim p \vee \sim q)$ and q. But then Jane would possess an attitude (or combination of attitudes) that she herself believes is rationally forbidden in her situation, which would violate the Akratic Principle.[56]

The top-down position may also seem to run directly afoul of the Fixed Point Thesis. Absent any cultural or authoritative testimony, it would be rationally forbidden for Jane to believe both $\sim(\sim p \vee \sim q)$ and $\sim q$. Top-down seems to license Jane to believe that that combination of beliefs is permissible, so top-down seems to make it rationally permissible for Jane to have a false belief about what's rational.

Yet the point is a delicate one. The top-down theorist holds that an agent's evidence about what is rationally forbidden or required of her affects what is indeed forbidden or required. On the top-down position, Jane's combination of p/q beliefs would be forbidden on its own, but once her testimonial evidence is added that combination becomes rationally acceptable. Thus the belief Jane forms on the basis of testimony about what's rationally permissible for her *turns out to be true* given that testimony and the belief it generates. Jane's higher-order belief correctly describes the lower-order requirements of rationality on her, so there is no straightforward violation of the Fixed Point Thesis or the Akratic Principle.

Another angle on the same point: Of the three positions we've considered, only mismatch directly contravenes the duality phenomenon I highlighted in Section 1. Both bottom-up and top-down take rational requirements on consistency and inference to stand or fall with requirements on attitudes toward particular propositions. The proposition that Jane's combination of p/q attitudes is rationally permissible is a dual of that permission itself. On the bottom-up reading, both the combination and her belief about that

$\sim q$. Both the top-down position and the bottom-up position would deny that this overall state is rationally permissible.

56 Acknowledging this tension, Weatherson offers his Kantians argument against the Akratic Principle so he can defend a mismatch position. Ralph Wedgwood's views are also interesting on this front, and have been evolving—Wedgwood (2012) defends a mismatch view, despite the fact that Wedgwood (2007) embraced a version of the Akratic Principle! (Thanks to Ralph Wedgwood for correspondence on this point.)

combination are rationally impermissible. On the top-down reading there are circumstances in which Jane's belief about the combination is rationally permissible, but in those circumstances the combination is permissible as well. Only the mismatch position suggests that Jane could be permitted to believe that a belief combination is required (or permitted) while that combination is in fact forbidden.

So the top-down position does not *directly* conflict with the Fixed Point Thesis in the way mismatch does. Yet I believe that top-down is ultimately inconsistent with that thesis as well. This is because any top-down view is committed to the possibility of an agent's being rationally permitted to believe something false about what's rationally required—if not in her own current situation, then in another. To see why, imagine Jane's case happens in two stages. At first she has no testimony about combinations of p/q beliefs, and simply believes both $\sim(\sim p \vee \sim q)$ and $\sim q$. At this point both bottom-up and top-down agree that her overall state is rationally flawed. Then Jane receives authoritative testimony that this combination of attitudes is rationally permitted, and comes to believe that she is permitted to possess the combination. According to the top-down position, at the later stage this claim about what's permitted is true, and Jane's overall state contains no rational flaws.

But what about Jane's beliefs at the later stage concerning what was rationally permissible for her at the earlier stage? I will argue that according to the top-down theorist, there will be cases in which it's rationally permissible for Jane to believe that at the earlier stage (before she received any authoritative testimony) it was rationally permissible for her to believe both $\sim(\sim p \vee \sim q)$ and $\sim q$. Since that's an a priori falsehood about what rationality requires (even by the top-down theorist's lights), the top-down position violates the Fixed Point Thesis.

Why must the top-down theorist permit Jane such a belief about her earlier situation? One reason is that the top-down view is motivated by the thought that the right kind of upbringing or testimony can make it rational for an agent to believe anything about what's rationally permissible. Suppose the authorities simply came to Jane and told her that believing both $\sim(\sim p \vee \sim q)$ and $\sim q$ was *permissible for her all along*. The top-down view of testimony and its higher-order influence suggests that under the right conditions it could be rational for Jane to believe this.

Even more damning, I think the top-down theorist *has* to take such higher-order beliefs to be permissible for Jane in order to read her story as he does. In our original two-stage version of the story, in which Jane first believes both $\sim(\sim p \vee \sim q)$ and $\sim q$ and then receives testimony making that combination of beliefs rationally permissible, what is the content of that testimony supposed to be? Does the authority figure come to Jane and say, "Look, the combination of beliefs about p and q you have right now is logically inconsistent, and so is rationally impermissible—until, that is, you hear this testimony and believe it, which will make your combination of beliefs rationally okay"? The top-down theorist doesn't imagine Jane's rational indoctrination proceeding via this sort of mystical bootstrapping. Instead, the top-down theorist imagines that Jane's miseducation about what's rationally permissible (whether it happens in stages or before she forms her fateful p/q beliefs) is a process whereby Jane comes to be misled about what's been rationally permissible for her all along. Even if Jane's beliefs about what's permissible in her own situation are accurate, her beliefs about what's rationally permissible in other situations (including perhaps her own former situation) are false, and are therefore forbidden by the Fixed Point Thesis.

The top-down theorist thinks the right higher-order beliefs can make any attitude combination permissible. But top-down still wants to be a *normative* position, so it has rules for which situational components (such as elements of the agent's evidence) permit which higher-order beliefs. As we saw in the Self-Undermining Argument, these rules come with restrictions to keep from undermining themselves. Once we recognize the possibility of looping, those restrictions broaden to forbid any false belief about what's rational in one's own situation or in others. Properly understood, the top-down position's own strictures make the position untenable.[57] Rational rules form an inviolate core of the theory of rationality; they limit what you can rationally be permitted to believe, even in response to authoritative testimony.

6. PEER DISAGREEMENT

The best objection I know to the Fixed Point Thesis concerns its consequences for peer disagreement. To fix a case before our minds, let's suppose Greg and Ben are epistemic peers in the sense that they're equally good at drawing rational conclusions from their evidence. Moreover, suppose that as part of their background evidence Greg and Ben both know that they're peers in this sense. Now suppose that at t_0 Greg and Ben have received and believe the same total evidence E relevant to some proposition h, but neither has considered h and so neither has adopted a doxastic attitude toward it. For simplicity's sake I'm going to conduct this discussion in evidentialist terms (the arguments would go equally well on other views), so Greg's and Ben's situation with respect to h is just their total relevant evidence E. Further suppose that for any agent who receives and believes total relevant evidence E, and who adopts an attitude toward h, the only rationally permissible attitude toward h is belief in it. Now suppose that at t_1 Greg realizes that E requires believing h and so believes h on that basis, while Ben mistakenly concludes that E requires believing $\sim h$ and so (starting at t_1) believes $\sim h$ on

[57] What if we went for a top-down position all the way up—a view on which what's rationally permissible for the agent to believe at higher orders in light of her evidence depends only on what the agent *believes* that evidence permits her to believe, and so on? Such a view would still need normative rules about what counts as correctly applying the agent's beliefs about what's permissible, and those rules could be fed into the Self-Undermining Argument. This point is similar to a common complaint against Quinean belief holism and certain versions of coherentism; even a Quinean web needs rules describing what it is for beliefs at the center to mesh with those in the periphery.

that basis. (To help remember who's who: *Greg* does a *good* job rationally speaking, while *Ben* does *badly*.)

At t_1 Greg and Ben have adopted their own attitudes toward *h* but each is ignorant of the other's attitude. At t_2 Greg and Ben discover their disagreement about *h*. They then have identical total evidence E', which consists of *E* conjoined with the facts that Greg believes *h* on the basis of *E* and Ben believes $\sim h$ on the basis of *E*. The question is what attitude Greg should adopt toward *h* at t_2.

A burgeoning literature in epistemology[58] examines this question of how peers should respond to disagreements in belief. Meanwhile peer disagreement about what to *do* (or about what intentions are required in a particular situation) is receiving renewed attention in moral theory.[59] I'll focus here on epistemological examples concerning what to believe in response to a particular batch of evidence, but my arguments will apply equally to disagreements about the intentions rationally required by a situation. To make the case even more concrete, I will sometimes suppose that in our Greg-and-Ben example *E* entails *h*. We might imagine that Greg and Ben are each solving an arithmetic problem, *E* includes both the details of the problem and the needed rules of arithmetic, and Ben makes a calculation error while Greg does not.[60] The arithmetic involved will be sufficiently obvious but not too obvious to fall into the "sweet spot" described in the previous section, so Ben's miscalculation is a genuine rational error. While the disagreement literature has certainly not confined itself to entailment cases, as far as I know every player in the debate is willing to accept entailments as a fair test of his or her view.

I will focus primarily on two responses to peer disagreement cases. The Split the Difference view (hereafter SD) holds that Greg, having recognized that an epistemic peer drew the opposite conclusion from him about *h*, is rationally required to suspend judgment about *h*.[61] The Right Reasons view (hereafter RR) says that since Greg drew the rationally required conclusion about *h* before discovering the disagreement, abandoning his belief in *h* at t_2 would be a rational mistake.

Ironically, a good argument for RR can be developed from what I think is the best argument *against* RR. The anti-RR argument runs like this: Suppose for *reductio* that RR is correct and Greg shouldn't change his attitude toward *h* in light of the information that his peer reached a different conclusion from the same evidence. Now what if Ben was an epistemic superior to Greg,

58 Besides the specific sources I'll mention in what follows, Feldman and Warfield (2010) and Christensen and Lackey (2013) are collections of essays exclusively about peer disagreement.

59 Of course, discussions of moral disagreement are as old as moral theory itself. The most recent round of discussion includes Setiya (2013), Enoch (2011: Ch. 8), McGrath (2008), Crisp (2007), Sher (2007), Wedgwood (2007: Sect. 11.3), and Shafer-Landau (2003).

60 This is essentially the restaurant-bill tipping example from Christensen (2007).

61 SD is distinct from the "Equal Weight View" defended by Elga (2007; 2010). But for cases with particular features (including the case we are considering), Equal Weight entails SD. Since SD can be adopted without adopting Equal Weight more generally, I will use it as my target here.

someone who Greg knew was much better at accurately completing arithmetic calculations? Surely Greg's opinion about *h* should budge a bit once he learns that an epistemic superior has judged the evidence differently. Or how about a hundred superiors? Or a thousand? At some point when Greg realizes that his opinion is in the minority amongst a vast group of people who are very good at judging such things, rationality must require him to at least suspend judgment about *h*. But surely these cases are all on a continuum, so in the face of just one rival view—even a view from someone who's just an epistemic peer—Greg should change his attitude toward *h* somewhat, *contra* the recommendation of RR.

Call this the Crowdsourcing Argument against RR.[62] It's a bit tricky to make out when we're working in a framework whose only available doxastic attitudes are belief, disbelief, and suspension of judgment—that framework leaves us fewer gradations to make the continuum case that if Greg should go to suspension in the face of some number of disagreeing experts then he should make at least *some* change in response to disagreement from Ben. But no matter, for all I need to make my case is that there's some number of epistemic superiors whose disagreement with Greg would make it rationally obligatory for him to suspend judgment about *h*. Because if you believe that, you must believe that there is some further, perhaps much larger number of epistemic superiors whose disagreement would make it rationally obligatory for Greg to believe $\sim h$. If you like, imagine the change happens in two steps, and with nice round numbers. First Greg believes *h* on the basis of *E*, and believes he is rationally required to do so. He then meets a hundred experts who believe $\sim h$ on the basis of *E*. At this point Greg suspends judgment about *h*. Then he meets another nine hundred experts with the same opinion, and finally caves. Respecting their expertise, he comes to believe $\sim h$.[63]

Once we see the full range of effects the SDer thinks expert testimony can have on Greg, we realize that the SD defender is essentially a top-down theorist. And so his position interacts with the Fixed Point Thesis in exactly the way we saw in the previous section. On the one hand, SD does not produce an immediate, direct violation of the thesis. SD says that at t_2, after Greg meets Ben, the required attitude toward *h* for Greg is suspension. We stipulated in our case that Greg's original evidence *E* requires belief in *h*, but Greg's total evidence at t_2 is now E'—it contains not only *E* but also evidence about what Ben believes. At t_2 Greg may not only suspend on *h* but also believe that

[62] You might think of Crowdsourcing as an argument *for* SD, but in fact it is merely an argument *against* RR. Kelly (2010: pp. 137ff.) makes exactly this argument against RR, then goes on to endorse a Total Evidence View concerning peer disagreement that is distinct from SD. Whether Crowdsourcing is an argument *for* anything in particular won't matter in what follows—though we should note that Kelly explicitly endorses the claim I just made that many-superior and single-peer cases lie on a continuum.

[63] The moral disagreement literature repeatedly questions whether there *are* such things as "moral experts" (see e.g. Singer (1972) and McGrath (2008)). If there aren't, this argument may need to be made for practical disagreement cases by piling millions and millions of disagreeing peers upon Greg instead of just one thousand superiors.

suspension is required in his current situation. But since his situation at t_2 contains total evidence E' instead of just E, he doesn't believe anything that contradicts the truths about rationality we stipulated in the case.

Nevertheless, we can create trouble for SD by considering Greg's later-stage beliefs about what was rationally permissible earlier on. If you have the intuition that got Crowdsourcing going to begin with, that intuition should extend to the conclusion that faced with enough opposing experts, Greg could be rationally permitted to believe not only $\sim h$ but also that $\sim h$ was rationally obligatory on E. Why is this conclusion forced upon the SD defender? Again, for two reasons. First, we can stipulate that when the mathematical experts talk to Greg they tell him not only that they believe $\sim h$, but also that they believe $\sim h$ is entailed by E. (It's our example—we can stipulate that if we like!) It would be implausible for the SDer to maintain that Greg must bow to the numerical and mathematical superiority of the arithmetic experts in adopting the outcome of their calculation, but not in forming his beliefs about whether that outcome is correct.

Second, Greg's adopting higher-order beliefs from the experts was probably what the SD defender was envisioning already. When Greg and Ben meet, they have a disagreement not just about whether h is true, but also about whether h was the right thing to conclude from E. SDers often argue that this higher-order disagreement should make Greg doubt whether he performed the calculation correctly (after all, Ben is just as good at figuring these things out as he), and ultimately lead him to suspend judgment on h. Similarly, when the thousand experts come to Greg and convince him to believe $\sim h$, it must be that they do so by telling him his original calculation was wrong. Contrary to what Greg originally thought (they say), E doesn't entail h; instead E entails $\sim h$, so that's what Greg ought to believe. The mathematicians aren't supposed to be experts on the rational influence of testimony; they aren't supposed to be making subtle arguments to Greg about what his total evidence will support after their interaction with him. They're supposed to be telling him something with mathematical content—the type of content to which their expertise is relevant.

And now SD has proved too much: By supposition, E entails h and therefore rationally requires belief in it. When the experts convince Greg that E entails $\sim h$, they thereby convince him that he was required to believe $\sim h$ all along—even before he encountered them. By the Fixed Point Thesis, Greg is now making a rational error in believing that E rationally requires belief in $\sim h$. So it is not rational for Greg to respect the experts in this way. By the continuum idea, it's not rational for Greg to suspend judgment in the face of fewer experts to begin with, or even to budge in the face of disagreement from Ben his peer.[64]

[64] Exactly how much of the Fixed Point Thesis do we need to get this result? As I see it, all we need is the Special Case Thesis plus the second generalization I described in Section 3. Belief in h is required on E, and after meeting the thousand mathematicians Greg believes that

We now have an argument from the Fixed Point Thesis to the Right Reasons view about peer disagreement. We argued for the Fixed Point Thesis from the Akratic Principle, so if the Akratic Principle is true then misleading evidence at higher levels about what attitudes are required at lower levels does not "trickle down" to permit attitudes that otherwise would have been forbidden. SD and the top-down position both fail because they are trickle-down theories. RR and the bottom-up position are correct: If one's initial situation requires a particular attitude, that attitude is still required no matter how much misleading evidence one subsequently receives about what attitudes were permitted in the initial situation.

I said that the best objection to the Fixed Point Thesis comes from its consequences for peer disagreement. Some epistemologists think that on an intuitive basis, Right Reasons (and therefore the Fixed Point Thesis) is simply getting peer disagreement wrong; Ben's general acuity should earn his beliefs more respect, even when he happens to have misjudged the evidence. While we'll return to this thought in Section 7, strictly speaking it isn't an *argument* against RR so much as a straightforward denial of the view. On the other hand, there are now a number of complex philosophical arguments available against RR: that its has deleterious long-term effects, that it leads to illicit epistemic "bootstrapping," etc. I think these arguments have been adequately addressed elsewhere.[65]

Yet there's an objection that immediately occurs to anyone when they first hear RR, an objection that I don't think has been resolved. One can't object to RR on the grounds that it will lead Greg to a conclusion forbidden by his initial evidence; by stipulation the view applies only when he's read that evidence right. But one might ask: How can Greg know that he's the one to whom the view applies—how can he know *he's* the one who got it right? This question may express a concern about guidance, about RR's being a principle an agent could actually apply. Or it may express a concern about Ben: Ben will certainly *think* he got things right initially, so his attempts to respect RR may lead him to

such belief is forbidden. *E* doesn't describe Greg's (entire) situation at the later stage, so we do need that second generalization. But that was the one we were able to establish in Section 4.

The Crowdsourcing continuum also shows another way to argue for the Special Case Thesis's first generalization from Section 3. Suppose we have a view that permits an agent to make rational-requirement errors other than errors in which he takes something to be forbidden that's required (the errors covered by the Special Case Thesis). Whatever kind of case motivates such permissions, we will be able to construct a more extreme version of that case in which the agent is indeed permitted to believe something's forbidden that's required. Facing just Ben, or just the first one hundred experts, didn't compel Greg into any errors covered by the Special Case Thesis (even with its second generalization). But by piling on more experts we could commit the SD defender to the kind of extreme mistake in which an agent inverts what's required and forbidden.

[65] I'm thinking especially of Elga's (2007) bootstrapping objection, which Elga thinks rules out any view other than SD. Kelly (2010: pp. 160ff.) shows that this objection applies only to a position on which both Greg *and* Ben should stick to their original attitudes (or something close to their original attitudes) once the disagreement is revealed. Thus bootstrapping is not an objection to RR or to Kelly's own Total Evidence View. (Though my "proves too much" objection to SD works against Total Evidence as well.)

form further unsupported beliefs (or at least to resist giving in to Greg when he should).

Here I think it helps to consider an analogy. Suppose I defend the norm, "If you ought to ϕ, then you ought to perform any available ψ necessary for ϕ-ing." There may be many good objections to this norm, but here's a bad objection: "If I'm trying to figure out whether to ψ, how can I tell whether I ought to ϕ?" The norm in question is a conditional—it only applies to people meeting a certain condition. It is not the job of this norm to tell you (or help you figure out) whether you meet that condition. Similarly, it's no objection to the norm to say that if someone mistakenly thinks he ought to ϕ (when really he shouldn't), then his attempts to follow this norm may lead him to perform a ψ that he really shouldn't either. The norm says how agents should behave when they *actually* ought to ϕ, not when they *think* they ought to.

RR is a conditional, describing what an agent is rationally required to do upon encountering disagreement *if* he drew the conclusion required by his evidence at an earlier time. It isn't RR's job to describe what Greg's initial evidence E requires him to believe; we have other rational rules (of entailment, of evidence, of perception, etc.) to do that. It also is no objection to RR that if Ben mistakenly thinks he meets its antecedent, his attempts to follow RR may lead him to adopt the wrong attitude toward h at t_2. In describing the case we stipulated that Ben was rationally required to believe h on the basis of E at t_1; Ben made a *rational error* when he concluded $\sim h$ instead. Any mistakes Ben then makes at t_2 from misapplications of RR are parasitic on his original t_1 miscalculation of what E rationally requires. It shouldn't surprise us that an agent who initially misunderstands what's rationally required may go on to make further rational mistakes.

Perhaps the objection to RR involves a Cognitive Reach concern: it's unreasonable to require Greg to stick to his beliefs at t_2 when it may not be obvious or accessible to him that he was the one who got things right. My response here is the same as it was to Cognitive Reach concerns about internalism and the Special Case Thesis.[66] The objection is motivated by the thought that in order for an attitude to be rationally required of an agent, the relevant relation between that attitude and the agent's situation must be sufficiently obvious or accessible. We stipulated in our example that at t_1 Greg and Ben are rationally required to believe h on the basis of E. In order for that to be true, the relevant relation between h and E (in the imagined case, an entailment) must be sufficiently obvious or accessible to both parties at t_1—it lands in our "sweet spot." That obviousness or accessibility doesn't disappear when Greg gains more evidence at t_2; adding facts about what Ben believes doesn't keep Greg from recognizing h's entailment by E. So the facts needed for Greg to

66 Once again (see note 39), I think the intuitive worry under consideration is available to both internalists and externalists in epistemology. Internalists are more likely to put the objection in terms of accessibility, while externalists are more likely to complain of insufficient obviousness.

determine what RR requires of him are still sufficiently obvious and accessible to him at t_2.

One might think that the extra information about Ben's beliefs contained in E' *defeats* what Greg knew at t_1—the extra evidence somehow destroys the all-things-considered justification Greg had for believing h at t_1. But that's just what's at issue between the RR-theorist and the SD-theorist: the former thinks E' still rationally requires Greg to believe E, while the latter does not. That E' contains defeaters for E's justification of h cannot be *assumed* in arguments between the two views.

7. CONCLUSION: ASSESSING THE OPTIONS

This essay began with logical omniscience. Examining formal epistemologists' struggles to remove logical omniscience requirements from their theories, we uncovered a duality phenomenon: any rational requirement—whether it be a requirement on beliefs or intentions, whether it be a requirement of attitudinal consistency or a constraint on inference—comes with particular propositions toward which agents are required (or forbidden) to adopt particular attitudes. Some of those propositions are propositions about rationality itself. The Fixed Point Thesis reveals that wherever there is a rational requirement, rationality also requires agents not to get the facts about that requirement wrong. This thesis concerns actual attitudes held by actual agents, not just agents who have been idealized somehow; it remains true whatever constraints we place on how many attitudes an agent can assign or how obvious a relation must be to generate rational requirements.

I established the Fixed Point Thesis through two arguments (No Way Out and Self-Undermining), each of which uses only the Akratic Principle as a premise. I then showed that the Fixed Point Thesis has surprising consequences for agents' responses to information about what's rational. If an agent has correctly determined what attitudes her situation requires, rationality forbids changing those attitudes when she receives apparent evidence that she's made the determination incorrectly. Applied to peer disagreement cases, this implies the Right Reasons view on which an agent who's adopted the attitude required by her evidence is required to maintain that attitude even after learning that others have responded differently.

To my mind the strongest objection to the Fixed Point Thesis is not to offer some recondite philosophical argument but simply to deny its implications for disagreement on intuitive grounds. It feels preposterous to hold that in the Crowdsourcing case Greg is required to stick to the (admittedly correct) conclusion of his calculations in the face of a thousand acknowledged mathematical experts telling him he's wrong.[67] If this is what the Akratic Principle requires, then perhaps we should drop that principle after all.[68]

[67] A similar intuitive point against the Fixed Point Thesis can be made using Elga's (ms) hypoxia case. (See also Christensen (2010).) Everything I say about Crowdsourcing in what follows applies equally well to hypoxia and similar examples.

[68] Thanks to Stew Cohen and Russ Shafer-Landau for discussion of this option.

Unfortunately, dropping the Akratic Principle is no panacea for counterintuitive cases; Horowitz (2013) describes a number of awkward examples confronted by Akratic Principle deniers. Dropping the principle also has difficult dialectical consequences for defenders of Split the Difference (or a compromise like Kelly's Total Evidence View).[69] The mismatch theorist holds that in Crowdsourcing Greg is required to agree with the experts in his higher-order views—that is, he is required to believe along with them that he should believe $\sim h$—but should nevertheless maintain his original, first-order belief in h. The usual reply to this suggestion is that such a response would put Greg in a rationally unacceptable akratic overall state. But this reply is unavailable if one has dropped the Akratic Principle. Without the Akratic Principle, Split the Difference is unable to defend itself from the mismatch alternative, on which agents are required to conform their explicit beliefs about what's rational to the views of peers and experts but those beliefs have negligible further effects.

More broadly, I think it's a mistake to assess the Akratic Principle by counting up counterintuitive cases on each side or by treating it and Split the Difference as rational rules on an intuitive par. The Akratic Principle is deeply rooted in our understanding of rational consistency and our understanding of what it *is* for a concept to be normative.[70] Just as part of the content of the concept *bachelor* makes it irrational to believe of a confirmed bachelor that he's married, the normative element in our concept of rationality makes it irrational to believe an attitude is rationally forbidden and still maintain that attitude. The rational failure in each case stems from some attitudes' not being appropriately responsive to the contents of others. This generates the Moore-paradoxicality Feldman notes in defending his Akratic Principle for belief.

While the Akratic Principle therefore runs deep, Split the Difference is grounded in an intuition that can be explained away. I've already suggested that this intuition is a mistaken overgeneralization of the rational significance we assign to testimony in normal situations. And we have a principled explanation for why that generalization gives out where it does. The blank check seemingly written by the rational role of testimony turns out to undermine itself. To maintain rationality's normativity—to enable it to draw *boundaries*—we must restrict rational rules from permitting false beliefs about themselves and each other.

We can also channel some of the intuitive push behind Split the Difference into other, nearby views. For example, we might concede that the mathematical experts' testimony diminishes Greg's *amount* of evidence—or even amount of *justification*—for his conclusion. (Though the Fixed Point Thesis

[69] Christensen (2013) has a particularly good discussion of SD's dependence on the Akratic Principle. See also Weatherson (2013).

[70] A longstanding philosophical tradition questions whether *akrasia* is even *possible*; I am aware of no philosophical tradition questioning whether it's possible to maintain one's views against the advice of experts.

will never allow testimony to swing Greg's total evidence around and all-things-considered support the opposite conclusion.) I would be happy to admit an effect like this in the mirror-image case: If the thousand experts had all told Greg he was absolutely correct, that feels like it would enhance his belief's epistemic status somehow.[71]

If you are convinced that the Akratic Principle should be maintained but just can't shake your Crowdsourcing intuitions, a final option is to hold that Crowdsourcing (and peer disagreement in general) presents a rational dilemma.[72] One might think that in the Crowdsourcing case, Greg's evidence renders rationally flawed any overall state that doesn't concede anything to the experts, while the Fixed Point Thesis draws on Akratic Principle considerations to make rationally flawed any overall state that concedes something to the experts. The result is that no rationally flawless overall state is available to Greg in the face of the experts' testimony, and we have a rational dilemma.

Some philosophers deny the existence of rational dilemmas;[73] they will reject this option out of hand. But a more subtle concern is why we went to all this trouble just to conclude that peer disagreement is a rational dilemma. After all, that doesn't tell us what Greg should *do* (or should *believe*) in the situation. We've returned to a concern about the significance of evaluations of rational flawlessness, especially when those evaluations don't straightforwardly issue in prescriptions.

Here I should emphasize again that the evaluations we've been considering are evaluations of *real* agents' overall states, not the states of mythical ideal agents. How can it be significant to learn that such an agent's state is rationally flawed? Consider Jane again, who believes $\sim q$ and $\sim(\sim p \vee \sim q)$ while thinking that belief-combination is rationally permissible. Having rejected the top-down view, we can confirm that Jane's overall state is rationally flawed. While that confirmation doesn't automatically dictate what Jane should believe going forward, it certainly affects prescriptions for Jane's beliefs. If the top-down theorists were right and there were no rational flaws in Jane's overall state, there would be no pressure for her to revise her beliefs and so no possibility of a prescription that she make any change.

When it comes to rational dilemmas, it can be very important to our prescriptive analysis to realize that a particular situation leaves no rationally flawless options—even if that doesn't immediately tell us what an agent should do in the situation. A number of epistemologists[74] have recently analyzed cases

71 For a detailed working-out of the justification-levels line, see Eagle (ms). Other alternatives to Split the Difference are available as well. van Wietmarschen (2013), for instance, suggests that while Greg's *propositional* justification for h remains intact in the face of Ben's report, his ability to maintain a *doxastically* justified belief in h may be affected by the peer disagreement.

72 Although Christensen (2013) employs different arguments than mine (some of which rely on intuitions about cases I'm not willing to concede), he also decides that the Akratic Principle is inconsistent with conciliationist views on peer disagreement. Christensen's conclusion is that peer disagreements create rational dilemmas.

73 See e.g. Broome (2007).

74 Such as Elga (ms), Hasan (ms), Weatherson (ms), Schechter (2013), Chalmers (2012: Ch. 2), Christensen (2010), and farther back Foley (1990) and Fumerton (1990).

in which an agent is misled about or unsure of what rationality requires in her situation (without having interacted with any peers or experts). Some have even proposed amendments to previously accepted rational principles on the grounds that those principles misfire when an agent is uncertain what's required.[75] Meanwhile practical philosophers[76] have considered what happens when an agent is uncertain which intentions are required by her situation. Many of these discussions begin by setting up a situation in which it's purportedly rational for an agent to be uncertain—or even make a mistake—about what rationality requires. As in peer disagreement discussions, authors then eliminate various responses the agent might have to her situation by pointing out that those responses violate putative rational rules (logical consistency of attitudes, probabilistic constraints on credences, versions of the Akratic Principle, etc.).

But now suppose that the moment the agent makes a mistake about what rationality requires (or even—if logical omniscience requirements are correct—the moment she assigns less than certainty to particular kinds of a priori truths), the agent has already made a rational error. Then it is no longer decisive to point out that a particular path the agent might take while maintaining the mistake violates some rational rule, *because no rationally flawless options are available to an agent who persists in such an error.* If we view a particular situation as a rational dilemma, determining the right prescription for an agent in that situation shifts from a game of avoiding rational-rule violations to one of making tradeoffs between unavoidable violations. That's a very different sort of normative task,[77] and the first step in engaging the norms-of-the-second-best involved in sorting out a rational dilemma is to realize that you're in one.[78]

Finally: To conclude that peer disagreements are rational dilemmas is not to deny the Fixed Point Thesis. The thesis holds that no situation rationally permits an overall state containing a priori false beliefs about what situations rationally require. It is consistent with this thesis that in some situations *no* overall state is rationally permissible—in some situations no rationally flawless state is available. So to insist that Greg is in a rational dilemma would not undermine any conclusion I have drawn in this essay.[79] We would still

[75] See, for example, criticisms of Rational Reflection in Christensen (2010) and Elga (2013), and criticisms of single-premise closure in Schechter (2013).

[76] Including Sepielli (2009), Wedgwood (2007: Sect. 1.4), and Feldman (ms).

[77] Compare Rawls's (1971) distinction between ideal and nonideal theory.

[78] Imagine someone ultimately develops a robust theory of the second best: some normative notion and set of rules for that notion that determine how one should make tradeoffs and what one should do when caught in a rational dilemma. Will those rules forbid states in which an agent believes the normative notion forbids an attitude yet maintains that attitude anyway? If so, we have a version of the Akratic Principle for that notion, and our arguments begin all over again....

[79] It wouldn't even undermine the Right Reasons position. I have tried to define Right Reasons very carefully so that it indicates a rational *mistake* if Greg *abandons* his belief in h at t_2—making RR consistent with the possibility that Greg is in a rational dilemma at t_2. If we carefully define Split the Difference in a parallel way, then if peer disagreement poses a

have my central claim that mistakes about rationality are mistakes of rationality; we would simply be admitting that those mistakes can sometimes be avoided only by offending rationality in other ways. As long as it's a rational mistake to think or behave as one judges one ought not, it will also be a rational mistake to make false judgments about what's rational.[80]

REFERENCES

Adler, J. E. (2002). *Belief's Own Ethics*. Cambridge, MA: MIT Press.

Arpaly, N. (2000). On acting rationally against one's best judgment. *Ethics* 110, 488–513.

Audi, R. (1990). Weakness of will and rational action. *Australasian Journal of Philosophy* 68, 270–81.

Balcerak Jackson, M. and B. Balcerak Jackson (2013). Reasoning as a source of justification. *Philosophical Studies* 164, 113–26.

Bergmann, M. (2005). Defeaters and higher-level requirements. *The Philosophical Quarterly* 55, 419–36.

Bjerring, J. C. (2013). Impossible worlds and logical omniscience: An impossibility result. *Synthese* 190, 2505–24.

Brandom, R. B. (1994). *Making It Explicit*. Cambridge, MA: Harvard University Press.

Broome, J. (1999). Normative requirements. *Ratio* 12, 398–419.

Broome, J. (2007). Wide or narrow scope? *Mind* 116, 359–70.

Brunero, J. (2013). Rational *akrasia*. *Organon F* 20, 546–66.

Chalmers, D. J. (2012). *Constructing the World*. Oxford: Oxford University Press.

Cherniak, C. (1986). *Minimal Rationality*. Cambridge, MA: The MIT Press.

Christensen, D. (2007). Epistemology of disagreement: The good news. *Philosophical Review* 116, 187–217.

Christensen, D. (2010). Rational reflection. *Philosophical Perspectives* 24, 121–40.

rational dilemma both RR and SD are true! Yet I don't think this is the reading of SD that most of its defenders want. They tend to write as if splitting the difference with Ben squares Greg entirely with rationality's demands, leaving him in a perfectly permissible, rationally flawless state. *That* position on peer disagreement contradicts the Fixed Point Thesis, and the Akratic Principle.

[80] For assistance, discussion, and feedback on earlier versions of this essay I am grateful to John Bengson, Selim Berker, J. C. Bjerring, Darren Bradley, Michael Caie, David Christensen, Stewart Cohen, Christian Coons, Daniel Greco, Ali Hasan, Shyam Nair, Ram Neta, Sarah Paul, Miriam Schoenfield, Russ Shafer-Landau, Roy Sorensen, Hank Southgate, Ralph Wedgwood, and Roger White; audiences at Rutgers University, the Massachusetts Institute of Technology, the University of Pittsburgh, Harvard University, Washington University in St. Louis, the University of Arizona, the third St. Louis Annual Conference on Reasons and Rationality, and the fifth annual Midwest Epistemology Workshop; and the students in my Spring 2011 Objectivity of Reasons seminar and my Spring 2012 Epistemology course at the University of Wisconsin-Madison. I am also grateful for two helpful referee reports for *Oxford Studies in Epistemology* from John Broome and John Gibbons.

Christensen, D. (2013). Epistemic modesty defended. In D. Christensen and J. Lackey (eds.), *The Epistemology of Disagreement: New Essays*, pp. 77–97. Oxford: Oxford University Press.
Christensen, D. and J. Lackey (eds.) (2013). *The Epistemology of Disagreement: New Essays*. Oxford: Oxford University Press.
Coates, A. (2012). Rational epistemic akrasia. *American Philosophical Quarterly* 49, 113–24.
Cresswell, M. J. (1975). Hyperintensional logic. *Studia Logica: An International Journal for Symbolic Logic* 34, 25–38.
Crisp, R. (2007). Intuitionism and disagreement. In M. Timmons, J. Greco, and A. R. Mele (eds.), *Rationality and the Good: Critical Essays on the Ethics and Epistemology of Robert Audi*, pp. 31–9. Oxford: Oxford University Press.
Descartes, R. (1988/1641). Meditations on first philosophy. In *Selected Philosophical Writings*, pp. 73–122. Cambridge: Cambridge University Press. Translated by John Cottingham, Robert Stoothoof, and Dugald Murdoch.
Eagle, A. (ms). The epistemic significance of agreement. Unpublished manuscript.
Eells, E. (1985). Problems of old evidence. *Pacific Philosophical Quarterly* 66, 283–302.
Elga, A. (2007). Reflection and disagreement. *Noûs* 41, 478–502.
Elga, A. (2010). How to disagree about how to disagree. In R. Feldman and T. A. Warfield (eds.), *Disagreement*, pp. 175–86. Oxford: Oxford University Press.
Elga, A. (2013). The puzzle of the unmarked clock and the new rational reflection principle. *Philosophical Studies* 164, 127–39.
Elga, A. (ms). Lucky to be rational. Unpublished manuscript.
Enoch, D. (2011). *Taking Morality Seriously: A Defense of Robust Realism*. Oxford: Oxford University Press.
Feldman, F. (ms). What to do when you don't know what to do. Unpublished manuscript.
Feldman, R. (2005). Respecting the evidence. *Philosophical Perspectives* 19, 95–119.
Feldman, R. and T. A. Warfield (eds.) (2010). *Disagreement*. Oxford: Oxford University Press.
Field, H. (2005). Recent debates about the a priori. In T. S. Gendler and J. Hawthorne (eds.), *Oxford Studies in Epistemology*, Volume 1, pp. 69–88. Oxford: Oxford University Press.
Foley, R. (1990). Fumerton's puzzle. *Journal of Philosophical Research* 15, 109–13.
Fumerton, R. (1990). *Reasons and Morality: A Defense of the Egocentric Perspective*. Ithaca, NY: Cornell University Press.
Gaifman, H. (2004). Reasoning with limited resources and assigning probabilities to arithmetical statements. *Synthese* 140, 97–119.
Garber, D. (1983). Old evidence and logical omniscience in Bayesian confirmation theory. In J. Earman (ed.), *Testing Scientific Theories*, pp. 99–132. Minneapolis: University of Minnesota Press.
Gibbons, J. (2006). Access externalism. *Mind* 115, 19–39.
Hasan, A. (ms). A puzzle for analyses of rationality. Unpublished manuscript.
Hegel, G. (1975). *Natural Law*. Philadelphia, PA: University of Pennsylvania Press. Translated by T. M. Knox.

Horowitz, S. (2013). Epistemic akrasia. *Noûs*. Published online first.
Ichikawa, J. and B. Jarvis (2013). *The Rules of Thought*. Oxford: Oxford University Press.
Kant, I. (1974). *Logic*. New York: The Bobbs-Merrill Company. Translated by Robert S. Hartman and Wolfgang Schwarz.
Kelly, T. (2010). Peer disagreement and higher-order evidence. In R. Feldman and T. A. Warfield (eds.), *Disagreement*, pp. 111–74. Oxford: Oxford University Press.
McGrath, S. (2008). Moral disagreement and moral expertise. In R. Shafer-Landau (ed.), *Oxford Studies in Metaethics*, Volume 3, pp. 87–108. Oxford: Oxford University Press.
Rawls, J. (1971). *A Theory of Justice*. Cambridge, MA: Harvard University Press.
Scanlon, T. M. (1998). *What We Owe to Each Other*. Cambridge, MA: Harvard University Press.
Scanlon, T. M. (2003). Metaphysics and morals. *Proceedings and Addresses of the American Philosophical Association* 77, 7–22.
Schechter, J. (2013). Rational self-doubt and the failure of closure. *Philosophical Studies* 163, 428–52.
Schroeder, M. (2008). Having reasons. *Philosophical Studies* 139, 57–71.
Sepielli, A. (2009). What to do when you don't know what to do. *Oxford Studies in Metaethics* 4, 5–28.
Setiya, K. (2013). *Knowing Right from Wrong*. Oxford: Oxford University Press.
Shafer-Landau, R. (2003). *Moral Realism: A Defence*. Oxford: Oxford University Press.
Sher, G. (2007). But I could be wrong. In R. Shafer-Landau (ed.), *Ethical Theory: An Anthology*, pp. 94–102. Oxford: Blackwell Publishing Ltd.
Singer, P. (1972). Moral experts. *Analysis* 32, 115–17.
Smithies, D. (2012). Moore's paradox and the accessibility of justification. *Philosophy and Phenomenological Research* 85, 273–300.
Titelbaum, M. G. (2014). How to derive a narrow-scope requirement from wide-scope requirements. *Philosophical Studies*. Published online first.
van Wietmarschen, H. (2013). Peer disagreement, evidence, and well-groundedness. *The Philosophical Review* 122, 395–425.
Weatherson, B. (2013). Disagreements, philosophical and otherwise. In D. Christensen and J. Lackey (eds.), *The Epistemology of Disagreement: New Essays*, pp. 54–76. Oxford: Oxford University Press.
Weatherson, B. (ms). Do judgments screen evidence? Unpublished manuscript.
Wedgwood, R. (2007). *The Nature of Normativity*. Oxford: Oxford University Press.
Wedgwood, R. (2012). Justified inference. *Synthese* 189, 273–95.
Weiner, M. (2007). More on the self-undermining argument. Blog post archived at <http://www.mattweiner.net/blog/archives/000781.html>.
Williams, B. (1981). Internal and external reasons. In *Moral Luck*, pp. 101–13. Cambridge: Cambridge University Press.
Williamson, T. (2000). *Knowledge and its Limits*. Oxford: Oxford University Press.
Williamson, T. (2011). Improbable knowing. In T. Dougherty (ed.), *Evidentialism and its Discontents*, pp. 147–64. Oxford: Oxford University Press.

10. An Inferentialist Conception of the A Priori

Ralph Wedgwood

When philosophers explain the distinction between the a priori and the a posteriori, they usually characterize the a priori negatively, as involving a kind of justification that does *not* come from experience. But this only raises the question: If we ever do have a priori justification for anything, where *does* this justification come from? One answer often given by those who believe that we sometimes have a priori justification is that it comes from a special faculty of 'intuition' or 'rational insight'. But without any further explanation of how this faculty of intuition or insight operates, appealing to this special faculty seems to be no more than a *label* for the problem rather than a solution to it.

For these reasons, it seems that we need a new conception of the distinction between the a priori and the a posteriori. I shall propose such a new conception here. Most discussions of the a priori have focused either on the difference between a priori *knowledge* and empirical knowledge, or else on the difference between *beliefs* that are justified a priori and beliefs that are justified a posteriori. In developing my proposal, I shall take a different approach. I shall start by initially focusing, not on knowledge or on the justification of beliefs, but instead on the justification of *inferences* and *inferential dispositions*; only after explaining how the distinction works in this inferential case shall I turn to the justification of beliefs.

1. THE NATURE OF INFERENCE

The first task before me, then, is to explain how I conceive of the phenomenon of inference.[1] The most important fact about inferences is that an inference is typically not an attitude towards a single proposition, but rather an attitude towards what I shall call an 'argument'—where an argument is a structured entity, composed out of propositions. The simplest arguments consist of a pair of items: a set of propositions—the argument's *premises*—and a further proposition—the argument's *conclusion*. More complex arguments may include *sub-arguments*, either instead of or in addition to a set of premises, as well as an ultimate conclusion for the argument as a whole.

[1] I have given a sketch of this conception of inference elsewhere; see in particular Wedgwood (2012).

This way of conceiving of arguments is familiar from systems of natural deduction in logic. Every instance of the rules of inference that are recognized in such systems would count as an argument in this sense.[2] For example, every instance of *modus ponens* (also known as 'if'-elimination) consists of a set of premises—two premises of the form 'If p, then q' and p—and a conclusion—the corresponding proposition q. Similarly, in every instance of *conditional proof* (also known as 'if'-introduction), instead of a set of premises, there is a sub-argument, from a premise p to a conclusion q, and as the conclusion of the whole argument, the corresponding conditional proposition 'If p, then q'.

Within this framework, there is no reason for us not to allow that in some cases, the set of propositions that constitutes the argument's premises may be simply the *empty set*. The idea of such zero-premise inferences is familiar from natural-deduction systems of logic, and it is a simple generalization of the basic idea of arguments as abstract structures built up out of propositions.

As I said above, an inference is a kind of attitude towards an argument. Strictly speaking, however, there are two importantly different forms that the phenomenon of inference can take. In one form, an inference is a *mental event*, which occurs at a particular time; in the other form, an inference is an *enduring mental state*, which is stored in the thinker's memory as a relatively permanent feature of that thinker's system of beliefs. The mental event of inference can be thought of as an event of *forming* or *coming to have* the corresponding enduring mental state of inference. Just to have some terminology for marking this distinction, I shall call the enduring mental state the state of '*accepting*' the relevant argument, and I shall call the corresponding mental event the event of '*drawing*' the relevant inference. However, to keep things simple, for the rest of this discussion, I shall mostly ignore the mental event of drawing an inference, and concentrate on the enduring state of 'accepting' an argument instead.

To 'accept' an argument, as I am using the term here, need not involve accepting either the premises or the conclusion of the argument. It is simply to accept the argument itself as a whole—which is something that you might do even if you did not believe either the argument's premises or its conclusion, but were simply *supposing* the argument's premises, purely 'for the sake of argument'. Some philosophers might be tempted to say that accepting an argument in this way is simply to *believe* that the argument's premises *entail* the conclusion, or at least that the conclusion is made *probable* by the premises. But then it would be impossible to accept an argument if one lacked such a belief—as many unsophisticated thinkers would have to, on account of their not even possessing the concept of *entailment* or *probability*. Since it seems to be possible for thinkers to accept an argument even if they lack a belief of this kind, I shall propose a more general conception of what it is to 'accept an argument' here.

According to the conception that I propose, to 'accept' an argument is to have a *conditional* belief in the argument's conclusion—conditionally on the

[2] For an example of such a system of natural deduction, see Tennant (1990).

assumptions of the argument's premises and sub-arguments. In the simple case of a single-premise argument, to accept the argument would be to have a conditional belief in the conclusion, conditionally on the assumption of the premise. This sort of conditional belief is not an attitude towards a single proposition *q*: it is intrinsically an attitude towards a *pair* of propositions *p* and *q*—the attitude of conditionally believing *q*, given the assumption of *p*. One can have such a conditional belief in *q* given *p* even if one does not have an unconditional belief either in *p* or in *q*. One might simply be supposing the premise *p* purely for the sake of argument, and accepting the conclusion *q* purely conditionally on the assumption of that premise.

One way of thinking of such conditional beliefs is as follows. To accept *q* conditionally on the assumption of *p* is, roughly, to engage in a kind of *make-believe*: one in effect *simulates* the state of believing the proposition *p*, and also simulates the adjusted belief-system that one is committed to having in the event of one's learning *p*; one conditionally accepts *q* on the assumption of *p* just in case this adjusted system of beliefs contains a belief in *q*.[3] According to the proposal that I am making here, to accept a simple single-premise argument is to have a conditional belief of this sort.

In addition to these simple cases of single-premise arguments, however, there can be other more complex cases as well. In some cases, there are *several* premises rather than just a single premise: in these cases, to accept the argument is to have a conditional belief in the conclusion, conditionally on the assumption of all the argument's *premises*. In other cases, the argument's premise-set is the *empty set*: in these cases, to accept the argument is simply to have an unconditional belief in the argument's conclusion.

Finally, to explain what it is to accept an argument that involves *sub-arguments*, we need to make sense of a sort of supposition or assumption that has a whole argument, rather than a single proposition (or even a set of propositions), as its object. As I have suggested, to suppose a proposition *p* ('purely for the sake of argument') is roughly like engaging in a kind of *make-believe*—in effect, a state of *simulating* the state of really believing the proposition *p*. To suppose a whole argument is, in a parallel way, like the state of simulating the state of *accepting* the argument, in the sense that I have explained. In this way, it seems that we can make sense of the idea of conditionally believing a conclusion, conditionally on the assumption of a sub-argument. In general, then, to accept an argument *A* is conditionally to believe *A*'s conclusion, conditionally on the assumptions of *A*'s premises and sub-arguments.

On some views, it is part of the nature or essence of the attitude of belief that whenever an agent believes a proposition *p*, the belief is *correct* if and only if the proposition *p* is *true*.[4] There is a natural way of extending this view to give an account of the nature of the state of accepting an argument. According to this extension of the view, the attitude of *accepting* an argument

[3] I cannot give a full account of the nature of these conditional beliefs here; but for an example of such an account, see Edgington (1995).

[4] I have defended a view of this kind myself; see Wedgwood (2013).

is correct if and only if the argument is *truth-preserving*. In the simplest cases, an argument without sub-arguments counts as truth-preserving if and only if it is not the case that all the premises are true while the conclusion is not true. In more complicated cases, an argument counts as truth-preserving if and only if it is not the case that all the argument's premises are true and all its sub-arguments are truth-preserving, while the conclusion is not true.

It seems plausible to me that all these beliefs and conditional beliefs come in *degrees*. That is, one can believe a conclusion, conditionally on the assumptions of various premises and sub-arguments, to varying degrees. In some cases, one might be conditionally *certain*—that is, one might have the highest possible degree of conditional belief in a conclusion, given the assumptions of the relevant premises and sub-arguments. When one is conditionally certain of the conclusion in this way, one is in effect treating the argument as though it were deductively valid. (This would be a way of 'treating arguments as deductively valid' that is available to agents who are incapable of thinking explicitly about 'deductive validity' because they do not even possess the concept of validity.)

In other cases, however, one might have a significantly lower level of conditional belief in an argument's conclusion, given the assumptions of the argument's premises and sub-arguments. This is in effect to treat the argument as though it were not deductively valid, but at most inductively strong. For most of this discussion, however, I shall focus only on cases where accepting the argument involves being conditionally certain of the conclusion, conditionally on the assumptions of the argument's premises and sub-arguments; I shall touch only briefly on cases where one's acceptance of the argument involves a lower degree of conditional belief in the conclusion, conditionally on the argument's premises and sub-arguments.

If beliefs come in degrees, then these degrees range over the whole gamut from complete confidence, through increasing degrees of uncertainty and doubt, all the way to total *disbelief*. This makes it easy for us to make sense of what we might call *anti-inferences* and *anti-arguments*. An anti-argument is just like an argument, except that accepting an anti-argument involves conditionally *disbelieving* its conclusion, conditionally on the assumptions of its premises and sub-arguments. Presumably, the correctness of an anti-inference is a matter, not of the anti-inference's being truth-preserving, but rather of its being *truth-excluding*, in the sense that if the premises are true, then the conclusion is false. I shall explore some such anti-arguments later on, when considering the distinctive inferential role of *negation*.

2. RATIONAL INFERENTIAL DISPOSITIONS

Suppose that a thinker accepts a certain argument. Considering her acceptance of the argument retrospectively, we could ask: Was the thinker's acceptance of that argument rational or justified?[5] Was she reasoning rationally in accepting that argument?

In one sense, you might rationally accept a complicated argument, merely because a respected authority figure asserts that the argument is valid. In that case, the argument itself does not represent the inferential process that explains your acceptance of it, because you accept the argument as a result of accepting a *further* argument—in effect, the argument from 'The authority figure asserts that the argument is valid' to the conclusion 'The argument is valid.' In what follows, I shall set such cases aside. I shall focus solely on cases where the thinker's acceptance of the argument is explained by the thinker's having attitudes towards each of the argument's sub-arguments, as well as towards the argument as a whole, and does not in this way depend on the thinker's acceptance of any other argument.

What is it for a thinker rationally to accept an argument in this sense, or for her acceptance of the argument to be (retrospectively) justified? I propose the following answer. A thinker's attitude of accepting an argument is (retrospectively) justified if and only if the thinker's having this attitude is the manifestation of a *rational inferential disposition* that the thinker has—a disposition that we could call a kind of rational inferential *ability* or *competence*.

The structure of each of these dispositions seems to be as follows. There is a certain range of arguments—typically, the arguments that exemplify a certain pattern or form of inference—such that, *ceteris paribus*, in normal cases in which the thinker *considers* an argument within this range, the thinker responds to considering this argument by *accepting* the argument.[6] For example, I believe that I have a disposition of this kind to accept instances of *modus ponens*. *Modus ponens* arguments are such that, *ceteris paribus*, in normal cases in which I consider one of these arguments, I respond by accepting the argument in question.

It seems to be indispensable to appeal to inferential dispositions of this sort in giving an account of rational or justified inferences, for the following reason. Every argument is in fact an instance of infinitely many argument-schemas. For example, every instance of *modus ponens*—indeed, every two-premise argument whatsoever—is an instance of the schema: '*p*, *q*, therefore *r*'. So it seems that it is not always the case that whenever you accept an argument that is in fact an instance of *modus ponens*, your acceptance of that argument is (retrospectively) justified. In particular, your acceptance of the

[5] This notion of a thinker's acceptance of an argument's being 'retrospectively' justified is the analogue, for inferences, of the notion of a belief's being *doxastically* justified. I shall consider when an inference counts as 'prospectively' justified—the analogue of the notion of a thinker's having *propositional* justification for believing a proposition—later on. (For recent discussion of the distinction between doxastic and propositional justification, see Turri 2010.)

[6] Since we are thinking of 'accepting' an argument as an enduring mental state here (rather than as the mental event of drawing an inference), we should also conceive of 'considering an argument' as the enduring state of having that argument 'within one's ken', as one of the arguments towards which one has any attitudes at all (rather than as a mental event of actively contemplating the argument).

argument will not count as retrospectively justified if it is in fact the manifestation of a crazy disposition, which is also manifested in your acceptance of all sorts of fallacious and invalid arguments, as well as in your acceptance of this particular argument (which happens to be an instance of *modus ponens*). In general, your acceptance of an argument will be retrospectively justified if and only if it is the manifestation of a *rational* inferential disposition.

Which inferential dispositions count as rational dispositions of this sort? In general, it seems that whether or not such an inferential disposition is rational depends, in part, on the *form* of the arguments that the disposition prompts the thinker to accept; that is, it depends on the pattern or argument-schema that these arguments exemplify.

There are two cases here. First, in the case of some argument-schemas, the thinker may have a rational disposition to accept *any* instance of those argument-schemas that she considers, without exception. These inferences would be in a sense *indefeasible*: no further factors can remove the rationality of accepting instances of these schemas. It may be, for example, that your disposition to accept instances of certain basic logical rules of inference (such as *modus ponens* or the like) is an indefeasibly rational disposition of this first kind. If that is right, then *no* instances of these basic rules of inference can be defeated, by any further features of the thinker's cognitive circumstances.

Secondly, in the case of certain other argument-schemas, special circumstances can arise which *defeat* certain instances of those other argument-schemas, and the thinker only has a rational disposition to accept the *undefeated* instances of those schemas. Still, even in this second kind of case, it is the *default* position that the thinker's rational disposition *normally* inclines the thinker to accept instances of these schemas. The circumstances that defeat certain instances of these argument-schemas are in some way (which we will not have time to explore here) special and unusual—in some way, these circumstances count as the exception to the normal rule.

We have already used the notion of a rational inferential disposition to characterize when a thinker's acceptance of an argument is retrospectively justified: your acceptance of an argument is *ex post* or retrospectively justified if and only if it is the manifestation of such a rational inferential disposition. Now that we also have articulated the notion of an 'undefeated instance' of an argument-schema, we can also characterize when a thinker has *ex ante* (or prospective) justification for accepting an argument: you have *ex ante* (prospective) justification for accepting an argument if and only if the argument is an undefeated instance of an argument-schema such that you have a rational inferential disposition to accept undefeated instances of that schema.[7]

[7] For this distinction between '*ex post*' (or retrospective) justifiedness and '*ex ante*' (or prospective) justification, see Goldman (1979). As I remarked in note 5, contemporary epistemologists typically mark this distinction by contrasting 'doxastic' and 'propositional' justification. However, as I have explained elsewhere (Wedgwood 2013), Goldman's terminology is preferable, because in fact this distinction is not limited to the justification of beliefs, but is exemplified throughout the whole normative domain.

However, even if the form of these argument-schemas is *part* of what determines whether a thinker's disposition to accept instances of these schemas counts as rational, it seems that some fact about the thinker at the relevant time must also be involved in determining whether or not this disposition is rational. This is because it appears that even if it is perfectly rational for one thinker to be disposed to accept instances of a certain argument-schema, it can *fail* to be rational for another thinker to have such a disposition. In short, two different thinkers—or even the same thinker at two different times—can differ with respect which argument-schemas it is rational for each thinker to be disposed to accept at each time.

So, it seems, there must be some fact about the thinker at the relevant time that determines *which* argument-schemas it is rational for the thinker to have a disposition to accept at that time. But what fact about the thinker is this?

3. TWO KINDS OF JUSTIFICATION FOR INFERENTIAL DISPOSITIONS

As I shall propose in this section, there are two importantly different kinds of case here. In some cases, an inferential disposition is necessarily rational for *any* thinker with certain basic cognitive capacities. In these cases, I propose, if a thinker has this inferential disposition, the rationality of the disposition is ultimately explained purely by the fact that the thinker has these basic cognitive capacities; as I shall explain, these inferential dispositions have a kind of basic a priori justification. In other cases, the rationality of an inferential disposition is not explained purely by the thinker's basic cognitive capacities in this way; these inferential dispositions do not have this kind of basic a priori justification.

Which basic cognitive capacities are relevant here? Taking different sets of basic cognitive capacities to be relevant could lead to different conceptions of a priori justification. One approach, which seems to lead to a plausible and theoretically interesting conception of what it is for the disposition to accept instances of a certain argument-schema to count as a priori justified, involves focusing on the capacities that are necessarily involved in even considering instances of this argument-schema. These capacities include, most prominently, the capacities for the *concepts* and *types of attitudes* that are involved in considering instances of this schema.

Several philosophers have suggested that the justification of certain inferences may be explained by what it is for thinkers to possess of some of the *concepts* that are deployed in the inference.[8] The possibility that I am exploring here is in effect a *generalization* of this suggestion. Specifically, I am exploring the possibility that the rationality of some inferential dispositions may be explained by our basic cognitive capacities—where this set of capacities includes, not just our possession of various concepts, but also our

[8] For proposals of this sort, see for example Boghossian (2003) and Peacocke (1992).

capacity for the various attitude-types that we are capable of (and perhaps other capacities as well).

If it is ever true that the rationality of an inferential disposition is explained by the basic cognitive capacities that the thinker has, it should be possible to fill in the details of this explanation. The proposal that I am making here does not depend on any particular way of filling in these details. Nonetheless, it may be helpful, to fix ideas, for me to give a sketch of how such explanations might work. Since the thinker's possession of the relevant concepts is among the basic cognitive capacities that I am taking to be relevant here, one way of filling in the details would be to adapt the suggestions that philosophers have made about how the rationality of certain inferential dispositions is explained by the thinker's possession of the relevant concepts.

For example, consider the disposition to accept instances of *modus ponens*. The capacities that are necessary for even considering instances of *modus ponens* include the capacity for thoughts of the form 'If *p*, then *q*': this is the capacity that many philosophers would call one's 'possession' of the 'concept "if"'. So the relevant capacities in this case include one's possession of the concept 'if'. Perhaps possessing the concept 'if' essentially involves having a disposition to accept instances of *modus ponens*, and this fact about the concept constrains the truth conditions of thoughts involving the concept 'if' in such a way as to ensure that all instances of *modus ponens* are truth-preserving. If this is the right account of what it is to possess the concept 'if', it may be the thinker's possession of this concept that explains the rationality of the thinker's being disposed to accept instances of *modus ponens*. Since the possession of this concept is one of the relevant cognitive capacities, this would be a case in which these cognitive capacities explain why it is rational for any thinker who has these capacities to have this inferential disposition.[9]

Another example might involve concepts—such as *negation* perhaps—such that possessing these concepts makes it rational to have certain dispositions with respect to certain patterns of *anti-inference*, as well as dispositions with respect to certain patterns of inference. Specifically, it may be that it is rational, for everyone who possesses the concept that is expressed by 'not', to have the following two dispositions: first, a disposition to have a conditional *disbelief* in *p* conditionally on the assumption of '$\neg p$'; and secondly, a disposition to have a conditional *belief* in '$\neg p$' conditionally on the assumption of any propositions $\{q_1, \ldots q_n\}$ and any *sub-anti-argument* acceptance of which would involve a conditional disbelief in *p* conditionally on the assumption of $\{q_1, \ldots q_n\}$. It may be that possession of the concept that is expressed by 'not' essentially involves having these dispositions, and this fact about the concept constrains the truth conditions of thoughts involving the concept in such as a way as to assume that all instances of these patterns of inference and

[9] This is the sort of answer that I have advocated elsewhere (see e.g. Wedgwood 2011).

anti-inference are correct.[10] This could be what explains the rationality of having these inferential and anti-inferential dispositions.

In addition to the concepts that are expressed by the logical constants, there may also be other concepts such that the thinker's possessing these concepts explains why it is rational for the thinker to be disposed to accept instances of certain argument-schemas. For example, perhaps your possession of the concept 'uncle' makes it rational for you to be disposed to accept arguments from any premise of the form '*x* is an uncle' to the corresponding conclusion '*x* is male'; and perhaps your possession of the concept 'knowledge' makes it rational for you to be disposed to accept arguments from any premise of the form '*x* knows *p*' to the corresponding conclusion *p*. In general, it may be plausible to take this view of many of the kinds of inference that have traditionally been regarded as conceptually or analytically valid.

It may be that these logical and conceptually valid rules of inference are *indefeasible* rules, in the sense that it is rational for any thinker who has the relevant capacities to be disposed to accept *any* instances of these rules, regardless of what background information the thinker may have. Even if there are indefeasible rules of this kind, there seem also to be some *defeasible* rules, such that it is rational for any thinker who is capable of considering instances of these rules to be disposed to accept *undefeated* instances of those rules, but not defeated instances of these rules. For example, one such defeasible rule of inference might be what we could call the 'rule of external-world introduction'—where in each instance of this rule, the premise is some proposition of the form 'It looks to me as though *p* is the case' and the conclusion is the proposition *p* itself. Clearly, it is not rational for every thinker who is capable of considering instances of the rule to be disposed to accept each instance of the rule that he or she considers—since there can be defeated instances of this rule, such as when the thinker has background information that he or she has just taken a powerful hallucinogen. But perhaps it is still rational for every thinker who has the relevant capacities to be disposed to accept the *normal, undefeated* instances of this rule.

For our purposes, however, the precise details of how the rationality of each of these inferential dispositions is explained by the fact that the thinker has these basic cognitive capacities do not matter. It also does not matter whether the explanation of the disposition's rationality is based on the thinker's possession of some concept, or on some other basic cognitive capacity instead. In general, there are many different ways in which the details of this general approach could be filled in; the account of the a priori that I am proposing here does not depend on these details. All that matters is that there are certain argument-schemas such that the rationality of the thinker's disposition to accept undefeated instances of these schemas is explained purely by the fact

[10] The 'correctness' of an anti-inference, as I suggested above, requires that the relevant argument is *truth-excluding* rather than truth-preserving. This proposal about negation is related to those of Rumfitt (2000) and Peacocke (1986), albeit with some minor differences that I cannot attempt to explore here.

that the thinker has the capacities that are necessary for even considering instances of those schemas. So long as this is the case with respect to some argument-schemas, then there are at least some examples of the first sort of justification that I am focusing on here.

For our purposes, the important point is that if there are any such inferential dispositions, these dispositions are rational for *all* thinkers who have these basic cognitive capacities, purely in virtue of the fact that they have these capacities. The central proposal that I am making in this paper is that these inferential dispositions have a certain basic kind of a priori justification.

There is an objection that many philosophers will be tempted to make here. How could the mere fact that a thinker has certain cognitive capacities *ever* explain why an inferential disposition is rational? Could it never happen that certain basic cognitive capacities require being disposed to make certain *invalid* or *irrational* inferences? For example, could there not be intrinsically defective concepts—such as, concepts that lead to paradox, perhaps, like the concepts of truth, or free will, or vague concepts? If there are defective concepts, then how could the thinker's possession of these concepts explain the rationality of *any* inferential disposition?

As a matter of fact, I am inclined to deny that there are any defective concepts in this sense. In general, my view is that basic cognitive capacities cannot require having any inferential dispositions unless those dispositions are essentially rational—that is, rational in all possible cases.[11] But for present purposes, I do not have to repeat the arguments that I have given for this view. This is because I am here simply proposing an account of what it is for an inferential disposition to be justified a priori. So the account that I am proposing here does not entail that the rationality of inferential dispositions is ever explained purely by the basic cognitive capacities that the thinker has. My account entails only that if the rationality of inferential dispositions is never explained purely by the thinker's basic cognitive capacities in this way, then no inferential disposition is justified a priori. This does not seem an implausible consequence of my proposal: if our basic cognitive capacities (including our possession of concepts and our capacities for the various attitude-types) never explain the rationality of any inferential disposition, then it is hard to see how any inferential disposition can be justified a priori.

At all events, even if there are cases of this first sort, in which the rationality of an inferential disposition is a priori, explained purely by the thinker's basic cognitive capacities, there may also be cases of the *second* sort as well. In cases of this second sort, the capacities that are necessarily involved in even considering the instances of a certain argument-schema do *not* suffice to explain why the thinker's disposition to accept instances of the schema is rational. Instead, there is something else, something extra, which is present in the thinker's

[11] For my arguments in favour of this view, and my defence of this view against a range of objections, see Wedgwood (2007).

mind—something that could have been absent even if the thinker had still had the capacity for considering instances of this argument-schema—that explains why the disposition to accept instances of the argument schema is rational.

For example, perhaps, you have a disposition to infer directly from the proposition that both hands on your watch are pointing straight up to the conclusion that it is 12 o'clock. In drawing this inference, you are manifesting a rational inferential disposition. But it seems plausible that the rationality of this inferential disposition is not explained purely by your having the capacities that are necessary for you even to consider instances of the relevant pattern of inference. Instead, the rationality of this disposition seems to be explained by your having a rational *background belief* to the effect that your watch is a reliable timepiece, and when a reliable timepiece's hands are both pointing straight up, that indicates that the time is 12 o'clock. When the rationality of an inferential disposition is explained in part by rational background beliefs in this way, I shall call it a 'non-basic' inferential disposition.

Again, the point that is crucial for our purposes here is not to investigate exactly what explains the rationality of inferential dispositions in cases of this second sort. The crucial point is simply that we can draw this distinction between two sorts of case: (a) cases in which the rationality of your having a certain inferential disposition is explained purely by your having basic cognitive capacities of the relevant kind; and (b) cases in which the explanation depends in addition on certain further factors that happen also to be present in your mind, over and above the mere fact that you have these cognitive capacities. My central proposal is that rational inferential dispositions of the first sort have a basic kind of a priori justification, while inferential dispositions of the second 'non-basic' sort do not have this basic kind of a priori justification.

Some philosophers might regard my usage of these terms 'a priori' and 'a posteriori' as eccentric. But in fact, it seems to me, my usage of these terms is entirely appropriate. First, it is consistent with the meanings of these two phrases in Latin: the phrase 'a priori' means 'from what comes beforehand' and 'a posteriori' means 'from what comes later'. The basic cognitive capacities necessary for even considering instances of the relevant argument-schemas—the capacities that explain the rationality of inferential dispositions of the first kind—in a clear sense 'come before' the additional further factors—like the thinker's background beliefs—which are involved in explaining the rationality of inferential dispositions of the second kind.

Secondly, this way of understanding the distinction is in harmony with one of the main ways in which Kant describes the a priori, as what our cognitive capacities somehow 'supply out of ourselves' (*was unser eigenes Erkenntnisvermögen ... aus sich selbst hergibt*).[12] If the rationality of your having a certain inferential disposition is explained purely by your having certain

[12] See the *Critique of Pure Reason*, Introduction to the Second Edition (B 1).

basic cognitive capacities, then in a good sense these capacities can be thought of as 'supplying' that inferential disposition 'out of themselves'.

Of course, this talk of our cognitive capacities' 'supplying' something 'out of themselves' is a metaphor. But it is a natural metaphor to use; and other metaphors that seem equally natural here also chime with traditional ways of describing the a priori. Thus, we could talk of the distinction between what is also 'built into' the basic cognitive capacities of the mind itself, and what we have access to as a result of taking in further information over and above what is already built into these capacities. What is a priori 'flows from' resources that are in some way already within these capacities of the mind itself, while what is a posteriori flows from information that is only contingently present in minds that possess those basic capacities.

One aspect of this picture of the a priori that might surprise some philosophers is that it gives no special role to sensory experience. I have not characterized the basic kind of a priori justification that I am concerned with here by claiming that it is 'independent' of sensory experience or the like.[13] I have characterized it as a kind of justification that depends purely on these basic cognitive capacities. Presumably, however, the relevant facts about my sensory experiences—the facts that could play a justificatory role in relation to some of my inferential dispositions—are not guaranteed to be present in my mind purely by my possession of these basic cognitive capacities. So, if one of my inferential dispositions is justified in part by such facts about my sensory experiences, its justification is not a priori in the basic way that I have described. However, there could in principle be other cases, in which the extra factor that plays a crucial role in justifying the inferential disposition is not a fact about the thinker's sensory experiences, but a fact about some other mental phenomena that happen to be present in the thinker's mind. So long as this fact is not guaranteed to be present by the thinker's possession of these basic cognitive capacities, this would not be a case of a priori justification (of the basic kind that I have characterized here).

So far, I have only offered a proposal about the justification of inferential dispositions. In the next section, I shall explain how to use this proposal to develop a conception of the distinction between a priori and a posteriori justified beliefs.

4. FROM A PRIORI RATIONAL INFERENTIAL DISPOSITIONS TO A PRIORI JUSTIFIED BELIEFS

I have already made a proposal, in section 2, about how these rational inferential dispositions are connected to justified inferences. The proposal was simple: an inference is *ex post* (or retrospectively) justified if and only if it is the manifestation of a rational or justified inferential disposition; and there is

[13] In this way, my characterization differs from those of such philosophers as BonJour (1999) or Burge (1993).

ex ante (or prospective) justification for a thinker to accept an argument if and only if the argument is an undefeated instance of an argument-schema such that the thinker has a rational disposition to accept undefeated instances of that schema.

This suggests an equally simple account of when inferences are a priori justified: an inference is retrospectively a priori justified if and only if it is the manifestation of an a priori justified disposition; and there is prospective a priori justification for a thinker to accept an argument if and only if the argument is an undefeated instance of a schema such that the thinker has an a priori justified disposition to accepted undefeated instances of the schema.

Admittedly, every manifestation of these dispositions is caused by a contingent factor that happens to be present in the thinker's mind—namely, the fact of the thinker's *considering* the relevant argument; and this factor is not guaranteed to be present purely because of the thinker's having the relevant cognitive capacities. However, it still seems plausible that the manifestations of these dispositions are justified a priori. The thinker's considering an argument does not involve taking in any new information from the outside world, over and above what is already built into the capacities that are necessary for considering the argument. Moreover, one's considering the argument is not necessary for one to have *ex ante* or prospective justification for accepting the argument: in this sense, the fact of one's considering the argument is not a reason that supports or justifies accepting the argument; it is just a way in which the mind can be caused to make explicit to itself what is already implicit within it.

So, I shall suppose that whenever one manifests an a priori justified inferential disposition of this sort, one's acceptance of the particular argument that one is considering is also justified a priori. For example, suppose that one considers an instance of disjunction-introduction, of the form 'p; so, p or q', and responds by accepting this argument—that is, by conditionally believing the conclusion 'p or q' given the assumption of the premise p. Then one's acceptance of this argument is justified a priori.

How can we get from being a priori justified in *accepting* certain *arguments* to being a priori justified in *believing* certain *propositions*? One very simple way in which this might happen is if the justified inferential dispositions that the thinker has include dispositions to accept certain *zero-premise* arguments. (For example, one justified inferential disposition might be a disposition to accept any argument whose conclusion is a proposition of the form 'p or $\neg p$', even if the argument has no premises at all.) In these cases, acceptance of one of these zero-premise arguments is already an outright belief in the conclusion of the argument.

However, there is also a general connection between justified inferences and justified beliefs in the case of arguments that have assumptions—including both sub-arguments and premises. In general, the connection seems to be this. Suppose that a thinker is justified in accepting an argument that involves a certain assumption A—regardless of whether this assumption A is a premise or a sub-argument—and suppose that the thinker is also justified in accepting this assumption *A*. Then the thinker is also justified in believing the conclusion of the whole argument, not just conditionally on the assumption of *all* of the argument's sub-arguments and premises, but conditionally on a *smaller* set of assumptions—namely, a set of assumptions that includes all of the argument's assumptions other than *A*.[14]

Thus, even if all of the thinker's justified inferential dispositions are dispositions to accept arguments that involve either premises or sub-arguments or both, it is still possible for the thinker to get justified beliefs out of justified inferences. The thinker simply needs to be justified in accepting an argument that has a sub-argument, while simultaneously being justified in accepting that sub-argument.

The simplest illustration of this involves the kind of suppositional reasoning that is often called 'conditional proof'. To illustrate this point, suppose that (as I suggested in the previous section) it is rational for anyone who possesses the concept *uncle* to be disposed to accept any argument from a premise of the form '*x* is an uncle' to the corresponding conclusion '*x* is male'. Then you might consider the argument from 'Ralph is an uncle' to 'Ralph is male', and, through manifesting this disposition, you might accept this argument. Then, by means of an instance of conditional proof in which this argument is a sub-argument, you could respond to your accepting this argument by having an outright belief in the proposition 'If Ralph is an uncle, Ralph is male.'

In all these cases, then, a priori justified beliefs simply correspond to a priori justified inferences of a certain distinctive sort. Specifically, they correspond to a priori justified inferences that, in the terminology of many natural-deduction systems of logic, have no 'undischarged assumptions'.[15] In the terminology that I have been employing here, these inferences consist in the acceptance of arguments that have no premises, but at most certain sub-arguments. Whenever a thinker is a priori justified in accepting an argument of this sort, and also a priori justified in accepting all of the argument's sub-arguments, she is a priori justified in believing the argument's conclusion.

The general connection between justified inferences and justified beliefs that I proposed above entails the following familiar idea: if the thinker is justified in accepting an argument, and also simultaneously justified in accepting of all of the argument's assumptions with the highest level of confidence, then the thinker is also justified in believing the conclusion of the whole argument. So, it seems, when the thinker is a priori justified in accepting an argument, and is also a priori justified in accepting all the argument's assumptions, then the thinker is a priori justified in believing the argument's conclusion as well.

[14] Strictly, as I have stated it here, this connection requires that the thinker must be justified in accepting the assumption *A* with maximum confidence. Where *A* is a sub-argument, this connection is in effect the idea of 'discharging' the assumptions of the sub-argument's premises. For more details, see Tennant (1990, 56).

[15] This point is clearly articulated by Peacocke (1993)—although Peacocke combines this point with a number of other theses that I am not defending here.

There are other ways in which we can now explain a priori justified beliefs as well. For example, suppose that in addition to having justified inferential dispositions to accept certain arguments, the thinker also has a justified disposition to reason in accordance with a certain *rule of proof*.[16] Roughly, manifesting such a disposition will involve responding to the fact that one has justified beliefs in certain propositions, and is simultaneously considering a certain further proposition, by having a belief in that further proposition as well. For example, suppose that you are a priori justified, as a result of a certain sort of suppositional reasoning, in believing a proposition *p*; then it may be rational for you to respond to having this sort of justified belief in *p* by also believing the proposition that *p is necessarily true*. If the rationality of your being disposed to reason in accordance with this rule of proof is explained purely by the basic cognitive capacities that you have, then this disposition is justified a priori; given that your belief in *p* is itself justified a priori, your belief in the proposition that *p* is necessarily true, which you have through following this rule of proof, is also justified a priori. So this could be a further way of having a priori justified beliefs.

Once we can have a priori justified beliefs in any of these ways, then it may be possible for these a priori justified beliefs to form part of the body of background beliefs that explains the justification of other beliefs and dispositions. In particular, as we have seen, such background beliefs may explain the justification of certain non-basic inferential dispositions. In some cases, it may be that none of the thinker's a posteriori justified background beliefs are relevant to the justification of one of these non-basic inferential dispositions, but some of the thinker's a priori justified beliefs are relevant. In this case, one may regard this non-basic inferential disposition as also in a non-basic way justified a priori.

The most plausible example of this case would involve a logician who has a disposition to accept instances of certain extremely complicated and sophisticated patterns of argument, where the rationality of the logician's accepting these arguments is not explained purely by the logician's possession of the capacities that are necessary for considering those arguments, but rather by the fact that the logician has an a priori justified belief that arguments of that form are valid. Then the logician's disposition to accept these sophisticated arguments might be an example of a non-basic inferential disposition that counts as a priori justified in this way.

5. FEATURES AND ADVANTAGES OF THIS APPROACH

One significant advantage of this approach is that we have not merely characterized the a priori negatively, in terms of where a priori justification does

[16] The terms 'rule of inference' and 'rule of proof' were first introduced by Smiley (1963, 114). For a classic discussion of this distinction (in which these two types of rules were labelled 'schematic' and 'thematic' rules respectively), see Geach (1980, 109–10).

not come from; we have characterized it positively, in terms of where it *does* come from—namely, the rational dispositions that are built into these basic cognitive capacities.

Moreover, by characterizing the a priori in this way, we also did not have to appeal to 'intuitions' or 'rational insight' as a fundamental explanatory notion in the theory.[17] This is an advantage because without further explanation, this appeal to 'intuitions' or 'rational insight' has always seemed deeply mysterious. Indeed, we can now explain intuitions away, in a sense. Such intuitions may arise from the inferential dispositions that are rational for us in virtue of our possessing certain basic capacities (like our possession of various concepts, or our capacity for various types of attitudes). For example, as I have already suggested, we can use the framework that I have articulated here to explain why we have the intuition that if someone knows a proposition p, then p is true, or the intuition that if a person is an uncle, then that person is male. This seems to explain why so many intuitions—like the examples that I have just given here—take the form of conditionals, or of universal generalizations of conditionals.

Some philosophers might object: in what way is it more intelligible to appeal to the rational acceptance of arguments that have no premises than to rational intuitions? But as we have seen, these zero-premise inferences are not a special *sui generis* phenomenon; they exist only as part of a system of inferences that also includes inferences that have premises—and every theory needs to explain how such inferences can be rational. So the inferentialist approach that I am proposing does provide a more illuminating account of a priori justification than the invocation of 'rational intuitions'.

One might wonder: is this conception of the a priori committed to a questionable kind of *foundationalism*? In appealing to a particular instance of a certain pattern of inference to explain why a certain belief is a priori justified, I might be thought to imply either or both of the following two archetypal foundationalist theses: first, the thesis that the prospective or 'propositional' justification of any a priori justified belief depends purely on the availability of a way of inferring the relevant proposition from a suitably 'privileged' set of premises (where the *empty* set of premises could clearly be regarded as 'privileged' in the relevant way), and does not depend on any holistic coherence relations that that belief stands in to the totality of the thinker's beliefs; secondly, such a belief can be retrospectively or 'doxastically' justified only if it is *based on* the thinker's in some sense 'carrying out' that inference.

Fortunately, the proposal outlined here is not committed to either of these controversial theses. First, the contrast that the first thesis rests on, between (i) the availability of a way of inferring a conclusion p from the empty set of premises and (ii) the holistic coherence of believing p with the thinker's other

[17] For an example of a philosopher who takes the notion of an 'intuition' as fundamental to the a priori, see Bealer (2000); for a philosopher who appeals to 'rational insight', see BonJour (1999).

beliefs, may in fact be a false dichotomy. It may be that the arguments that the thinker is a priori justified in accepting are part of what determines what it is for the thinker's beliefs to count as 'coherent'. For example, the fact that there is an indefeasible argument of this sort from a premise *p* to a conclusion q_1 may be at least part of what explains why it would be incoherent for the thinker's belief-system to include a higher degree of confidence in *p* than in q_1; and the fact that there is an indefeasible argument of this sort that has no premises at all for the conclusion q_2 may be part of what explains why it would be incoherent to have any degree of belief that falls short of maximum level of confidence in q_2.[18]

Secondly, my picture does not imply that doxastically justified beliefs must be 'based on' the thinker's 'carrying out' any inference (at least not if this is understood as a conscious mental process of some kind). They simply imply that doxastically justified beliefs must be the manifestations of rational dispositions; and the relevant disposition may just be a disposition to respond to 'considering' the relevant proposition—that is, to having the proposition 'in one's ken' in the relevant way—by having the kind of doxastic attitude towards the proposition that rational coherence requires. This is, in effect, a disposition to conform to the requirements of coherence among one's doxastic attitudes in a non-accidental way. So this conception of doxastic justification does not obviously incur any commitments to any controversial sort of foundationalism.

One of the most famous controversies in philosophy concerns whether *mathematical* reasoning is justified a priori. Can the approach that is being proposed here shed any light on this controversy?

With respect to mathematics, let us just focus on mathematical proofs. Are we justified a priori in accepting such proofs? It will be clear that we are justified a priori if a *logicist* programme, of the sort developed by Crispin Wright (1983), is correct. As Neil Tennant (1987, 275–300) has shown, this sort of logicist programme can be put into an inferentialist form, according to which possession of arithmetical concepts involves the capacity for reasoning in accordance with rules of inference that allow all the axioms of Peano arithmetic to be derived from the empty set of assumptions.

Within the framework that is being proposed here, what is distinctive of the logicist programme is that it explains the rationality of accepting all the axioms and arguments that are distinctive of arithmetic on the basis of what is involved in our possession of arithmetical *concepts*. In that sense, the logicist programme implies that all the truths of arithmetic are *conceptual* truths. However, within my framework, our possession of concepts is not the only basic cognitive capacity that can explain the a priori justification of an inferential disposition. In principle, other sorts of basic cognitive capacities may also be relevant. Perhaps, for example, in the case of mathematics, the

18 For this picture of the relation between rational inference and rational coherence, see Wedgwood (2012).

relevant capacities consist of our general ability for understanding certain sort of *structures*, including the structures of *merely possible* states of affairs, along with the ability for recognizing the features that are *shared* by various different sorts of structures.[19] Even if these cognitive capacities do not consist simply in our possession of any particular concepts, it might be that it is our possession of these cognitive capacities that explains the rationality of our accepting the axioms and inferences of mathematics. If so, then even if mathematical truths are not all conceptual truths, our justification for accepting mathematical proofs would still be a priori.[20]

6. GENERALIZING BEYOND INFERENTIAL DISPOSITIONS

In section 4, I mentioned that there might be a priori justified dispositions to reason in accordance with rules of proof as well as rules of inference. A disposition to reason in accordance with a certain rule of proof is not strictly speaking an inferential disposition, although it is undoubtedly a broadly doxastic disposition (in the sense of a disposition to have a certain belief in response to having certain other beliefs). In this sense, my conception of the a priori is not inferentialist in the strong sense that it implies that *every* case of a priori justification involves a priori justified *inferential* dispositions: other doxastic dispositions besides inferential dispositions—like dispositions to reason in accordance with a certain rule of proof—may also count as a priori justified.

Indeed, doxastic dispositions of many different kinds may be justified a priori—that is, justified purely because of the thinker's possession of basic cognitive capacities of the relevant sort. For example, the dispositions for taking one's sensory experiences and apparent episodic memories at face value may also be justified a priori in this way. However, while these dispositions may be justified a priori, these doxastic dispositions are not themselves sources of any a priori justified beliefs. The disposition to take one's sensory experiences at face value will only lead one to have a belief in a proposition p if the proposition p is part of the content of one's sensory experiences; and so what makes it the case that one is justified in believing p as a result of manifesting this disposition is never just one's possession of the relevant basic cognitive capacities, but always also the fact of one's having appropriate sensory experiences as well. Hence such beliefs are only ever justified empirically, not a priori.

In general, then, my account of the a priori is inferentialist in a somewhat weaker sense: according to this account, a priori justified inferences are

[19] This suggestion is offered on behalf of those who take a 'structuralist' view of mathematics; for discussions of structuralism in the philosophy of mathematics, see especially Hellmann (2005) and McBride (2005).

[20] Similar points may hold about our justification for *philosophical* beliefs. But I cannot investigate these difficult meta-philosophical issues here.

involved in every case of a priori justified *belief*, and so also in every case of a priori knowledge. Even reasoning in accordance with a rule of proof, as we have in effect already seen, will only yield an a priori justified belief if all the beliefs to which one responds in reasoning in accordance with that rule are themselves a priori justified beliefs. So, a priori justified beliefs which manifest a disposition for reasoning in accordance with this rule of proof depend on the thinker's having other a priori justified beliefs as well. It seems then that even in these cases there must be some a priori justified beliefs that arise in some way other than through following such rules of proof. Presumably, these other a priori justified beliefs arise from a priori justified inferences in the way that I described in section 4.

Indeed, it is not just doxastic phenomena that can be justified a priori. We can make sense of the idea that dispositions of *practical* thought might also count as justified a priori. A disposition of practical thought would count as justified a priori just in case it is a rational disposition, and the explanation of its rationality is based purely on one's possession of certain basic mental capacities (not on any further facts about one's mental life that are not guaranteed to be present purely by one's possession of these capacities.)

So, consider our dispositions to form intentions or choices on the basis of our beliefs, and perhaps on the basis of our desires or preference as well. For example, consider the disposition to have the intention to perform an act *A* in response to the belief that this act *A* is what one ought to do in the relevant situation. This disposition might be rational or justified purely a priori—that is, purely in virtue of what is built into the relevant capacities.

There also seems to be a kind of suppositional *practical* reasoning that is possible here. One can suppose, purely for the sake of argument, that doing *B* is what one ought to do; and then one can have a purely *conditional intention* to do *B*, conditionally on the assumption that *B* is what one ought to do. Arguably, this conditional intention is also justified a priori—it is a rational conditional intention to have, and its rationality is explained in a fundamentally analogous way to the rationality of a priori justified beliefs.

In this way, then, the general theory of the a priori that I am advocating here is not really 'inferentialist': it is a theory that fundamentally appeals to the way in which our basic mental capacities can ground the rationality of various dispositions of thought. Nonetheless, when it comes to a priori justified beliefs (and to a priori knowledge), this approach gives a fundamental explanatory role to the notion of a priori justified inference.[21]

[21] An earlier version of this paper was presented at a Kline workshop on a priori knowledge at the University of Missouri in March 2013. I am grateful to the members of that audience—and especially to my commentator, Matthew McGrath—and also to two anonymous referees for *Oxford Studies in Epistemology*, for helpful comments.

REFERENCES

Bealer, George (2000). 'A Theory of the A Priori', *Pacific Philosophical Quarterly* 81 (1): 1–30.

Boghossian, Paul (2003). 'Blind Reasoning', *Aristotelian Society Supplementary Volume* 77: 225–48.

BonJour, Laurence (1999). *In Defense of Pure Reason* (Cambridge: Cambridge University Press).

Burge, Tyler (1993). 'Content Preservation', *Philosophical Review* 102 (no. 4, October): 457–88.

Edgington, Dorothy (1995). 'On Conditionals', *Mind* 104 (414): 235–329.

Geach, P. T. (1980). *Reference and Generality*, 3rd edition (Ithaca, NY: Cornell University Press).

Goldman, Alvin (1979). 'What Is Justified Belief?', in G. S. Pappas, ed., *Justification and Knowledge* (Dordrecht: Reidel), 1–24.

Hellmann, Geoffrey (2005). 'Structuralism', in Scott Shapiro, ed., *The Oxford Handbook of Philosophy of Mathematics and Logic* (Oxford: Oxford University Press), 536–62.

McBride, Fraser (2005). 'Structuralism Reconsidered', in Scott Shapiro, ed., *The Oxford Handbook of Philosophy of Mathematics and Logic* (Oxford: Oxford University Press), 563–89.

Peacocke, Christopher (1986). 'Understanding Logical Constants: A Realist's Account', *Proceedings of the British Academy* 73: 153–99.

Peacocke, Christopher (1992). *A Study of Concepts* (Cambridge, MA: MIT Press).

Peacocke, Christopher (1993). 'How Are A Priori Truths Possible?' *European Journal of Philosophy* 1 (2): 175–99.

Rumfitt, Ian (2000). '"Yes" and "No"', *Mind* 109: 781–823.

Smiley, T. J. (1963). 'Relative Necessity', *Journal of Symbolic Logic* 28: 113–34.

Tennant, Neil (1987). *Anti-realism and Logic* (Oxford: Clarendon Press).

Tennant, Neil (1990). *Natural Logic*, revised edition (Edinburgh: Edinburgh University Press).

Turri, John (2010). 'On the Relationship between Propositional and Doxastic Justification', *Philosophy and Phenomenological Research* 80: 312–26.

Wedgwood, Ralph (2007). 'Normativism Defended', in *Contemporary Debates in the Philosophy of Mind*, ed. Brian P. McLaughlin and Jonathan Cohen (Oxford: Blackwell), 85–101.

Wedgwood, Ralph (2011). 'Primitively Rational Belief-Forming Processes', in *Reasons for Belief*, ed. Andrew Reisner and Asbjørn Steglich-Petersen (Cambridge: Cambridge University Press), 180–200.

Wedgwood, Ralph (2012). 'Justified Inference', *Synthese* 189 (no. 2, November): 273–95. DOI: 10.1007/s11229-011-0012-8.

Wedgwood, Ralph (2013). 'The Right Thing to Believe', in Timothy Chan, ed., *The Aim of Belief* (Oxford: Clarendon Press), 123–39.

Wright, Crispin (1983). *Frege's Conception of Numbers as Objects* (Aberdeen: Aberdeen University Press).

INDEX

The manufacturer's authorised representative in the EU for product safety is Oxford University Press España S.A. of El Parque Empresarial San Fernando de Henares, Avenida de Castilla, 2 - 28830 Madrid (www.oup.es/en or product.safety@oup.com). OUP España S.A. also acts as importer into Spain of products made by the manufacturer.
Printed and bound by CPI Group (UK) Ltd, Croydon, CR0 4YY
14/09/2026
02216307-0013